Trefftz and Fundamental Solution-Based Finite Element Methods

Edited by

Qing-Hua Qin
Department of Engineering
Shenzhen MSU-BIT University
518172 Shenzhen
China

Trefftz and Fundamental Solution-Based Finite Element Methods

Editor: Qing-Hua Qin

ISBN (Online): 978-981-4998-54-3

ISBN (Print): 978-981-4998-55-0

ISBN (Paperback): 978-981-4998-56-7

Bentham Science Publishers Pte. Ltd.
80 Robinson Road #02-00
Singapore 068898
Singapore
Email: subscriptions@benthamscience.net

BENTHAM
SCIENCE

CONTENTS

PREFACE

Articles on Trefftz and fundamental solution-based finite element methods (FEM), also known as boundary type FEM, have been published widely in journals attracting the attention of both researchers and engineers with backgrounds in materials science, mechanical engineering, applied physics, mechanics of solids, and applied mathematics. There is, however, not a single-volumed document that can provide both comprehensive and technically detailed coverage of this field that has been available up to the present. Now it seems timely to collect the research results and to present a collective approach of these useful but scattered data, while this data should be made available to professional researchers, scientists, engineers and students in materials engineering and applied mechanics, *e.g.*, physicists, mechanical and materials scientists.

Unique in its direct and comprehensive approach, the aim of this book is to fill the gap mentioned above and present fundamental knowledge of boundary type FEM and comprehensive treatment of its algorithms. The volume summarizes the current state of practice and presents the most recent research outcomes in the subject. It details the development of each of the ideas and techniques beginning with fundamental principles, and new concepts are illustrated with simple examples wherever possible.

This book will be more focused on fundamental aspects of Trefftz FEM and fundamental solution based-FEM. The book will be written for scientists and engineers working in industry and academia. A major distinction in comparison with other books will be that it will provide a comprehensive, up-to-date perspective from well-known, distinguished scientists and experts in the field. Indeed, there is no such book out there at present and we believe this book has significant potential to make a real difference.

Trefftz and fundamental solution-based FEM is a broad field, and, practically speaking, this book can only cover a fraction of the many relatively advanced topics. It consists of two major parts: Part I Trefftz FEM and Part II fundamental solution-based FEM. In the first part, Chapter 1 gives an introduction and overview of the hybrid-Trefftz FEM. Basic concept and general element formulations of the method are described. Chapters 2 to 4 discuss selected topics of non-homogeneous parabolic problems, thermal analysis of composites, and heat conduction in nonlinear functionally graded materials. In part II of the book, following a brief summary of the fundamental solution based-FEM, Chapters 6 and 7 discuss axisymmetric potential problems and Rotordynamic Response of Tapered Composites. Finally, Chapters 8 and 9 cover n-sided polygonal hybrid finite elements and piezoelectric materials.

The editor is grateful to the contributors for their timely contribution and patience while the chapters were being processed and reviewed. We are greatly indebted to reviewers for their generous help and time given to review the chapters. Finally, the editor must thank the Bentham Science editor and staff for their support in bringing out this book.

Qing-Hua Qin
Department of Engineering
Shenzhen MSU-BIT University
Shenzhen 518172
China

List of Contributors

Ana Coutinho	Faculdade de Ciências e Tecnologia, Universidade Nova de Lisboa, 1649004 Lisboa, Portugal
Changyong Cao	For Soft Machines & Electronics, Michigan State University, East Lansing, MI 48824, USA
Hui Wang	School of Civil Engineering and Architecture, Hainan University, Haikou, China
Ildi Cismasiu	Faculdade de Ciências e Tecnologia, Universidade Nova de Lisboa, 1649004 Lisboa, Portugal
Ionut Dragos Moldovan	CERIS, Instituto Superior Técnico, Universidade de Lisboa, 1649004 Lisboa, Portugal
Junchen Zhou	School of Mechanical and Automotive Engineering, Shanghai University of Engineering Science, Shanghai, China
Keyong Wang	School of Mechanical and Automotive Engineering, Shanghai University of Engineering Science, Shanghai, China
Qiang Xi	Center for Numerical Simulation Software in Engineering and Sciences, College of Mechanics and Materials, Hohai University Nanjing 211100, P. R. China
Qing-Hua Qin	Department of Engineering, Shenzhen MSU-BIT University, 518172 Shenzhen, China
Rajamohan Ganesan	Concordia Centre for Composites (CONCOM), Department of Mechanical, Industrial and Aerospace Engineering, Concordia University, Montreal, Canada
Wan-Qing Lin	College of Civil Engineering, Henan University of Technology, Zhengzhou, China
Wen-Zhi Xu	Center for Numerical Simulation Software in Engineering and Sciences, College of Mechanics and Materials, Hohai University Nanjing 211100, P. R. China
Ze She	School of Mechanical and Automotive Engineering, Shanghai University of Engineering Science, Shanghai, China
Zhuo-Jia Fu	Center for Numerical Simulation Software in Engineering and Sciences, College of Mechanics and Materials, Hohai University Nanjing 211100, P. R. China

CHAPTER 1

Introduction to Hybrid-Trefftz Finite Element Method

Qing-Hua Qin[1]*

[1]*Department of Engineering, Shenzhen MSU-BIT University, 518172 Shenzhen, China*

Abstract: This chapter presents a review on the hybrid-Trefftz (HT) finite element method (FEM) and its applications in applied mechanics and mechanical engineering. First, fundamental issues on the hybrid Trefftz FE approach are described, including elements associated with special-purposed functions, generalized variational functionals, T-complete functions, internal fields and boundary fields in an element. Then, the Trefftz FEM of contact problems, two-dimensional elastic problems, elastoplasticity, and piezoelectricity are described. Mathematical expressions for the cases mentioned are derived by means of T-complete solutions and a modified variational functional. In the case of plane elasticity and elastoplastic problems, exact solutions derived in a well-established way from complex variable equations are used for the intra-element fields, and an iterative form of the fundamental equations is employed in the case of elastoplasticity. The creation of force-displacement equations from the variational functional is also addressed. Finally, some conclusions are presented and some open questions in this area are discussed.

Keywords: Contact problem, Elastoplasticity, FEM, Piezoelectricity, Plane elasticity, Trefftz function, Variational principle.

INTRODUCTION

The HT FEM, using a set of FEs with the Trefftz functional as interpolation function, is a hybrid FEM, including the adoption of an inter-element frame field to link the intra-displacement fields of the elements. It is noted that, apart from closed-form solutions that are available for only a few problems with simple boundary conditions and geometries [1-10], the FEM has provided an alternative approach to approximate the solutions of engineering problems in the past decades [11-14].

Historically, the development of powerful FEM has mainly focused on four different categories [11, 12, 15-18]. The first is the so-called traditional FE

***Corresponding author Qing-Hua Qin:** Department of Engineering, Shenzhen MSU-BIT University, 518172 Shenzhen, China; E-mail: qinghua.qin@smbu.edu.cn formulation [11]. It is based on a suitable polynomial interpolation function which has already been used to analyse various physical problems. With this method, the solution domain is divided into a number of cells or elements, and material properties are defined at element level. The second is the natural-mode FEM initiated by Argyris *et al.* [17]. The natural FEM presents a significant alternative to the conventional FEM, with ramifications for all aspects of structural analysis. It makes a distinction between the constitutive and geometric parts of element tangent stiffness, which can lead effortlessly to the nonlinear effects associated with large displacements. The third is the hybrid-Trefftz FEM [19-22]. Unlike the conventional and natural FEMs, the Trefftz method combines the advantages of the FEM [11, 23] and the boundary element method (BEM) [1, 24-27]. In contrast to the first two methods, the internal fields of the Trefftz FEM, chosen in such a way that satisfy, a priori, the governing field equations, have easily been represented as the sum of a suitably truncated Trefftz function series (T-complete set) of regular homogeneous solutions, multiplied by undetermined coefficients and a particular integral of non-homogeneous equations. The theoretical fundamentals of T-complete sets have been presented mainly in [28-31], in which the authors named such systems as C-complete systems. According to Zienkiewicz's suggestion, Herrera and colleagues changed this description to the T(Trefftz)-complete system of solutions, in honour of the inventor of such non-singular solutions. As such, the terminology 'TH-families' is usually used when referring to systems of functions which satisfy the criteria originated by Herrera [4]. Inter-element continuity is guaranteed by using a special variational functional and a frame field defined on inter-element boundary. The process of generating element formulation leads to the standard linear stiffness equation, with a positive definite element stiffness matrix. Clearly, while the traditional FE formulation may be assimilated to a special form of the Rayleigh-Ritz approach, the HT FEM has a close relationship with the hybrid-Trefftz method [32]. The final form is the hybrid FEM based on the fundamental solution, discussed in detail in Chapter 5 [33].

Since the first attempt to generate a general-purpose Trefftz formulation in 1977 [12], the Trefftz-element concept has become increasingly popular, attracting a growing number of researchers into this field [34-43]. To date, Trefftz elements have been successfully applied to problems of two-dimensional elasticity [42, 44-47], fracture mechanics [48-51], Kirchhoff thin plates [12, 22, 52-54], moderately thick Reissner-Mindlin plates [55-58], thick plates [59-63], general 3-D solid mechanics [42, 64], axisymmetric solid mechanics [65, 66], potential problems [67-70], shells [71, 72], elastodynamic problems [22, 38, 73], micromechanics analysis [74, 75], heat conduction analysis [20, 76-78], geometrically nonlinear plates [53, 62, 63, 79], materially nonlinear elasticity [80-83], contact problems [84-86], and piezoelectric materials [51, 87-90]. Moreover, the concept of special-purpose functions has shown great efficiency in handling various geometries or local effects and load-dependent singularities (*e.g.*, cracks, obtuse or reentrant corners, elliptic holes, patch or concentrated loads [91, 92]). Further, the idea of developing p- or h-versions of Trefftz finite elements has already been identified as particularly efficient from the viewpoints of computation [63]. Huang and Li [93] developed an Adini's element coupled with the Trefftz finite method, which is suitable for modelling locally singular problems. The first monograph for describing in detail the HT FEM and its applications in practical engineering was published in 2000 [19]. Besides, a wealthy source of information pertaining to the HT FEM exists in a wide range of general or special review-type articles such as those of Herrera [94], Jirousek and Zielinski [95], Kita and Kamiya [96], and Qin [19, 97].

Following the introduction above, this chapter consists of six sections. Fundamental concepts and general method of element formulations, which include descriptions of practical problems, Trefftz functions, two groups of independently assumed displacement fields, and special constructed variational functions, are described in the section "BASIC FORMULATIONS FOR TREFFTZ FEM." The section "PLANE ELASTICITY" focuses on the fundamentals of Trefftz FEs for linear two-dimensional elastic problems based on Trefftz functions and the specially constructed variational principle appearing in the previous section. It details the method of deriving element stiffness equations, Trefftz functions, and special-purpose functions accounting for local effects. Applications of Trefftz FEs to contact problems, elastoplastic bodies, and piezoelectricity are described in the subsequent three sections.

BASIC FORMULATIONS FOR TREFFTZ FEM

Some basic concepts are discussed in this section, emphasizing specially constructed variational principles, Trefftz functions, and elemental stiffness matrix. The descriptions are based on the works of Jirousek and Zielinski [95], Qin [97], and Qin and Wang [21]. A Cartesian coordinate system is employed, the position of a field point is represented by $\mathbf{x}$ (or x_i), and both traditional Cartesian notation (x, y, z) and conventional indicial notation (x_i) are utilised. Einstein summation convention is applied to repetitive indexes. Bold face letters represent vectors, tensors, or their matrix representations.

Basic Relationships in Engineering Problems

Many physical problems in practical engineering are boundary value problems. A given numerical solution to any of these problems must satisfy the governing equations, boundary conditions, and so on. For an engineering problem, its physical behaviour is generally governed by the following differential equations:

$$\mathbf{L}\boldsymbol{\sigma} + \bar{\mathbf{b}} = 0 \qquad \text{(partial differential equation)} \tag{1}$$

$$\boldsymbol{\sigma} = \mathbf{D}\boldsymbol{\varepsilon} \qquad \text{(constitutive law)} \tag{2}$$

$$\boldsymbol{\varepsilon} = \mathbf{L}^T\mathbf{u} \qquad \text{(generalized geometrical relationship)} \tag{3}$$

with the boundary conditions:

$$\mathbf{u} = \bar{\mathbf{u}} \qquad \text{(on } \Gamma_u \text{, essential boundary condition)} \tag{4}$$

$$\mathbf{t} = \mathbf{A}\boldsymbol{\sigma} = \bar{\mathbf{t}} \qquad \text{(on } \Gamma_t \text{, natural boundary condition)} \tag{5}$$

where the variables $\mathbf{u}$, $\boldsymbol{\sigma}$, $\boldsymbol{\varepsilon}$, and $\bar{\mathbf{b}}$ are vectors of displacements, stresses, strains, and body forces, **L**, **A**, and **D** stand for differential operator matrix, transformation matrix, and constitutive coefficient matrix respectively, including the external normal unit vector of the boundary. In the form of Trefftz FE, Eqs (1)-(5) are completed by enforcing following inter-element continuity requirements:

$$\mathbf{u}_e = \mathbf{u}_f \qquad \text{(on } \Gamma_e \cap \Gamma_f \text{, conformity),} \tag{6}$$

$$\mathbf{t}_e + \mathbf{t}_f = 0 \qquad \text{(on } \Gamma_e \cap \Gamma_f \text{, traction reciprocity),} \tag{7}$$

where '*e*' and '*f*' stand for any two neighbouring elements. With suitably defined matrices **L**, **D** and **A**, one can describe a particular physical problem through the general relationships (1)-(7). The first step in a FE analysis therefore decides what kind of problem is at hand. This decision is based on the assumptions used in the theory of physical and mathematical approaches to the solution of specific problems. Some typical problems encountered involve: (a) piezoelectricity, (b) heat conduction, (c) contact problem, (d) two-dimensional elasticity, (e) elastoplasticity, (f) thin plate bending, (g) thick plate, and (h) three-dimensional (3D) elastic problem. For illustration, let us consider a plane stress problem. For this special problem, we have

$$\mathbf{u} = \{u\ v\}^T,\ \ \bar{\mathbf{b}} = \{\bar{b}_x\ \bar{b}_y\}^T,\ \boldsymbol{\varepsilon} = \{\varepsilon_{xx}\ \varepsilon_{yy}\ \ 2\varepsilon_{xy}\}^T,\ \ \boldsymbol{\sigma} = \{\sigma_{xx}\ \sigma_{yy}\ \sigma_{xy}\}^T,$$

$$\mathbf{v} = \{u\ v\}^T,\quad \mathbf{L} = \begin{bmatrix} \partial/\partial x & 0 & \partial/\partial y \\ 0 & \partial/\partial y & \partial/\partial x \end{bmatrix},$$

$$\mathbf{D} = \frac{E}{1-\mu^2}\begin{bmatrix} 1 & \mu & 0 \\ \mu & 1 & 0 \\ 0 & 0 & \dfrac{1-\mu}{2} \end{bmatrix},\quad \mathbf{A} = \begin{bmatrix} n_x & 0 & n_y \\ 0 & n_y & n_x \end{bmatrix},\quad \mathbf{t} = \mathbf{A}\sigma = \{t_1, t_2\}^T \qquad \textbf{(8)}$$

where u, v and $\bar{b}_i$ are, respectively, displacements in x_1 and x_2 directions and the vector of body forces, σ_{ij} and ε_{ij} are stresses and strains respectively; μ and E are Poisson's ratio and Young's modulus, n_i are the external normal unit vector, and t_i are the surface traction vector.

Assumed Fields

The basic idea of the Trefftz FEM is to establish a FE formulation whereby inter-element continuity is applied on a nonconforming internal field constructed to a priori satisfy the field equation of the problem under consideration [21]. In other words, as an alternative to the Rayleigh-Ritz approach as a basis for a finite element method, the formulation here is based on the Trefftz approach [32]. With this method, the solution domain Ω is divided into elements, and the assumed intra-element fields over each element "*e*" are:

$$\mathbf{u} = \breve{\mathbf{u}} + \sum_{i=1}^{m} \mathbf{N}_i \mathbf{c}_i = \breve{\mathbf{u}} + \mathbf{N}\mathbf{c} \qquad \textbf{(9)}$$

where $\mathbf{N}_i$ and $\breve{\mathbf{u}}$ are given functions and $\mathbf{c}_i$ is a coefficient vector to be determined. If the field equations are written as:

$$\Re\mathbf{u}(\mathbf{x}) = \bar{\mathbf{b}}(\mathbf{x}), \qquad (\mathbf{x} \in \Omega_e) \qquad \textbf{(10)}$$

where $\Re$ represents the differential operator, $\mathbf{x}$ stands for the position coordinates, the overhead bar denotes the specified quantities and Ω_e represents for the eth element sub-domain, then $\breve{\mathbf{u}} = \breve{\mathbf{u}}(\mathbf{x})$ and $\mathbf{N}_i = \mathbf{N}_i(\mathbf{x})$ in Eqn (9) must be chosen such that

$$\Re\breve{\mathbf{u}} = \bar{\mathbf{b}} \quad \text{and} \quad \Re\mathbf{N}_i = 0, \qquad (i = 1, 2, \ldots, m) \tag{11}$$

everywhere in Ω_e. The unknown coefficient $\mathbf{c}$ can be calculated from the conditions on the external boundary and/or the continuity conditions on the inter-element boundary. Thus various Trefftz-element models can be obtained by using different approaches to enforce these conditions. In the majority of the approaches, a hybrid technique is usually used, whereby the elements are linked through an auxiliary conforming displacement frame which has the same form as in the conventional FE method. This means that, in the Trefftz FE approach, a conforming potential (or displacement in solid mechanics) field should be independently defined on the element boundary to enforce the potential continuity between elements and also to link the coefficient $\mathbf{c}$, appearing in Eq. (9), with the nodal displacement $\mathbf{d}$ (=$\{d\}$). The frame is defined as:

$$\tilde{\mathbf{u}}(\mathbf{x}) = \tilde{\mathbf{N}}(\mathbf{x})\mathbf{d}, \qquad (\mathbf{x} \in \Gamma_e) \tag{12}$$

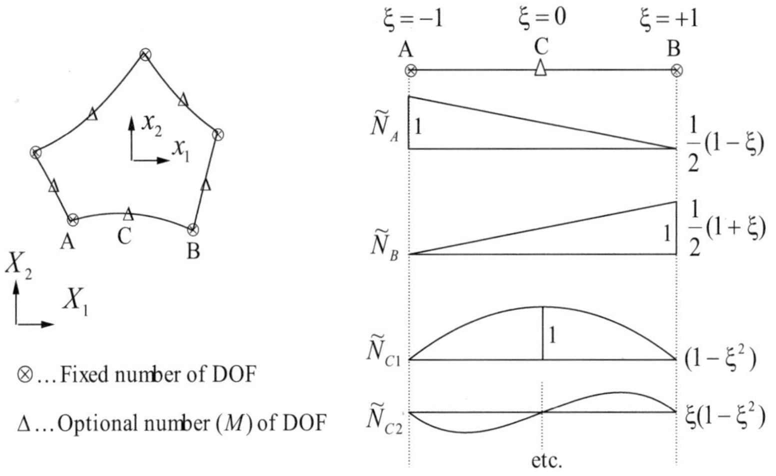

Fig. (1). Configuration of the Trefftz-element model.

where the symbol "~" is used to specify that the field is defined on the element boundary only, $\mathbf{d}$=$\mathbf{d}(\mathbf{c})$ stands for the vector of the nodal displacements, which are the final unknowns of the problem, Γ_e represents the boundary of element e, and $\tilde{\mathbf{N}}$ is a matrix of the related shape functions. Fig. (**1**) displays a typical example of the shape functions used here.

T-complete Functions

As an important function in deriving Trefftz FE formulation T-complete functions (or Trefftz functions) are discussed in this section. Firstly, we illustrate how to derive Trefftz functions and developing suitable criteria for completeness. The general procedures of the derivation can be found in the work of Colton [98] and Herrera [99]. For the simplicity of illustration, consider the following equation:

$$\nabla^2 u = 0 \tag{13}$$

where $\nabla^2 = \partial^2 / \partial x^2 + \partial^2 / \partial y^2$ is the 2D Laplace operator. Its Trefftz functions are a series of functions satisfying Eq. (13) and these functions are complete in the sense of containing all possible solutions in the domain under consideration. It is obvious that the following functions satisfies Eq. (13):

$$1, r\cos\theta, r\sin\theta, \cdots, r^m \cos m\theta, r^m \sin m\theta, \cdots, \tag{14}$$

where r and θ are two polar coordinates. As a result, the T-complete set, denoted by $\mathbf{T}$, is written as:

$$\mathbf{T} = \{1, r^m \cos m\theta, r^m \sin m\theta\} = \{T_i\} \tag{15}$$

Modified Variational Functionals

The Trefftz FE formulation for the engineering problem (1)-(7) can be established by means of variational principle [21]. Noted that the stationary conditions of the traditional variational functional may not satisfy the inter-element continuity condition which is required in Trefftz finite element analysis, some modified variational functionals have been constructed and used in the literature to establish the Trefftz FE formulation. We here present three of these which have already been used in engineering analysis.

The modified variational functional used throughout this paper is [21]:

$$\Pi_m = \sum_e \Pi_{me} = \sum_e \{\Pi_e + \int_{\Gamma_{te}} (\bar{\mathbf{t}} - \mathbf{t})\tilde{\mathbf{u}} ds - \int_{\Gamma_{Ie}} \mathbf{t}\tilde{\mathbf{u}} ds\} \quad \textbf{(16)}$$

$$\Psi_m = \sum_e \Psi_{me} = \sum_e \{\Psi_e + \int_{\Gamma_{ue}} (\bar{\mathbf{u}} - \tilde{\mathbf{u}})\mathbf{t} ds - \int_{\Gamma_{Ie}} \mathbf{t}\tilde{\mathbf{u}} ds\} \quad \textbf{(17)}$$

where

$$\Pi_e = \iint_{\Omega_e} \Pi(\boldsymbol{\sigma}) d\Omega - \int_{\Gamma_{ue}} \mathbf{t}\bar{\mathbf{u}} ds \quad \textbf{(18)}$$

$$\Psi_e = \iint_{\Omega_e} (\Psi(\boldsymbol{\varepsilon}) - \bar{\mathbf{b}}\mathbf{u}) d\Omega - \int_{\Gamma_{te}} \bar{\mathbf{t}}\tilde{\mathbf{u}} ds \quad \textbf{(19)}$$

with

$$\Pi(\boldsymbol{\sigma}) = \frac{1}{2}\boldsymbol{\sigma}^T \mathbf{C}\boldsymbol{\sigma}, \quad \Psi(\boldsymbol{\varepsilon}) = \frac{1}{2}\boldsymbol{\varepsilon}^T \mathbf{D}\boldsymbol{\varepsilon} \quad \textbf{(20)}$$

where $\mathbf{C}=\mathbf{D}^{-1}$ and Eq. (1) are satisfied due to the use of Trefftz function.. Here, the terminology "modified principle" refers to add some modified terms to a conventional functional for the construction of a special variational functional to account for additional requirements including the condition defined in Eqs (6) and (7).

The boundary Γ_e of a particular element consists of the following parts:

$$\Gamma_e = \Gamma_{ue} \cup \Gamma_{te} \cup \Gamma_{Ie} \quad \textbf{(21)}$$

where

$$\Gamma_{ue} = \Gamma_u \cap \Gamma_e, \quad \Gamma_{te} = \Gamma_t \cap \Gamma_e, \quad \textbf{(22)}$$

and Γ_{Ie} denotes the inter-element boundary of the element '*e*'. The derivation of the HT FEM using the functional (16) or (17) was discussed by Qin [21].

Derivation of Element Stiffness Matrix

The FE displacement-force equation can be produced by setting $\delta\Pi_{me} = 0$ or $\delta\Psi_{me} = 0$. By virtue of the solution properties of the intra-element trial functions, the functional Π_{me} in Eq. (16) can be simplified to:

$$\Pi_{me} = \frac{1}{2}\int_{\Omega_e} \mathbf{u}\bar{\mathbf{b}} d\Omega + \frac{1}{2}\int_{\Gamma_e} \mathbf{t}\mathbf{u} ds + \int_{\Gamma_{te}} (\bar{\mathbf{t}} - \mathbf{t})\tilde{\mathbf{u}} ds - \int_{\Gamma_{Ie}} \mathbf{t}\tilde{\mathbf{u}} ds - \int_{\Gamma_{ue}} \mathbf{t}\bar{\mathbf{u}} ds . \tag{23}$$

Substituting the expressions given in Eqs (9) and (12) into (16) and using Eqs (2), (3) and (5) produces:

$$\Pi_{me} = -\frac{1}{2}\mathbf{c}^T \mathbf{H}\mathbf{c} + \mathbf{c}^T \mathbf{S}\mathbf{d} + \mathbf{c}^T \mathbf{r}_1 + \mathbf{d}^T \mathbf{r}_2 + \text{terms without } \mathbf{c} \text{ or } \mathbf{d} \tag{24}$$

in which the matrices **H**, **S** and the vectors $\mathbf{r}_1$, $\mathbf{r}_2$ are all known [21].

To enforce inter-element continuity on the common element boundary, the unknown vector **c** should be expressed in terms of nodal DOF (degrees of freedom) **d**. An optional relationship between **c** and **d** in the sense of variation can be obtained from:

$$\frac{\partial \Pi_{me}}{\partial \mathbf{c}^T} = -\mathbf{H}\mathbf{c} + \mathbf{S}\mathbf{d} + \mathbf{r}_1 = 0 . \tag{25}$$

This leads to:

$$\mathbf{c} = \mathbf{G}\mathbf{d} +, \mathbf{g} \tag{26}$$

where $\mathbf{G} = \mathbf{H}^{-1}\mathbf{S}$ and $\mathbf{g} = \mathbf{H}^{-1}\mathbf{r}_1$, and then straightforwardly yields the expression of Π_{me} in terms only of **d** and other known matrices:

$$\Pi_{me} = \frac{1}{2}\mathbf{d}^T \mathbf{G}^T \mathbf{H}\mathbf{G}\mathbf{d} + \mathbf{d}^T(\mathbf{G}^T \mathbf{H}\mathbf{g} + \mathbf{r}_2) + \text{terms without } \mathbf{d} \tag{27}$$

Therefore, the element stiffness matrix equation can be obtained by taking the vanishing variation of the functional Π_{me} as:

$$\frac{\partial \Pi_{me}}{\partial \mathbf{d}^T} = 0 \Rightarrow \quad \mathbf{K}\mathbf{d} = \mathbf{P} \tag{28}$$

where $\mathbf{K}=\mathbf{G}^T\mathbf{H}\mathbf{G}$ and $\mathbf{P}=-\mathbf{G}^T\mathbf{H}\mathbf{g}-\mathbf{r}_2$ are, respectively, the element stiffness matrix and the equivalent nodal flow vector. The expression (28) is the elemental stiffness matrix equation for Trefftz FE analysis.

PLANE ELASTICITY

This section discusses HT FE theory in linear plane elastici problem. The linear displacement-strain relationship is here assumed [100-102] and developments of the HT FE formulation in two-dimensional elasticity are reviewed.

In this case, the intra-element field (9) becomes:

$$\mathbf{u} = \begin{Bmatrix} u_1 \\ u_2 \end{Bmatrix} = \begin{Bmatrix} \breve{u}_1 \\ \breve{u}_2 \end{Bmatrix} + \sum_{j=1}^{m} \mathbf{N}_j \mathbf{c}_j = \breve{\mathbf{u}} + \mathbf{Nc}, \tag{29}$$

where $\mathbf{c}_j$ are undetermined coefficients and the known coordinate functions $\breve{\mathbf{u}}$ and $\mathbf{N}_j$ are respectively particular integrals and homogeneous solutions to the equations:

$$\mathbf{LDL}^T \breve{\mathbf{u}} + \bar{\mathbf{b}} = 0 \qquad (\text{on } \Omega_e), \tag{30}$$

and

$$\mathbf{LDL}^T \mathbf{N}_j = 0 \qquad (\text{on } \Omega_e), \tag{31}$$

where **D, L**, and $\bar{\mathbf{b}}$, are defined in Eq. (8) for the case of plane stress. In the case of plane strain, E and μ above are replaced by:

$$E^* = \frac{E}{1-\mu^2},\ \mu^* = \frac{\mu}{1-\mu}. \tag{32}$$

If the body forces ($\bar{b}_1$ and $\bar{b}_2$) are constant, the associated particular solution can betaken as:

$$\breve{\mathbf{u}} = -\frac{1+\mu}{E} \begin{Bmatrix} \bar{b}_1 y^2 \\ \bar{b}_2 x^2 \end{Bmatrix}. \tag{33}$$

The distribution of the frame (12) can now be written as:

$$\tilde{u}_1 = \tilde{N}_A \tilde{u}_{1A} + \tilde{N}_B \tilde{u}_{1B} + \sum_{i=1}^{M} \delta^{i-1} \tilde{N}_{Ci} a_{Ci}, \tag{34}$$

$$\widetilde{u}_2 = \widetilde{N}_A \widetilde{u}_{2A} + \widetilde{N}_B \widetilde{u}_{2B} + \sum_{i=1}^{M} \delta^{i-1} \widetilde{N}_{Ci} b_{Ci} , \tag{35}$$

along a given side, saying A-C-B, of an element (Fig. **1**), where the difinitions of $\widetilde{N}_A, \widetilde{N}_B$ and $\widetilde{N}_{Ci}$ are listed in Fig. (**1**), δ is a number equal to either -1 or 1, in accordance with the orientation of the given side A-C-B Fig. (**1**) in the global system (X_1, X_2):

$$\delta = \begin{cases} +1 & \text{if } X_{1B} - X_{1A} \le X_{2B} - X_{2A}, \\ -1 & \text{otherwise.} \end{cases} \tag{36}$$

A series of Trefftz function $\mathbf{N}_j$ can be obtained in a well-established way from Muskhelishvili's complex variable procedure [102]. They are [44]:

$$2G\mathbf{N}_{ej} = \begin{Bmatrix} \operatorname{Re} Z_{1k} \\ \operatorname{Im} Z_{1k} \end{Bmatrix} \text{ with } Z_{1k} = i\kappa z^k + kiz\bar{z}^{k-1} , \tag{37}$$

$$2G\mathbf{N}_{ej+1} = \begin{Bmatrix} \operatorname{Re} Z_{2k} \\ \operatorname{Im} Z_{2k} \end{Bmatrix} \text{ with } Z_{2k} = \kappa z^k - kz\bar{z}^{k-1} , \tag{38}$$

$$2G\mathbf{N}_{ej+2} = \begin{Bmatrix} \operatorname{Re} Z_{3k} \\ \operatorname{Im} Z_{3k} \end{Bmatrix} \text{ with } Z_{3k} = i\bar{z}^k , \tag{39}$$

$$2G\mathbf{N}_{ej+3} = \begin{Bmatrix} \operatorname{Re} Z_{4k} \\ \operatorname{Im} Z_{4k} \end{Bmatrix} \text{ with } Z_{4k} = -\bar{z}^k . \tag{40}$$

The related stress field is given by means of the constitutive relation (2):

$$\boldsymbol{\sigma} = \begin{Bmatrix} \sigma_{11} \\ \sigma_{22} \\ \sigma_{12} \end{Bmatrix} = \breve{\boldsymbol{\sigma}} + \sum_{j=1}^{m} \mathbf{T}_j \mathbf{c}_j = \breve{\boldsymbol{\sigma}} + \mathbf{Tc} , \tag{41}$$

while the special solution $\breve{\boldsymbol{\sigma}}$ can be conveniently obtained by setting $\breve{\boldsymbol{\sigma}} = \mathbf{D}\mathbf{L}^T\breve{\mathbf{u}}$. Derivation of the element force-displacement equation is carried out through use of the functional:

$$\Pi_{me} = \frac{1}{2}\int_{\Omega_e} (\mathbf{L}^T\mathbf{u})^T \mathbf{D}\mathbf{L}^T\mathbf{u}d\Omega - \int_{\Gamma_{eu}} \tilde{\mathbf{t}}\bar{\mathbf{u}}ds + \int_{\Gamma_{e\sigma}} (\bar{\mathbf{t}} - \mathbf{t})\tilde{\mathbf{u}}ds - \int_{\Gamma_{el}} \mathbf{t}\tilde{\mathbf{u}}ds. \quad \textbf{(42)}$$

We now turn our attention to discussion of two representative special-purpose element models. First, we consider a concentrated load acting at a point of any element (Fig. **2**). Singularities produced by the load can be accounted for accurately by augmenting the particular solution $\breve{\mathbf{u}}_e$ with a suitable singular solution $\hat{\mathbf{u}}_e$. For an isolated force in an infinite plane, for example (Fig. **2**), the plane stress solution [103] yields the displacements:

$$\hat{u}_1 = \frac{1+\mu}{4\pi E}P_1\left[(1+\mu)\frac{x_1^2}{r^2} - \frac{3-\mu}{2}\ln\frac{r^2}{l^2}\right] + \frac{(1+\mu)^2}{4\pi E}P_2\frac{x_1x_2}{r^2}, \quad \textbf{(43)}$$

$$\hat{u}_2 = \frac{(1+\mu)^2}{4\pi E}P_1\frac{x_1x_2}{r^2} + \frac{1+\mu}{4\pi E}P_2\left[(1+\mu)\frac{x_2^2}{r^2} - \frac{3-\mu}{2}\ln\frac{r^2}{l^2}\right] \quad \textbf{(44)}$$

where $l > 0$ is an arbitrary constant employed to offer a reference frame, $r^2 = x_1^2 + x_2^2$, and P_1 and P_2 are the external loads shown in Fig. **2**.

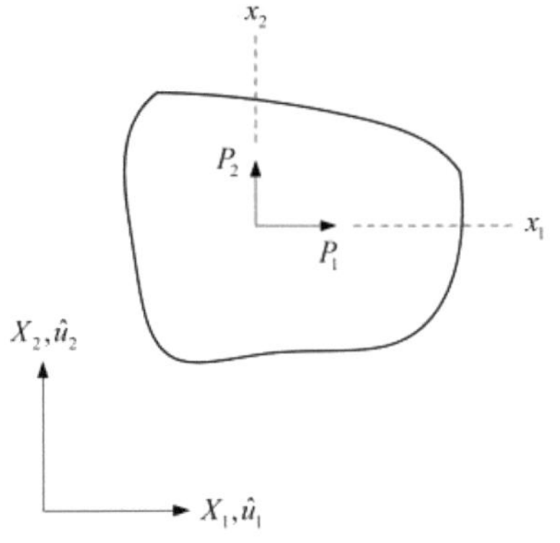

Fig. (2). Isolated concentrated loads in infinite plane.

A singular corner is another typical special-purpose element model (Fig. **3**). of the associated Trefftz functions satisfying the free stress conditions along the sides of a notch can be generated using the Williams'.eigenfunctions [104]. These Trefftz functions can be used to form special-purpose elements with singular corners. They are written as:

$$2Gu_1 = a\sum_n \mathrm{Re}\{\left(\frac{r}{a}\right)^{\lambda_n} \beta_n[(\kappa+\lambda_n \cos 2\alpha + \cos 2\lambda_n\alpha)\cos\lambda_n\theta$$
$$-\lambda_n \cos(\lambda_n - 2)\theta] - \left(\frac{r}{a}\right)^{\zeta_n} \eta_n[(\kappa+\zeta_n \cos 2\alpha \tag{45}$$
$$-\cos 2\zeta_n\alpha)\sin\zeta_n\theta - \zeta_n \sin(\zeta_n - 2)\theta]\},$$

$$2Gu_2 = a\sum_n \mathrm{Re}\{\left(\frac{r}{a}\right)^{\lambda_n} \beta_n[(\kappa-\lambda_n \cos 2\alpha - \cos 2\lambda_n\alpha)\sin\lambda_n\theta$$
$$+\lambda_n \sin(\lambda_n - 2)\theta] + \left(\frac{r}{a}\right)^{\zeta_n} \eta_n[(\kappa-\zeta_n \cos 2\alpha \tag{46}$$
$$+\cos 2\zeta_n\alpha)\cos\zeta_n\theta + \zeta_n \cos(\zeta_n - 2)\theta]\},$$

where a is defined by:

$$a = \sum_{i=1}^{N} \frac{(x_{1i}^2 + x_{2i}^2)^{1/2}}{N}, \tag{47}$$

in which N is the number of nodes in the corresponding element, β_n and η_n are real unknown constants, α and θ are listed in Fig. (**3**), while λ_n and ζ_n are eigenvalues having a real part >= 1/2. They are the solutions of the following equations:

$$\sin 2\lambda_n\alpha = -\lambda_n \sin 2\alpha, \tag{48}$$

for symmetric (tension) loading, and:

$$\sin 2\zeta_n\alpha = \zeta_n \sin 2\alpha, \tag{49}$$

for antisymmetric (pure shear) loading.

In addition to the high efficiency in solving singular corner problems, the great advantage of the presented special-purpose elements is the possibility of evaluating straightforwardly the stress intensity factors K_I (opening mode) and K_{II} (sliding mode) from the internal parameters β_1 and η_1:

$$K_I = \sqrt{2\pi}\lambda_1 a^{1-\lambda_1}(\lambda_1 + 1 - \lambda_1 \cos 2\alpha - \cos 2\lambda_1 \alpha)\beta_1, \tag{50}$$

$$K_{II} = \sqrt{2\pi}\zeta_1 a^{1-\zeta_1}(\zeta_1 - 1 - \zeta_1 \cos 2\alpha + \cos 2\zeta_1 \alpha)\eta_1. \tag{51}$$

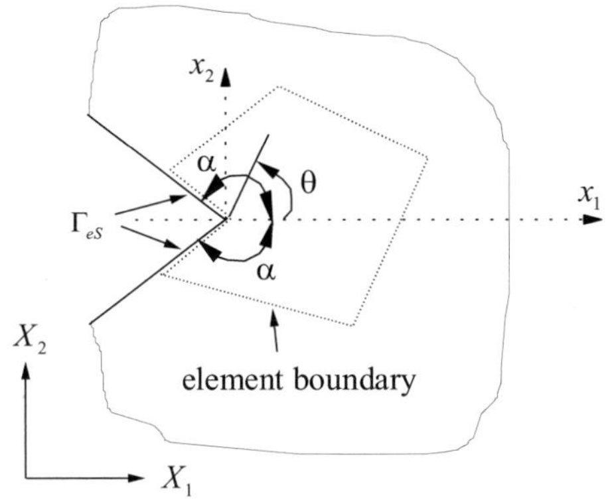

Fig. (3). Singular V-notched element.

ELASTOPLASTICITY

This section is concerned with application of the Trefftz FEM to the solution of elastoplastic problems [83]. As described in [83], a modified variational functional is introduced and used to derive hybrid-Trefftz finite element formulation for elastoplasticity of bulky solids. The formulation is applicable to either strain hardening or elastic perfectly plastic materials. A solution algorithm based on the initial stress formulation is implemented into the new element model.

Basic Governing Equations of Elastoplasticity

Consider an elastoplastic solid, occupying a two-dimensional arbitrarily shaped domain Ω bounded by its boundary Γ. In the "small" deformation range, Eqs (1)-(5) are now expressed in the incremental form as [83]:

$$[\mathbf{L}]^T \{\dot{\boldsymbol{\sigma}}\} + \{\dot{\mathbf{b}}\} = 0 \qquad \text{in } \Omega \tag{52}$$

$$\{\dot{\boldsymbol{\varepsilon}}\} = [\mathbf{L}]\{\dot{\mathbf{u}}\} \qquad \text{in } \Omega \tag{53}$$

$$\{\dot{\boldsymbol{\sigma}}\} = [\mathbf{D}_{ep}]\{\dot{\boldsymbol{\varepsilon}}\} \qquad \text{in } \Omega \tag{54}$$

$$\{\dot{\mathbf{u}}\} = \{\dot{\bar{\mathbf{u}}}\} \qquad \text{on } \Gamma_u \tag{55}$$

$$\{\dot{\mathbf{t}}\} = [\dot{\boldsymbol{\sigma}}]\{\mathbf{n}\} = \{\dot{\bar{\mathbf{t}}}\} \qquad \text{on } \Gamma_t \tag{56}$$

where

$$\{\dot{\mathbf{u}}\} = \{\dot{u}_1, \dot{u}_2\}^T, \quad \{\dot{\boldsymbol{\varepsilon}}\} = \{\dot{\varepsilon}_{11}, \dot{\varepsilon}_{22}, 2\dot{\varepsilon}_{12}\}^T, \quad \{\dot{\boldsymbol{\sigma}}\} = \{\dot{\sigma}_{11}, \dot{\sigma}_{22}, \dot{\sigma}_{12}\}^T, \tag{57}$$

$$\{\dot{\bar{\mathbf{u}}}\} = \{\dot{\bar{u}}_1, \dot{\bar{u}}_2\}^T, \quad \{\dot{\bar{\mathbf{t}}}\} = \{\dot{\bar{t}}_1, \dot{\bar{t}}_2\}^T, \quad \{\dot{\mathbf{b}}\} = \{\dot{b}_1, \dot{b}_2\}^T, \quad \{\mathbf{n}\} = \{n_1, n_2\}^T \tag{58}$$

$$[\mathbf{L}] = \begin{bmatrix} \frac{\partial}{\partial x_1} & 0 \\ 0 & \frac{\partial}{\partial x_2} \\ \frac{\partial}{\partial x_2} & \frac{\partial}{\partial x_1} \end{bmatrix}, \ [\dot{\boldsymbol{\sigma}}] = \begin{bmatrix} \sigma_{11} & \sigma_{12} \\ \sigma_{12} & \sigma_{22} \end{bmatrix}, \tag{59}$$

$$[\mathbf{D}_{ep}] = [\mathbf{D}_e] - [\mathbf{D}_e]\left\{\frac{\partial F}{\partial\{\boldsymbol{\sigma}\}}\right\}\left\{\frac{\partial F}{\partial\{\boldsymbol{\sigma}\}}\right\}^T [\mathbf{D}_e]\left[A + \left\{\frac{\partial F}{\partial\{\boldsymbol{\sigma}\}}\right\}^T [\mathbf{D}_e]\left\{\frac{\partial F}{\partial\{\boldsymbol{\sigma}\}}\right\}\right]^{-1}, \tag{60}$$

$$[\mathbf{D}_e] = \frac{E}{1-\mu^2}\begin{bmatrix} 1 & \mu & 0 \\ \mu & 1 & 0 \\ 0 & 0 & \frac{1-\mu}{2} \end{bmatrix} \tag{61}$$

for plane stress, and

$$[\mathbf{D}_e] = \frac{E(1-\mu}{(1+\mu)(1-2\mu)}\begin{bmatrix} 1 & \frac{\mu}{1-\mu} & 0 \\ \frac{\mu}{1-\mu} & 1 & 0 \\ 0 & 0 & \frac{1-2\mu}{2(1-\mu)} \end{bmatrix} \tag{62}$$

for plain strain, in which $\dot{\sigma}_{ij}$ and $\dot{\varepsilon}_{ij}$ are stresses and strains, $\dot{b}_i$ body forces, $\dot{u}_i$ displacements, and $\dot{\bar{u}}_i$ and $\dot{\bar{t}}_j$ prescribed boundary displacements and tractions. The yield function F and parameter [**A**] can be found in [83]. The strain $\{\dot{\boldsymbol{\varepsilon}}\}$ can be decomposed into its elastic and plastic parts:

$$\{\dot{\boldsymbol{\varepsilon}}\} = \{\dot{\boldsymbol{\varepsilon}}^e\} + \{\dot{\boldsymbol{\varepsilon}}^p\} \tag{63}$$

With expression (63), the stress strain equation can be written as:

$$\{\dot{\boldsymbol{\sigma}}\} = [\mathbf{D}_e](\{\dot{\boldsymbol{\varepsilon}}\} - \{\dot{\boldsymbol{\varepsilon}}^p\}) = \{\dot{\boldsymbol{\sigma}}^e\} + \{\dot{\boldsymbol{\sigma}}^p\} \tag{64}$$

where $\{\dot{\boldsymbol{\sigma}}^e\}$ and $\{\dot{\boldsymbol{\sigma}}^p\}$ are defined by:

$$\{\dot{\boldsymbol{\sigma}}^e\} = [\mathbf{D}_e]\{\dot{\boldsymbol{\varepsilon}}\}, \quad \{\dot{\boldsymbol{\sigma}}^p\} = -[\mathbf{D}_e]\{\dot{\boldsymbol{\varepsilon}}^p\} \tag{65}$$

Substituting Eq. (64) into (52) and using Eq. (53), the governing differential equation (52) can be rewritten as [105]:

$$[\mathbf{L}]^T[\mathbf{D}_e][\mathbf{L}]\{\dot{\mathbf{u}}\} + \{\dot{\mathbf{b}}^{ep}\} = 0 \qquad \text{in } \Omega \tag{66}$$

and the natural boundary (56) is replaced by:

$$\{\dot{\mathbf{t}}_0\} = [\dot{\boldsymbol{\sigma}}^e]\{\mathbf{n}\} = ([\dot{\boldsymbol{\sigma}}] - [\dot{\boldsymbol{\sigma}}^p])\{\mathbf{n}\} = \{\dot{\bar{\mathbf{t}}}\} - [\dot{\boldsymbol{\sigma}}^p]\{\mathbf{n}\} = \{\dot{\bar{\mathbf{t}}}_0\} \tag{67}$$

where $\{\dot{\mathbf{b}}^{ep}\} = \{\dot{\mathbf{b}}\} + [\mathbf{L}]^T\{\dot{\boldsymbol{\sigma}}^p\}$. The plastic part of stresses $\{\dot{\boldsymbol{\sigma}}^p\}$ is described in [83].

Moreover, in the Trefftz FE approach, Eqs (52)-(67) should be completed by adding the following inter-element continuity requirements:

$$\{\dot{\mathbf{u}}\}_e = \{\dot{\mathbf{u}}\}_f, \qquad (\text{on } \Gamma_e \cap \Gamma_f, \text{ conformity}), \tag{68}$$

$$\{\dot{\mathbf{t}}\}_e + \{\dot{\mathbf{t}}\}_f = 0, \qquad (\text{on } \Gamma_e \cap \Gamma_f, \text{ reciprocity}) \tag{69}$$

where '*e*' and '*f*' stand for any two neighbouring elements. Equations (52)-(69) are taken as the basis to establish the modified variational principle for Trefftz FE analysis in elastoplasticity.

Modified Variational Principles

The Trefftz FE equation for an elastoplastic solid can be established by the variational approach [19]. The modified variational functional for the case of an elastoplastic problem is based on the total potential energy as [21]:

$$\Psi_m = \sum_e \Psi_{me} = \sum_e \{\Psi_e + \int_{\Gamma_{ue}} (\{\dot{\bar{\mathbf{u}}}\}^T - \{\dot{\tilde{\mathbf{u}}}\}^T)\{\dot{\mathbf{t}}\}ds - \int_{\Gamma_{Ie}} \{\dot{\mathbf{t}}\}^T \{\dot{\tilde{\mathbf{u}}}\}ds\} \tag{70}$$

where

$$\Psi_e = \iint_{\Omega_e} (\Psi(\{\dot{\boldsymbol{\varepsilon}}\}) - \{\dot{\mathbf{b}}\}^T \{\dot{\mathbf{u}}\})d\Omega - \int_{\Gamma_{te}} \{\dot{\bar{\mathbf{t}}}\}^T \{\dot{\tilde{\mathbf{u}}}\}ds, \tag{71}$$

with

$$\Psi(\{\dot{\boldsymbol{\varepsilon}}\}) = \frac{1}{2}\{\dot{\boldsymbol{\varepsilon}}\}^T [\mathbf{D}_{ep}]\{\dot{\boldsymbol{\varepsilon}}\} \tag{72}$$

Assumed Fields

In the case of elastoplastic problems, Eq. (29) becomes:

$$\{\dot{\mathbf{u}}\} == \{\dot{\breve{\mathbf{u}}}\} + \sum_{i=1}^{m} \{\mathbf{N}_i\}^T \{\dot{\mathbf{c}}_i\} = \{\dot{\breve{\mathbf{u}}}\} + [\mathbf{N}]\{\dot{\mathbf{c}}\} \tag{73}$$

where $\{\dot{\mathbf{c}}_i\}$ stands for undetermined coefficient vector and $\{\dot{\breve{\mathbf{u}}}\}$ $(=\{\dot{\breve{u}}_1, \dot{\breve{u}}_2\}^T)$ and $[\mathbf{N}_i]$ are known functions. $\{\dot{\breve{\mathbf{u}}}\} = \{\dot{\breve{\mathbf{u}}}(\mathbf{x})\}$ and $\{\mathbf{N}_i\} = \{\mathbf{N}_i(\mathbf{x})\}$ in Eq. (73) should be chosen such that:

$$[\mathbf{L}]^T[\mathbf{D}][\mathbf{L}]\{\dot{\breve{\mathbf{u}}}\} + \{\dot{\mathbf{b}}^{ep}\} = 0 \text{ and } [\mathbf{L}]^T[\mathbf{D}][\mathbf{L}]\{\mathbf{N}_i\} = 0, \qquad (i = 1, 2, \ldots, m) \tag{74}$$

everywhere in Ω_e. A complete system of homogeneous solutions $\{\mathbf{N}_i\}$ is given in Eqs (37)-(40).

The particular solution $\{\dot{\breve{\mathbf{u}}}\}$ can be obtained by means of its source function. The source function corresponding to Eq. (66) was given in [106] as:

$$u_{ij}^{*}(r_{pq}) = \frac{1+\nu}{4\pi E}[-(3-\nu)\delta_{ij}\ln r_{pq} + (1+\nu)r_{pq,i}r_{pq,j}] \tag{75}$$

where $r_{pq} = [(x_q - x_p)^2 + (y_q - y_p)^2]^{1/2}$, $u_{ij}^{*}(r_{pq})$ denotes the ith component of displacement at the field point q of the solid under consideration when a unit point force is applied in the jth direction at the source point p. Using this source function, the particular solution can be expressed by:

$$\{\dot{\breve{\mathbf{u}}}\} = \begin{Bmatrix} \dot{\breve{u}}_1 \\ \dot{\breve{u}}_2 \end{Bmatrix} = \iint_{\Omega} \dot{b}_j^{ep} \begin{Bmatrix} u_{1j}^{*} \\ u_{2j}^{*} \end{Bmatrix} d\Omega \tag{76}$$

The unknown coefficient $\{\dot{\mathbf{c}}\}$ may be calculated from conditions on the external boundary and/or the continuity conditions on the inter-element boundary. Thus various Trefftz-element models can be obtained by using different approaches to enforce these conditions. In the majority of them a hybrid technique is used, whereby the elements are linked through an auxiliary conforming displacement frame which has the same form as in the conventional FE method. This means that in the Trefftz FE approach, a conforming displacement field should be independently defined on the element boundary to enforce the field continuity between elements and also to link the coefficient $\dot{\mathbf{c}}$, appearing in Eq. (73), with nodal displacement $\{\dot{\mathbf{d}}\}$. The frame is defined as:

$$\{\dot{\tilde{\mathbf{u}}}(\mathbf{x})\} = [\tilde{\mathbf{N}}(\mathbf{x})]\{\dot{\mathbf{d}}\}, \qquad (\mathbf{x} \in \Gamma_e) \tag{77}$$

where $\{\dot{\mathbf{d}}\} = \{\dot{\mathbf{d}}(\dot{\mathbf{c}})\}$ standing for the vector of the nodal displacements which are the final unknowns of the finite element formulation.

In the development of the present p-element, the following assumptions are adopted. First of all, the problem is assumed to be a plane strain (or plane stress) problem of elastoplasticity. Secondly, the element may be of a general quadrilateral shape or a triangular shape with two DOF (u_1, u_2) at each corner node (see Fig. **4**). Thirdly, to achieve higher order variations, an optional number of extra hierarchic modes is introduced along with the hierarchic DOF, a_{ci} for $\tilde{u}_1$, b_{ci} for $\tilde{u}_2$, which are conveniently associated with the mid-side node C (see Fig. **4**). Thus, along the side A-C-B of a particular element (see Fig. **4**), a simple interpolation of the frame displacement field can be given in the form:

$$\{\dot{\widetilde{\mathbf{u}}}_{AB}\} = \begin{Bmatrix} \dot{\widetilde{u}}_{1AB} \\ \dot{\widetilde{u}}_{2AB} \end{Bmatrix} = \begin{bmatrix} \widetilde{N}_1 & 0 & \widetilde{N}_2 & 0 \\ 0 & \widetilde{N}_1 & 0 & \widetilde{N}_2 \end{bmatrix} \{\dot{\mathbf{d}}_{AB}\} + \sum_{i=1}^{M} \gamma^{i-1} R_i \begin{Bmatrix} a_{ci} \\ b_{ci} \end{Bmatrix} \quad \mathbf{(78)}$$

where M is the order of the hierarchic DOF, and $\{\dot{\mathbf{d}}_{AB}\} = \{\dot{u}_{1A}\ \dot{u}_{2A}\ \dot{u}_{1B}\ \dot{u}_{2B}\}^T$.

$$\widetilde{N}_1 = \frac{1-\xi}{2}, \quad \widetilde{N}_2 = \frac{1+\xi}{2}, \quad R_i = \xi^{i-1}(1-\xi^2) \quad \mathbf{(79)}$$

and where ξ is defined in Fig. (**4**).

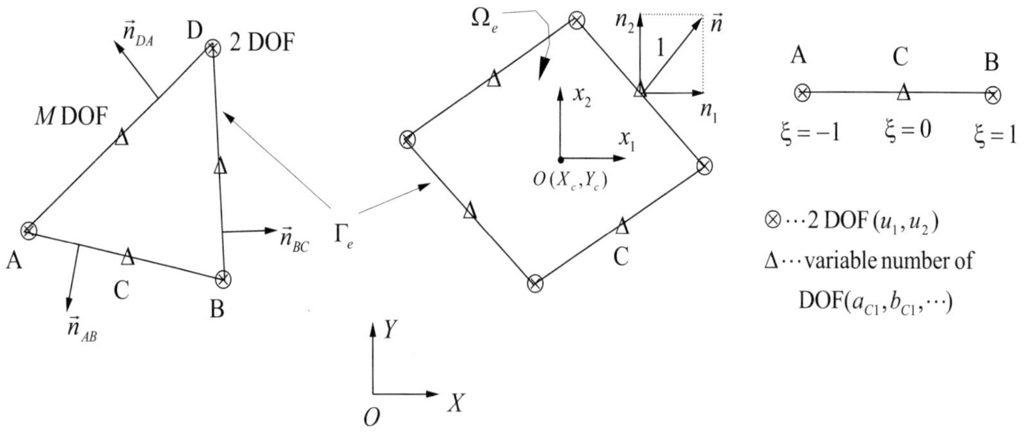

Fig. (4). The Trefftz p-element in plane elastoplasticity.

The coefficient γ is equal to +1 or -1 depending on the orientation of the side A-C-B (see Fig. **4**) in the global coordinate system (X,Y):

$$\gamma = \begin{cases} +1 \ \text{if}\ X_B - X_A \le Y_B - Y_A \\ -1 \ \text{if}\ X_B - X_A > Y_B - Y_A \end{cases} \quad \mathbf{(80)}$$

The purpose of using the coefficient γ is to ensure an unequivocal definition of the frame functions $\widetilde{\mathbf{u}}$ in terms of parameters a_{ci} and b_{ci} and common to two elements sharing the mid-side node C.

The tractions $\{\dot{\mathbf{t}}_0\} = \{\dot{t}_{10}, \dot{t}_{20}\}^T$ appearing in Eq. (67) can be derived from Eqs (56), (57), (73) and (80), and denote:

$$\{\dot{\mathbf{t}}_0\} = [\dot{\boldsymbol{\sigma}}^e]\{\mathbf{n}\} = [\mathbf{Q}]\{\dot{\mathbf{c}}\} + [\dot{\check{\mathbf{T}}}] \tag{81}$$

Element Stiffness Matrix

The element matrix equation can be generated by setting $\delta\Psi_{me} = 0$. To simplify the derivation, we first transform all domain integrals in Eq. (70) into boundary ones. In fact, by reason of the solution properties of the intra-element trial functions, the functional Ψ_{me} can be simplified to:

$$\begin{aligned}\Psi_{me} = &-\frac{1}{2}\int_{\Omega_e}\{\mathbf{b}\}^T\{\dot{\mathbf{u}}\}d\Omega + \frac{1}{2}\int_{\Gamma_e}\{\dot{\mathbf{t}}_0\}^T\{\dot{\mathbf{u}}\}ds - \int_{\Gamma_{te}}(\{\dot{\bar{\mathbf{t}}}_0\}^T\{\dot{\mathbf{u}}\}ds - \int_{\Gamma_{Ie}}\{\dot{\mathbf{t}}_0\}^T\{\dot{\tilde{\mathbf{u}}}\}ds \\ &- \int_{\Gamma_{ue}}\{\dot{\mathbf{t}}_0\}^T(\{\dot{\tilde{\mathbf{u}}}\} - \{\dot{\mathbf{u}}\})ds + \frac{1}{2}\int_{\Gamma_e}\{\dot{\mathbf{t}}^P\}^T\{\dot{\mathbf{u}}\}ds - \int_{\Gamma_{Ie}}\{\dot{\mathbf{t}}^P\}^T\{\dot{\tilde{\mathbf{u}}}\}ds + \int_{\Gamma_{ue}}\{\dot{\mathbf{t}}^P\}^T\{\dot{\mathbf{u}}\}ds \\ &- \int_{\Gamma_{te}}(\{\dot{\mathbf{t}}^P\}^T\{\dot{\mathbf{u}}\}ds\end{aligned} \tag{82}$$

Substituting the expressions given in Eqs (73) and (77) into (81) produces:

$$\Psi_{me} = \frac{1}{2}\{\dot{\mathbf{c}}\}^T[\mathbf{H}]\{\dot{\mathbf{c}}\} + \{\dot{\mathbf{c}}\}^T[\mathbf{S}]\{\dot{\mathbf{d}}\} + \{\dot{\mathbf{c}}\}^T\{\mathbf{r}_1\} + \{\dot{\mathbf{d}}\}^T\{\mathbf{r}_2\} + \text{terms without } \{\dot{\mathbf{c}}\} \text{ or } \{\dot{\mathbf{d}}\} \tag{83}$$

in which the matrices [**H**], [**S**] and the vectors $\{\mathbf{r}_1\}$, $\{\mathbf{r}_2\}$ are:

$$[\mathbf{H}] = \int_{\Gamma_e}[\mathbf{Q}]^T[\mathbf{N}]ds$$

$$[\mathbf{S}] = -\int_{\Gamma_{ue}}[\mathbf{Q}]^T[\tilde{\mathbf{N}}]ds - \int_{\Gamma_{Ie}}[\mathbf{Q}]^T[\tilde{\mathbf{N}}]ds$$

$$\begin{aligned}\{\mathbf{r}_1\} = &-\frac{1}{2}\int_{\Omega_e}[\mathbf{N}]^T\{\dot{\mathbf{b}}\}d\Omega + \frac{1}{2}\int_{\Gamma_e}[[\mathbf{Q}]^T\{\dot{\mathbf{u}}\} + [\mathbf{N}]^T(\{\dot{\check{\mathbf{T}}}\} + \{\dot{\mathbf{t}}^P\})]ds \\ &+ \int_{\Gamma_{ue}}[\mathbf{Q}]^T\{\dot{\bar{\mathbf{u}}}\}ds - \int_{\Gamma_{te}}[\mathbf{N}]^T\{\dot{\bar{\mathbf{t}}}\}ds\end{aligned}$$

$$\{\mathbf{r}_2\} = -\int_{\Gamma_{ue}} [\tilde{\mathbf{N}}]^T \{\dot{\bar{\mathbf{T}}}\} ds - \int_{\Gamma_{Ie}} [\tilde{\mathbf{N}}]^T (\{\dot{\bar{\mathbf{T}}}\} + \{\dot{\mathbf{t}}^P\}) ds \qquad \textbf{(84)}$$

where $[\dot{\bar{\mathbf{t}}}_0\} = \{\dot{\bar{t}}_{10}, \dot{\bar{t}}_{20}\}^T$ is the prescribed traction vector.

To enforce inter-element continuity on the common element boundary, the unknown vector $\{\dot{\mathbf{c}}\}$ should be expressed in terms of nodal DOF $\{\dot{\mathbf{d}}\}$. An optional relationship between $\{\dot{\mathbf{c}}\}$ and $\{\dot{\mathbf{d}}\}$ in the sense of variation can be obtained from:

$$\frac{\partial \Psi_{me}}{\partial \{\dot{\mathbf{c}}\}^T} = [\mathbf{H}]\{\dot{\mathbf{c}}\} + [\mathbf{S}]\{\dot{\mathbf{d}}\} + \{\mathbf{r}_1\} = 0. \qquad \textbf{(85)}$$

This leads to:

$$\{\dot{\mathbf{c}}\} = -[\mathbf{G}]\{\dot{\mathbf{d}}\} - \{\mathbf{g}\}, \qquad \textbf{(86)}$$

where $[\mathbf{G}] = [\mathbf{H}]^{-1}[\mathbf{S}]$ and $\{\mathbf{g}\} = [\mathbf{H}]^{-1}\{\mathbf{r}_1\}$, and then straightforwardly yields the expression of Ψ_{me} only in terms of $\{\dot{\mathbf{d}}\}$ and other known matrices:

$$\Psi_{me} = -\frac{1}{2}\{\dot{\mathbf{d}}\}^T[\mathbf{G}]^T[\mathbf{H}][\mathbf{G}]\{\dot{\mathbf{d}}\} + \{\dot{\mathbf{d}}\}^T(\{\mathbf{r}_2\} - [\mathbf{G}]^T\{\mathbf{r}_1\}) + \text{terms without } \{\dot{\mathbf{d}}\} \qquad \textbf{(87)}$$

Therefore, the element stiffness matrix equation can be obtained by taking the vanishing variation of the functional Ψ_{me} as:

$$\frac{\partial \Psi_{me}}{\partial \dot{\mathbf{d}}^T} = 0 \Rightarrow \quad \mathbf{K}\dot{\mathbf{d}} = \dot{\mathbf{P}} = \dot{\mathbf{P}}_0 + \mathbf{P}(\{\dot{\boldsymbol{\sigma}}^P\}) \qquad \textbf{(88)}$$

where $[\mathbf{K}]=[\mathbf{G}]^T[\mathbf{H}][\mathbf{G}]$ and $\{\mathbf{P}\}=-[\mathbf{G}]^T\{\mathbf{r}_1\}+\{\mathbf{r}_2\}$ are, respectively, the element stiffness matrix and the equivalent nodal force vector. The expression (88) is the elemental stiffness matrix equation for Trefftz FE analysis. $[\mathbf{K}]$ and $\{\dot{\mathbf{P}}_0\}$ can be calculated in the usual way, while $\{\mathbf{P}(\{\dot{\boldsymbol{\sigma}}^P\})\}$ contains unknown $\dot{\sigma}_{ij}^P$. An iterative procedure is thus required. The procedure is briefly described in the next section.

Moreover, to ensure good numerical conditioning during the inversion of matrix $[\mathbf{H}]$, the homogeneous solution $\{\mathbf{N}_i\}$ in Eq. (73) must be expressed in terms of suitably scaled local coordinates (x_1, x_2) originating at the element centroid (Fig. **4**):

$$x_1 = (X - X_c)/a, \quad x_2 = (Y - Y_c)/a \tag{89}$$

where X_c and Y_c stand for global coordinates at the centroid of the element and a is the average distance between the centroid and node i of the element:

$$a = \sum_{i}^{N} (x_{1i}^2 + x_{2i}^2)^{1/2} / N \tag{90}$$

and where N is the number of nodes for the element under consideration.

It should also be pointed out that the standard hybrid-Trefftz formulation implies that all terms representing the rigid body modes (three in-plane problems) have been discarded from the intra-element field to prevent the matrix [**H**] from being singular. However, after the finite element assembly has been solved for nodal displacements, those missing terms can easily be recovered. The intra-element field $\{\dot{\mathbf{u}}\}$ in an element may be augmented by the rigid body modes:

$$\{\dot{\mathbf{u}}_e\} = \{\dot{\breve{\mathbf{u}}}_e\} + [\mathbf{N}_e]\{\dot{\mathbf{c}}_e\} + \begin{bmatrix} 1 & 0 & x_2 \\ 0 & 1 & -x_1 \end{bmatrix} \{\dot{\mathbf{c}}_0\} \tag{91}$$

where 'e' stands for element e. Using a least-square procedure to match the nodal displacements $(\tilde{u}_{1i}, \tilde{u}_{2i})$ at corner nodes:

$$\sum_{i=1}^{N} [(\dot{\tilde{u}}_{1i} - \dot{u}_{1i})^2 + (\dot{\tilde{u}}_{2i} - \dot{u}_{2i})^2] = \min \text{ (with respect to } \{\dot{\mathbf{c}}_0\}) \tag{92}$$

Equation (92) provides

$$\{\dot{\mathbf{c}}_0\} = [\mathbf{R}]^{-1}\{\mathbf{r}\} \tag{93}$$

with

$$[\mathbf{R}] = \sum_{i=1}^{N} \begin{bmatrix} 1 & 0 & x_{2i} \\ 0 & 1 & -x_{1i} \\ x_{2i} & -x_{1i} & x_{1i}^2 + x_{2i}^2 \end{bmatrix}, \quad \{\mathbf{r}\} = \sum_{i=1}^{N} \begin{bmatrix} \dot{\tilde{u}}_{1i} - \dot{u}_{1i} \\ \dot{\tilde{u}}_{2i} - \dot{u}_{2i} \\ (\dot{\tilde{u}}_{1i} - \dot{u}_{1i})x_{2i} - (\dot{\tilde{u}}_{2i} - \dot{u}_{2i})x_{1i} \end{bmatrix} \tag{94}$$

Determination of Elastoplastic Status and Iterative Scheme

Determination of Elemental Stress Status

To determine the stress status of an element at any loading step, say the *i*th step, assuming that $\{\boldsymbol{\sigma}\}_i$ stands for the stress at the *i*th loading step, if $F(\{\boldsymbol{\sigma}\}_i) < 0$, the element is said to be in an elastic condition. Then, the stress increment at the *i*th loading step can be calculated by:

$$\{\boldsymbol{\sigma}\}_i = \{\boldsymbol{\sigma}\}_{i-1} + \{\dot{\boldsymbol{\sigma}}\}_i = \{\boldsymbol{\sigma}\}_{i-1} + [D_e]\{\dot{\boldsymbol{\varepsilon}}\}_i \tag{95}$$

However, if $F(\{\boldsymbol{\sigma}\}_{i-1}) < 0$ and $F(\{\boldsymbol{\sigma}\}_i) > 0$, the element is said to experience a transition from elastic to plastic conditions during the loading step. Let $\xi\{\dot{\boldsymbol{\sigma}}\}_i$ be the proportion of the stress increment at which the plastic behaviour is encountered [107]. Then *r* can be determined from:

$$F(\{\boldsymbol{\sigma}\}_{i-1} + \xi\{\dot{\boldsymbol{\sigma}}\}_i) = 0 \tag{96}$$

Solving Eq. (96) yields:

$$\xi = \frac{-B + \sqrt{B^2 - 4AC}}{2A} \tag{97}$$

where

$$\begin{aligned} A &= \frac{1}{2}[(\dot{s}_{11}^e)_i^2 + (\dot{s}_{11}^e)_i^2 + 2(\dot{s}_{12}^e)_i^2], \\ B &= (\dot{s}_{11}^e)_i(s_{11})_i + (\dot{s}_{22}^e)_i(s_{22})_i + 2(\dot{s}_{12}^e)_i(s_{12})_i \\ C &= (s_{11})_i^2 + (s_{22})_i^2 + 2(s_{12})_i^2 - \frac{2}{3}(\sigma_s)_i^2 \end{aligned} \tag{98}$$

Having determined the parameter ξ, the stress increment $\{\dot{\boldsymbol{\sigma}}\}_i$ is calculated from:

$$\{\dot{\boldsymbol{\sigma}}\}_i = (\xi[\mathbf{D}_e] + (1-\xi)[\mathbf{D}_{ep}])\{\dot{\boldsymbol{\varepsilon}}\}_i \tag{99}$$

Iterative Scheme

The steps of the iterative procedure in this analysis are described as follows:

At the first loading step, assume $\mathbf{P}(\{\dot{\boldsymbol{\sigma}}^p\}) = 0$. Then solve Eq. (88) for $\{\dot{\mathbf{d}}\}$ and calculate the parameter ξ in section 6.1, $\{\dot{\boldsymbol{\sigma}}^e\}_i$ and $\{\dot{\boldsymbol{\sigma}}^p\}_i$ using Eq. (65) as well as $\mathbf{P}(\{\dot{\boldsymbol{\sigma}}^p\})$.

Suppose that $\{\mathbf{d}\}^{(k)}$, $\{\boldsymbol{\sigma}\}^{e(k)}$ and $\{\boldsymbol{\sigma}\}^{p(k)}$ stand for the kth approximations, which can be obtained from the proceeding cycle of iteration. The $(k+1)^{\text{th}}$ solution may be calculated as:

(1) Apply a load incrementand take as initial values; $\{\mathbf{d}\}^{(k)}, \{\boldsymbol{\sigma}^e\}^{(k)}$ and $\{\boldsymbol{\sigma}^p\}^{(k)}$
(2) Enter the iterative cycle for i=1,2,.... Calculate the stress increments in all elements using the formulation in Appendix A. Calculate the total stress and compile a list of yielded elements. Calculate the correct stresses in the elastoplastic elements using Eq. (65));
(3) Calculate in Eq. (88) using the current value of, where the superscript $\mathbf{P}(\{\dot{\boldsymbol{\sigma}}^p\})$ [i-1] stands for the increments at the (i-1)th iterative cycle. Solve Eq. (88) for $\{\dot{\mathbf{d}}\}^{[i]}$; $\{\dot{\mathbf{d}}\}^{[i-1]}$ and $\{\dot{\boldsymbol{\sigma}}^p\}^{[i-1]}$
(4) $\eta_i = [(\{\dot{\mathbf{d}}\}^{[i]})^T \{\dot{\mathbf{d}}\}^{[i]} - (\{\dot{\mathbf{d}}\}^{[i-1]})^T \{\dot{\mathbf{d}}\}^{[i-1]}] / (\{\dot{\mathbf{d}}\}^{[i-1]})^T \{\dot{\mathbf{d}}\}^{[i-1]} \leq \eta$ If (η is a convergence tolerance), proceed to the next loading step and calculate

$$\begin{aligned} \{\mathbf{d}\}^{(k+1)} &= \{\mathbf{d}\}^{(k+1)} + \{\dot{\mathbf{d}}\}^{[i]} \\ \{\boldsymbol{\sigma}\}^{(k+1)} &= \{\boldsymbol{\sigma}\}^{(k)} + \{\dot{\boldsymbol{\sigma}}\}^{[i]} \\ \{\boldsymbol{\sigma}^p\}^{(k+1)} &= \{\boldsymbol{\sigma}^p\}^{(k)} + \{\dot{\boldsymbol{\sigma}}^p\}^{[i]} \end{aligned} \tag{100}$$

otherwise, go back to step (2).

It is noted that the required stiffness matrices appearing in Eq. (88) do not change through each step of computation. Hence, once the matrix [**K**] has been formed, they can be stored in the core and used in each cycle of iteration without any change. Obviously, this can save a large amount of computing time.

CONTACT PROBLEM

In this section, we address a direct constraint technique for solving elastic contact problems without friction. Beginning with a discussion on contact kinematics, we introduce a substructure interface model for establishing HT FEM formulation for contact problems. For efficiency, static condensation is employed to condense a large model down to a smaller one which involves nodes within the potential contact surface only. This model can reduce computational time and effort remarkably. Subsequently, the contact interface equation is constructed by introducing the contact conditions of compatibility and equilibrium. The discussion is based on the work in [85].

Contact Kinematics

Consider two linear elastic bodies involving contact, which are denoted in their reference configuration $\Omega^{(\beta)}$, β=1, 2 bounded by their boundaries $\Gamma^{(\beta)}$ (see Fig. **5**). The boundaries of each contacting body are composed of three disjoint parts: $\Gamma_u^{(\beta)}$, where displacement is prescribed, $\Gamma_t^{(\beta)}$, where traction conditions are imposed, and $\Gamma_c^{(\beta)}$, which is the potential contact surface taken to be sufficiently large to cover the actual contact area under loading.

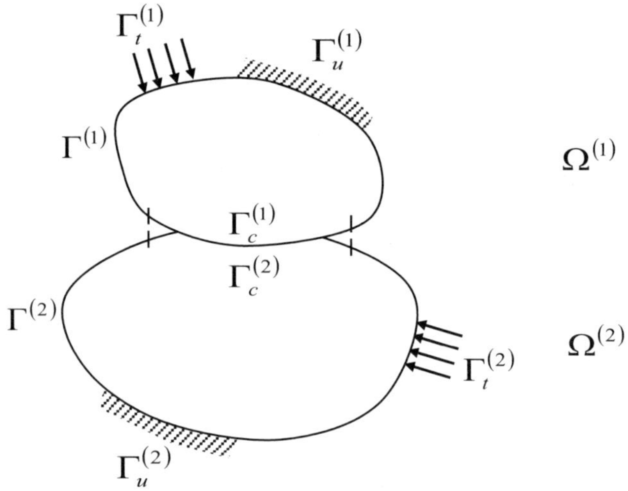

Fig. (5). Two bodies in contact.

The classical field equations of plane elasticity are governed by Eqs (1)-(5).

For contact problems, Eqs (1-5) should also be subject to the kinematical contact conditions. The potential contact region denoted by $\Gamma_c = \Gamma_c^{(1)} \cap \Gamma_c^{(2)}$ is where the two bodies meet, as shown in Fig. (**5**). It is worthy of note that the determination of the common normal unit vector **n** on the contact region Γ_c plays a significant role in the contact analysis. For this purpose, the expression of **n** is presented in the deformation configuration below (Eq. (103). The discretizations performed on both bodies in the potential contact zone match, so that the contact behaviour occurs just between each pair of opposite nodes $i^{(\beta)}$ (called the contact node pair, CNP for short), as shown in Fig. (**6**). Unit normal vectors, $\mathbf{n}_1$ and $\mathbf{n}_2$, can be constructed through each contact node $i^{(\beta)}$ on bodies 1 and 2, to the straight element sides $\Gamma^{(\beta)}_{c_{i\_-i}}$ respectively. Thus, the subtraction of $\mathbf{n}_2$ and $\mathbf{n}_1$ is then taken as an approximate common normal $\mathbf{n}_{21}$ [108].

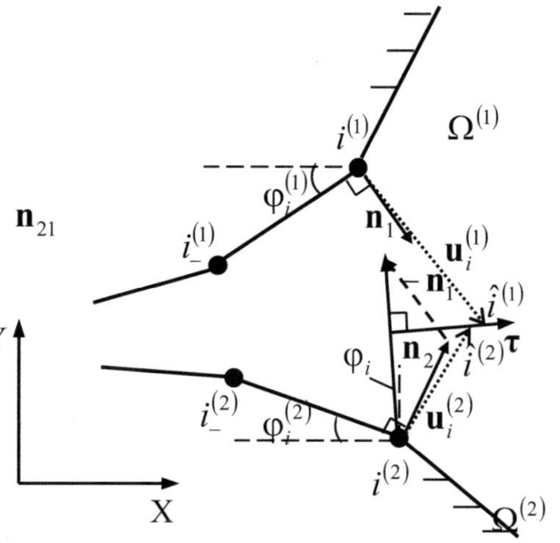

Fig. (6). Definition of contact coordinate system.

When the nodes $i^{(\beta)}$ on opposite surfaces come into contact under loading, their projections onto the common normal $\mathbf{n}_{21}$ should overlap. Obviously, the straight element sides $i_-^{(\beta)}$- $i^{(\beta)}$ (see Fig. **6**) pertinent to each particular body make the angles to the X axis equal to $\varphi_i^{(\beta)}$:

$$\varphi_i^{(\beta)} = \operatorname{arctg}\left|\frac{Y_i^{(\beta)} - Y_{i\_}^{(\beta)}}{X_i^{(\beta)} - X_{i\_}^{(\beta)}}\right| \tag{101}$$

Accordingly, the common normal $\mathbf{n}_{21}$ reads:

$$\begin{aligned}\mathbf{n}_{21} = \mathbf{n}_2 - \mathbf{n}_1 &= \sum_\beta (-1)^\beta \mathbf{n}_\beta \\ = \sum_\beta \cos\left(\frac{\pi}{2} + (-1)^{\beta-1}\varphi_i^{(\beta)}\right)\mathbf{e}_1 &+ \sin\left(\frac{\pi}{2} + (-1)^{\beta-1}\varphi_i^{(\beta)}\right)\mathbf{e}_2 \\ = 2\cos\frac{1}{2}\sum_\beta \varphi_i^{(\beta)}\left[\sum_\beta \cos\frac{1}{2}\left(\pi + \sum_\beta (-1)^{\beta-1}\varphi_i^{(\beta)}\right)\mathbf{e}_1 \right. &\left. + \sin\frac{1}{2}\left(\pi + \sum_\beta (-1)^{\beta-1}\varphi_i^{(\beta)}\right)\mathbf{e}_2\right]\end{aligned} \tag{102}$$

Then, the common normal unit vector $\mathbf{n}$ is conveniently given for CNP i by the expression:

$$\mathbf{n} = \frac{\mathbf{n}_{21}}{|\mathbf{n}_{21}|} = \cos\left(\frac{\pi}{2} + \frac{\varphi_i^{(2)} - \varphi_i^{(1)}}{2}\right)\mathbf{e}_1 + \sin\left(\frac{\pi}{2} + \frac{\varphi_i^{(2)} - \varphi_i^{(1)}}{2}\right)\mathbf{e}_2$$
$$= \cos\left(\frac{\pi}{2} + \varphi_i\right)\mathbf{e}_1 + \sin\left(\frac{\pi}{2} + \varphi_i\right)\mathbf{e}_2 = \cos\theta_i \mathbf{e}_1 + \sin\theta_i \mathbf{e}_2 \quad \textbf{(103)}$$

or in vector form:

$$\mathbf{n} = \{\cos\theta_i \quad \sin\theta_i\}^{\mathrm{T}} \quad \textbf{(104)}$$

Where $\mathbf{e}_1$ and $\mathbf{e}_2$ represent the unit vectors parallel to the X and Y axes respectively, and φ_i and θ_i stand for the angles from $\mathbf{n}$ to $\mathbf{e}_1$ and from $\mathbf{e}_2$ to $\mathbf{n}$ respectively.

The contact conditions normally considered for the frictionless problems studied herein can be grouped into compatibility conditions of normal displacements, and equilibrium conditions of forces, as shown here.

(a) Compatibility Condition:

$$\left(\mathbf{u}_i^{(1)} - \mathbf{u}_i^{(2)}\right)^{\mathrm{T}} \cdot \mathbf{n} + g_{0ni} = 0 \quad \textbf{(105)}$$

where the normal initial gap at CNP i, g_{0ni}, is defined in terms of the initial geometry, and

$$\mathbf{u}_i^{(\alpha)} = \{u_{iX} \quad u_{iY}\}^{\mathrm{T}} \quad \textbf{(106)}$$

$$g_{0ni} = \left(X_i^{(1)} - X_i^{(2)}\right)\cos\theta_i + \left(Y_i^{(1)} - Y_i^{(2)}\right)\sin\theta_i \quad \textbf{(107)}$$

Substitution of Eqs (106) and (107) into Eq. (105) leads to:

$$\left(u_{iX}^{(1)} - u_{iX}^{(2)}\right)\mathrm{tg}\,\varphi_i - u_{iY}^{(1)} + u_{iY}^{(2)} = \hat{g}_i \quad \textbf{(108)}$$

where

$$\hat{g}_i = Y_i^{(1)} - Y_i^{(2)} - \left(X_i^{(1)} - X_i^{(2)}\right)\mathrm{tg}\,\varphi_i \quad \textbf{(109)}$$

(b) Equilibrium Condition:

By the principle of action and reaction, the contact forces acting on CNP i must be equal and opposite, namely:

$$\mathbf{f}_i^{(1)} + \mathbf{f}_i^{(2)} = 0 \qquad \textbf{(110)}$$

and in terms of their components:

$$f_{iX}^{(1)} + f_{iX}^{(2)} = 0, \quad f_{iY}^{(1)} + f_{iY}^{(2)} = 0 \qquad \textbf{(111a, b)}$$

where

$$\mathbf{f}_i^{(\alpha)} = \{f_{iX} \quad f_{iY}\}^{\mathrm{T}} \qquad \textbf{(112)}$$

In addition, it is necessary that in the case of adhesive contact:

$$\mathbf{f}_i^{(1)} \cdot \boldsymbol{\tau} = \mathbf{f}_i^{(2)} \cdot \boldsymbol{\tau} = 0 \qquad \textbf{(113)}$$

or

$$f_{iX}^{(1)} \cos\varphi_i + f_{iY}^{(1)} \sin\varphi_i = 0, \quad f_{iX}^{(2)} \cos\varphi_i + f_{iY}^{(2)} \sin\varphi_i = 0 \qquad \textbf{(114a, b)}$$

where denotes the tangential direction at CNP i.

Substructure Interface Model

The Trefftz FEM used for the contact problem is the same as that in Eqs (28)-(41). In Eq. (29), it is important to note that the necessary but not sufficient condition for the element flexibility matrix **H** appearing in Eq. (29) to have full rank is:

$$m > \mathrm{NDOF} - 3 \qquad \textbf{(115)}$$

where m is the number of Trefftz functions (see Eq. (29)) and NDOF is the number of nodal DOF on the element. It has been verified from many practical applications that the choices of m=7, 9, or 11 and higher often yield robust and stable results. For the best compromise between accuracy and computational effort, however, the value of m=9 is selected in the present study. On the other hand, starting with k=1 and relations (38), the sequence generates an independent displacement pattern. As a consequence, the explicit 2×9 matrix $\mathbf{N}_e$ (see Eq. (29)) involving the 9 homogeneous solutions can be obtained as:

$$\mathbf{N}_e=\frac{1}{2G}\begin{bmatrix}(\kappa-1)x & y & -x & -2\kappa xy & (\kappa-2)x^2-(\kappa+2)y^2\\ (\kappa-1)y & x & y & (\kappa+2)x^2-(\kappa-2)y^2 & 2\kappa xy\end{bmatrix}$$

$$\begin{matrix}2xy & y^2-x^2 & (3-3\kappa)x^2y+(\kappa+3)y^3 & (\kappa-3)x^3-(3\kappa+3)xy^2\\ x^2-y^2 & 2xy & (\kappa+3)x^3-(3\kappa-3)xy^2 & (3\kappa+3)x^2y-(\kappa-3)y^3\end{matrix} \quad \textbf{(116)}$$

where "*e*" means the quantity associated with element *e*. Using Eqs (41) and (81), the corresponding matrices $\breve{\boldsymbol{\sigma}}_e$, $\mathbf{T}_e$, $\breve{\mathbf{t}}_e$, and $\mathbf{Q}_e$ can be given explicitly as:

$$\breve{\boldsymbol{\sigma}}_e=\begin{Bmatrix}0\\0\\ \bar{b}_x y+\bar{b}_y x\end{Bmatrix} \quad \textbf{(117a)}$$

$$\mathbf{T}_e=\begin{bmatrix}2 & 0 & -1 & -6y & 2x & 2y & -2x & -12xy & -12y^2\\ 2 & 0 & 1 & -2y & 6x & -2y & 2x & -12xy & 12x^2\\ 0 & 1 & 0 & 2x & -2y & 2x & 2y & 6(x^2+y^2) & 0\end{bmatrix} \quad \textbf{(117b)}$$

$$\breve{\mathbf{t}}_e=(\bar{b}_x y+\bar{b}_y x)\begin{Bmatrix}n_y\\ n_x\end{Bmatrix} \quad \textbf{(118a)}$$

$$\mathbf{Q}_e=\begin{bmatrix}2n_x & n_y & -n_x & 2n_y x-6n_x y & 2n_x x-2n_y y & 2n_y x+2n_x y\\ 2n_y & n_x & n_y & 2n_x x-2n_y y & 6n_y x-2n_x y & 2n_x x-2n_y y\end{bmatrix}$$

$$\begin{matrix}2n_y y-2n_x x & 6n_y(x^2+y^2)-12n_x xy & -12n_x y^2\\ 2n_y x+2n_x y & 6n_x(x^2+y^2)-12n_y xy & 12n_y x^2\end{matrix} \quad \textbf{(118b)}$$

In implementation of the HT FE model, the Trefftz dimensionless coordinate system:

$$\xi=\frac{x}{a},\ \eta=\frac{y}{a} \quad \textbf{(119)}$$

together with

$$a = \frac{1}{n}\sum_{i=1}^{n}\sqrt{x_i^2 + y_i^2} \tag{120}$$

must be exploited to replace *x, y* in matrices $\mathbf{H}_e$, $\mathbf{T}_e$, and $\mathbf{Q}_e$ to prevent the element flexibility matrix $\mathbf{H}_e$ from overflow. Here, *a* denotes the average distance between the element nodes and its centroid.

Using a similar technique to that in the conventional FE model, the overall frame function matrix $\tilde{\mathbf{N}}_e$ can be directly defined and assembled in positions appropriate to the node numbers of the element. For example, a simple interpolation function on the side 1-2 of a particular element (Fig. **1**) can be given in the form:

$$\tilde{\mathbf{N}}_{12} = \begin{bmatrix} \tilde{N}_1 & 0 & \tilde{N}_2 & 0 \\ 0 & \tilde{N}_1 & 0 & \tilde{N}_2 \end{bmatrix} \tag{121}$$

where

$$\tilde{N}_1 = \frac{1}{2}(1-\zeta),\ \tilde{N}_2 = \frac{1}{2}(1+\zeta) \tag{122}$$

To obtain the expression of $\tilde{\mathbf{N}}_e$, we introduce a new function $\psi_{ij}(\zeta)$[19] such that:

$$\psi_{ij}(\zeta) = \begin{cases} 1 & \text{when } \zeta \in \text{side } i-j \\ 0 & \text{otherwise} \end{cases} \tag{123}$$

and then $\tilde{\mathbf{N}}_e$ can be given as:

$$\tilde{\mathbf{N}}_e = \begin{bmatrix} \tilde{N}_1^* & 0 & \tilde{N}_2^* & 0 & \tilde{N}_3^* & 0 & \tilde{N}_4^* & 0 \\ 0 & \tilde{N}_1^* & 0 & \tilde{N}_2^* & 0 & \tilde{N}_3^* & 0 & \tilde{N}_4^* \end{bmatrix} \tag{124}$$

Where $\tilde{N}_1^* = \psi_{12}\tilde{N}_1 + \psi_{41}\tilde{N}_2$, $\tilde{N}_2^* = \psi_{23}\tilde{N}_1 + \psi_{12}\tilde{N}_2$, $\tilde{N}_3^* = \psi_{34}\tilde{N}_1 + \psi_{23}\tilde{N}_2$ and $\tilde{N}_4^* = \psi_{41}\tilde{N}_1 + \psi_{34}\tilde{N}_2$.

Substituting Eqs (9), (12), and (33) into the modified variational functional (123), we can obtain Eqs (24)-(28), in which:

$$\mathbf{K}_e = \mathbf{G}_e^{\mathrm{T}} \mathbf{H}_e^{-1} \mathbf{G}_e \quad \textbf{(125)}$$

is the element stiffness matrix, and

$$\mathbf{P}_e = \mathbf{G}_e^{\mathrm{T}} \mathbf{H}_e \mathbf{g}_e + \mathbf{h}_e \quad \textbf{(126)}$$

is the vector of the equivalent nodal forces. The auxiliary matrices $\mathbf{H}_e$, $\mathbf{G}_e$, $\mathbf{h}_e$, and $\mathbf{g}_e$ are explicitly expressed as:

$$\mathbf{H}_e = \int_{\Gamma_e} \mathbf{Q}_e^{\mathrm{T}} \mathbf{N}_e \, \mathrm{d}\Gamma = \int_{\Gamma_e} \mathbf{N}_e^{\mathrm{T}} \mathbf{Q}_e \, \mathrm{d}\Gamma \quad \textbf{(127a)}$$

$$\mathbf{G}_e = \int_{\Gamma_{el}+\Gamma_{et}} \mathbf{Q}_e^{\mathrm{T}} \widetilde{\mathbf{N}}_e \, \mathrm{d}\Gamma \quad \textbf{(127b)}$$

$$\mathbf{h}_e = -\frac{1}{2}\int_{\Omega_e} \mathbf{N}_e^{\mathrm{T}} \overline{\mathbf{b}}_e \, \mathrm{d}\Omega - \frac{1}{2}\int_{\Gamma_e} (\mathbf{Q}_e^{\mathrm{T}} \breve{\mathbf{u}}_e + \mathbf{N}_e^{\mathrm{T}} \breve{\mathbf{t}}_e) \mathrm{d}\Gamma + \int_{\Gamma_{eu}} \mathbf{Q}_e^{\mathrm{T}} \overline{\mathbf{u}} \, \mathrm{d}\Gamma \quad \textbf{(127c)}$$

$$\mathbf{g}_e = \int_{\Gamma_{et}} \widetilde{\mathbf{N}}_e^{\mathrm{T}} (\overline{\mathbf{t}}_e - \breve{\mathbf{t}}_e) \mathrm{d}\Gamma - \int_{\Gamma_{el}} \widetilde{\mathbf{N}}_e^{\mathrm{T}} \breve{\mathbf{t}}_e \, \mathrm{d}\Gamma \quad \textbf{(127d)}$$

Once the element stiffness matrix $\mathbf{K}_e$ and the equivalent nodal forces $\mathbf{P}_e$ for all the elements have been evaluated, the structural stiffness equation can be assembled in the matrix form analogous to that adopted in conventional FEM [11]:

$$\mathbf{K}\mathbf{d} = \mathbf{P} \quad \textbf{(128)}$$

Contact Interface Equation

In accordance with Eq. (36), the structural HT FE stiffness equation for each contacting body can be formulated as:

$$\mathbf{K}^{(\beta)} \mathbf{d}^{(\beta)} = \mathbf{P}^{(\beta)} + \mathbf{f}^{(\beta)} \quad \textbf{(129)}$$

where $\mathbf{f}^{(\beta)}$ are contact forces. In most contact problems, the contact region is small compared to the domain of the structure. Further, the subdomains $\Omega^{(\beta)}$ associated with the HT FE model can be viewed as substructures. Therefore, static condensation [109], which is used to condense the model down to a smaller one which involves nodes within the potential contact surfaces only, would be efficient for contact analysis. After introducing the prescribed displacements into Eq. (129) and splitting the unknowns into two parts, one part (denoted by subscript *c*) containing unknowns associated with the potential contact surfaces and the other part (denoted by subscript *r*) containing the remaining unknowns, we arrive from (129) at the following form in terms of submatrices for each contacting body:

$$\begin{bmatrix} \mathbf{K}_{rr}^{(\beta)} & \mathbf{K}_{rc}^{(\beta)} \\ \mathbf{K}_{cr}^{(\beta)} & \mathbf{K}_{cc}^{(\beta)} \end{bmatrix} \begin{Bmatrix} \mathbf{d}_{r}^{(\beta)} \\ \mathbf{d}_{c}^{(\beta)} \end{Bmatrix} = \begin{Bmatrix} \mathbf{P}_{r}^{(\beta)} \\ \mathbf{P}_{c}^{(\beta)} \end{Bmatrix} + \begin{Bmatrix} \mathbf{0} \\ \mathbf{f}_{c}^{(\beta)} \end{Bmatrix} \tag{130}$$

The solution of the first submatrix equation for $\mathbf{d}_{r}^{(\beta)}$ yields:

$$\mathbf{d}_{r}^{(\beta)} = \mathbf{K}_{rr}^{(\beta)-1}\left(\mathbf{P}_{r}^{(\beta)} - \mathbf{K}_{rc}^{(\beta)}\mathbf{d}_{c}^{(\beta)}\right) \tag{131}$$

Since $\mathbf{d}_{r}^{(\beta)}$ depends only indirectly on $\mathbf{f}_{c}^{(\beta)}$, it can be eliminated by substituting (131) into the second submatrix equation of (130). This leads to a much smaller system of equations:

$$\mathbf{K}^{*(\beta)}\mathbf{d}_{c}^{(\beta)} - \mathbf{f}_{c}^{(\beta)} = \mathbf{P}^{*(\beta)} \tag{132}$$

where the following notations are used:

$$\mathbf{K}^{*(\beta)} = \mathbf{K}_{cc}^{(\beta)} - \mathbf{K}_{cr}^{(\beta)}\mathbf{K}_{rr}^{(\beta)-1}\mathbf{K}_{cr}^{(\beta)\mathrm{T}} \tag{133a}$$

$$\mathbf{P}^{*(\beta)} = \mathbf{P}_{c}^{(\beta)} - \mathbf{K}_{cr}^{(\beta)}\mathbf{K}_{rr}^{(\beta)-1}\mathbf{P}_{r}^{(\beta)} \tag{133b}$$

Here, $\mathbf{K}^{*(\beta)}$ and $\mathbf{P}^{*(\beta)}$ are obtained at an intermediate step of the Gaussian elimination process, resulting in a significant reduction in computational effort. The advantage of static condensation is that only a smaller system of equations must be solved during the contact analysis. Also, the calculation of matrices $\mathbf{K}^{*(\beta)}$ and $\mathbf{P}^{*(\beta)}$

needs to be done only once, and their results can be used many times in the contact solution process.

The incorporation of the potential contact surfaces $\Gamma_c^{(\beta)}$ is of the form:

$$\begin{bmatrix} \mathbf{K}^{*(1)} & -\mathbf{I} & \mathbf{0} & \mathbf{0} \\ \mathbf{0} & \mathbf{0} & \mathbf{K}^{*(2)} & -\mathbf{I} \end{bmatrix} \begin{Bmatrix} \mathbf{d}_c^{(1)} \\ \mathbf{f}_c^{(1)} \\ \mathbf{d}_c^{(2)} \\ \mathbf{f}_c^{(2)} \end{Bmatrix} = \begin{Bmatrix} \mathbf{P}^{*(1)} \\ \mathbf{P}^{*(2)} \end{Bmatrix} \tag{134}$$

where n_c refers to the current number of CNPs along the contact zone, $\mathbf{K}^{*(\beta)}$ is $2n_c \times 2n_c$ matrix, $\mathbf{I}$ the identity $2n_c \times 2n_c$ matrix, and $\mathbf{P}^{*(\beta)}$, $\mathbf{d}_c^{(\beta)}$, $\mathbf{f}_c^{(\beta)}$ are $2n_c$ column vectors. So far, a unique solution to (134) cannot be obtained, because the number of unknowns is $8n_c$ while the number of equations is only $4n_c$. Therefore, it is necessary to provide $4n_c$ more contact conditions as additional equations to (134) for the solution. For conciseness, four of the independent compatibility and equilibrium conditions, for example, Eqs (108), (111a, b), and (114a), at each CNP may be rephrased in matrix notation as:

$$\begin{bmatrix} \operatorname{tg}\varphi_i & -1 & 0 & 0 & -\operatorname{tg}\varphi_i & 1 & 0 & 0 \\ 0 & 0 & 1 & 0 & 0 & 0 & 1 & 0 \\ 0 & 0 & 0 & 1 & 0 & 0 & 0 & 1 \\ 0 & 0 & \cos\varphi_i & \sin\varphi_i & 0 & 0 & 0 & 0 \end{bmatrix} \begin{Bmatrix} u_{iX}^{(1)} \\ u_{iY}^{(1)} \\ f_{iX}^{(1)} \\ f_{iY}^{(1)} \\ u_{iX}^{(2)} \\ u_{iY}^{(2)} \\ f_{iX}^{(2)} \\ f_{iY}^{(2)} \end{Bmatrix} = \begin{Bmatrix} \widehat{g}_i \\ 0 \\ 0 \\ 0 \end{Bmatrix} \tag{135}$$

After introducing the contact conditions (135) into (134), we obtain the resulting contact interface equation:

$$\mathbf{SX} = \mathbf{C} \tag{136}$$

where **S** is the contact coefficient matrix and **X** are the unknowns including contact forces and associated displacements. Since zero appears on the diagonal of **S,** which poses some difficulties in the solution process, the above system must be solved using Gaussian elimination with partial pivoting [110]. Then the contact forces and displacements are computed simultaneously. Accordingly, the contact stress for each particular CNP i is readily obtained:

$$\sigma_{in} = \left(f_{iY}^{(1)} \cos|\varphi_i| - f_{iX}^{(1)} \sin|\varphi_i|\right) / A_i \quad \textbf{(137)}$$

where A_i represents the contact sub-area dominated by CNP i.

Solution Strategy

The distinct characteristic of the present algorithm is that the problem can actually be analysed using only one total load step in the case of frictionless contact. In order to determine the actual contact boundary, the bisection method [110] is adopted in the iterative process. When $\left[f_n^{(1)}\right]_{A^{(1)}} > \text{TOL}$ and $\left[f_n^{(1)}\right]_{B^{(1)}} < -\text{TOL}$, where TOL designates a user-specified tolerance, the actual contact boundary is somewhere (denoted by $C^{(\beta)}$) between $A^{(\beta)}$ and $B^{(\beta)}$ as shown in Fig. (7). The left CNP may be considered the lower limit of the contact boundary and the right CNP is the upper limit. For the actual contact boundary, the position of $B^{(\beta)}$ can be adjusted following the bisection method and with Eq. (136) resolved until $\left[f_n^{(1)}\right]_{B^{(1)}} > -\text{TOL}$. The corresponding overall solution algorithm is summarized as follows:

Step 0: Initialize n_c and other information.
Step 1: Compute (105) for each CNP, and calculate (133) for each contacting body.
Step 2: Calculate (137a, b) for each contacting body.
Step 3: Construct (140) and solve it using Gaussian elimination with partial pivoting.
Step 4: Check the normal force of $f_{n_c}^{(1)}$ at the last CNP n_c.

IF $\left|f_{n_c}^{(1)}\right| < \text{TOL}$ THEN
go to Step 6
ELSE
IF $f_{n_c}^{(1)} > 0$ (compressive) THEN
$n_c = n_c + 1$, return to Step 2
ELSE
go to Step 5
END IF
END IF

Step 5: Adjust the position of the last CNP n_c using the bisection method, return to Step 1.
Step 6: Output the results.

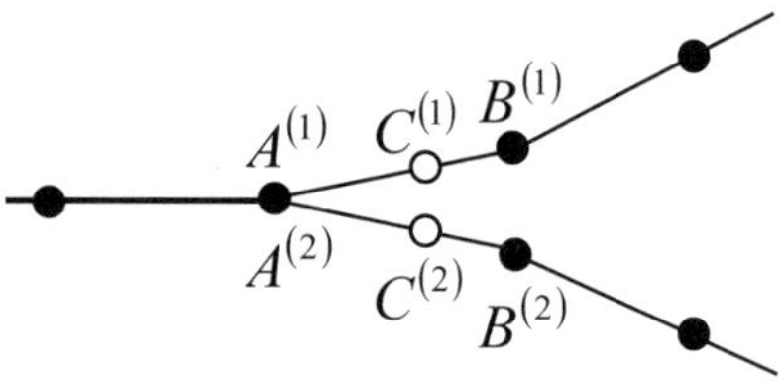

Fig. (7). Determination of actual contact boundary.

PIEZOELECTRIC PROBLEM

In the previous two sections, we have discussed applications of the Trefftz FEM for elastoplastic and contact problems. This approach can also be used for multifield problems such as piezoelectric materials. In the process, solutions in Stroh formalism for transversely isotropic piezoelectric materials are used for the intra-element displacement and electric potential (DEP) fields, together with an independent DEP frame function field along element boundaries. The formulation is based on a modified variational functional in which electric field and strain are taken as basic variables. The final unknowns are the parameters of the frame function field consisting of the usual DOF at corner nodes and an optional number of hierarchic DOF associated with the mid-side nodes. This section is presented based on the work in [90].

Governing Equations and Their Trefftz Functions

Basic Functions

In this section we recall briefly the governing equations of linear piezoelectricity. The summation convention is invoked over repeated indices. For convenience, matrices are represented by bold face letters and a comma after a variable implies differentiation with respect to Cartesian coordinates. Then, for a linear piezoelectric material, the differential governing equations of linear piezoelectric material in the Cartesian coordinates x_i (i=1, 2, 3) are given by:

$$\sigma_{ij,j} + \bar{b}_i = 0, \quad D_{i,i} + \bar{q}_b = 0 \qquad \text{in } \Omega \tag{138}$$

where σ_{ij} is the stress tensor, D_i is the electric displacement vector, $\bar{b}_i$ is the body force vector, $\bar{q}_b$ is the electric charge density, Ω is the solution domain, For an anisotropic piezoelectric material, the constitutive relation is [111]:

$$\sigma_{ij} = \frac{\partial H(\boldsymbol{\varepsilon},\mathbf{E})}{\partial \varepsilon_{ij}} = c^E_{ijkl}\varepsilon_{kl} - e_{kij}E_k, \quad D_i = -\frac{\partial H(\boldsymbol{\varepsilon},\mathbf{E})}{\partial E_i} = e_{ikl}\varepsilon_{kl} + \kappa^{\varepsilon}_{ik}E_k \tag{139}$$

where $H(\boldsymbol{\varepsilon},\mathbf{E}) = \frac{1}{2}c^E_{ijkl}\varepsilon_{ij}\varepsilon_{kl} - \frac{1}{2}\kappa^{\varepsilon}_{ij}E_iE_j - e_{kij}\varepsilon_{ij}E_k$, c^E_{ijkl} is the stiffness coefficient tensor for **E**=0, $\kappa^{\varepsilon}_{ij}$ the permittivity constant matrix for $\boldsymbol{\varepsilon}$=0, ε_{ij} and E_i are, respectively, the elastic strain tensor and the electric field intensity vector, e_{kij} is piezoelectric stress constants.

The relation between the strain tensor and the displacement, u_i, is given by:

$$\varepsilon_{ij} = \frac{1}{2}(u_{i,j} + u_{j,i}) \tag{140}$$

and the electric field components are related to the electric potential ϕ by:

$$E_i = -\phi_{,i} \tag{141}$$

The boundary conditions of the boundary value problem (138)-(141) can be given by:

$$u_i = \bar{u}_i \qquad \text{on } \Gamma_u \tag{142}$$

$$t_i = \sigma_{ij}n_j = \bar{t}_i \qquad \text{on } \Gamma_t \tag{143}$$

$$D_n = D_in_i = -\bar{q}_n = \bar{D}_n \qquad \text{on } \Gamma_D \tag{144}$$

$$\phi = \bar{\phi} \qquad \text{on } \Gamma_\phi \tag{145}$$

where $\bar{u}_i, \bar{t}_i, \bar{q}_n$ and $\bar{\phi}$ are, respectively, prescribed boundary displacement, traction vector, surface charge and electric potential, an overhead bar denotes prescribed value, $\Gamma=\Gamma_u+\Gamma_t=\Gamma_D+\Gamma_\phi$ is the boundary of the solution domain Ω.

Moreover, in the Trefftz finite element formulation, Eqs (138)-(145) should be completed by the following inter-element continuity requirements:

$$u_{ie} = u_{if}, \quad \phi_e = \phi_f, \qquad (\text{on } \Gamma_e \cap \Gamma_f, \text{ conformity}), \tag{146}$$

$$t_{ie} + t_{if} = 0, \quad D_{ne} + D_{nf} = 0 \qquad (\text{on } \Gamma_e \cap \Gamma_f, \text{ reciprocity}) \tag{147}$$

where '*e*' and '*f*' stand for any two neighbouring elements. Equations (138)-(147) are taken as the basis to establish the modified variational principle for the Trefftz finite element analysis of piezoelectric materials.

Trefftz Functions

The Trefftz function plays an important role in the derivation of Trefftz finite element formulation. A complete system of homogeneous solutions to equation (138) can be generated in a systematic way from the Stroh formalism technique [1]:

$$\mathbf{u} = 2\,\mathrm{Re}\{\mathbf{A}\langle f(z_\alpha)\rangle\mathbf{c}\} \tag{148}$$

where 'Re' stands for the real part of a complex number, **A** is the material eigenvector matrix. which has been well defined in the literature [1], $\langle f(z_\alpha)\rangle = \mathrm{diag}[f(z_1)\, f(z_2)\, f(z_3)\, f(z_4)]$ is a diagonal 4×4 matrix, while $f(z_i)$ is an arbitrary function with the argument $z_i = x_1 + p_i x_2$. p_i (i=1-4) are the material eigenvalues. Of particular interest is the complete set of polynomial solutions which may be generated by setting in Eq. (148) in turn

$$\left.\begin{aligned} f(z_\alpha) &= z_\alpha^k \\ f(z_\alpha) &= iz_\alpha^k \end{aligned}\right\} \; (k = 1,2,\cdots) \tag{149}$$

where $i = \sqrt{-1}$. Thus the Trefftz functions of equation (138) can be given from:

$$\mathbf{u} = \sum_{j=1}^{\infty} [\mathrm{Re}\{\mathbf{A}\langle z_\alpha^j\rangle\}\mathbf{a}_j + \mathrm{Re}\{\mathbf{A}\langle iz_\alpha^j\rangle\}\mathbf{b}_j] \tag{150}$$

Elemental Stiffness Formulation

Assumed Fields

The Trefftz finite element model is based on the assumption of two distinct DEP fields: the internal field **u** and the frame function $\tilde{\mathbf{u}}$ [19]. The field **u** fulfils identically the governing differential equations (138) and is assumed as:

$$\mathbf{u}=\begin{Bmatrix} u_1 \\ u_2 \\ \phi \end{Bmatrix}=\begin{Bmatrix} \breve{u}_1 \\ \breve{u}_2 \\ \breve{\phi} \end{Bmatrix}+\begin{Bmatrix} \mathbf{N}_1 \\ \mathbf{N}_2 \\ \mathbf{N}_3 \end{Bmatrix}\mathbf{c}=\breve{\mathbf{u}}+\sum_{j=1}\mathbf{N}_j\mathbf{c}_j=\breve{\mathbf{u}}+\mathbf{N}\mathbf{c} \qquad \textbf{(151)}$$

for a two-dimensional problem, where $\mathbf{c}_i$ stands for an undetermined coefficient, and $\breve{\mathbf{u}}(=\{\breve{u}_1,\breve{u}_2,\breve{\phi}\}^T)$ and $\mathbf{N}_j$ are known functions. If the governing differential equation (138) is rewritten in a general form

$$\Re\mathbf{u}(\mathbf{x})+\overline{\mathbf{b}}(\mathbf{x})=0, \qquad (\mathbf{x}\in\Omega_e) \qquad \textbf{(152)}$$

where $\Re$ stands for the differential operator matrix for Eq. (138), **x** for the position vector, $\overline{\mathbf{b}}$ $(=\{\{\overline{b}_1,\overline{b}_2,\overline{q}_b\}^T$ for the known right-hand side term, the overhead bar indicates the imposed quantities and Ω_e stands for the *e*th element sub-domain, then $\breve{\mathbf{u}}=\breve{\mathbf{u}}(\mathbf{x})$ and $\mathbf{N}=\mathbf{N}(\mathbf{x})$ in Eq. (151) must be chosen so that:

$$\Re\breve{\mathbf{u}}+\overline{\mathbf{b}}=0 \quad \text{and} \quad \Re\mathbf{N}=0 \qquad \textbf{(153)}$$

everywhere in Ω_e. Thus $\mathbf{N}_j$ in Eq. (151) can be formed by a suitably truncated complete system of (150). For example, we may set:

$$\mathbf{N}_{2j}=2\,\mathrm{Re}\{\mathbf{A}\langle z_\alpha^j\rangle\} \qquad \textbf{(154)}$$

$$\mathbf{N}_{2j+1}=2\,\mathrm{Re}\{\mathbf{A}\langle iz_\alpha^j\rangle\} \qquad \textbf{(155)}$$

The unknown coefficient **c** may be calculated from the conditions on the external boundary and/or the continuity conditions on the inter-element boundary. Thus, various Trefftz-element models can be obtained by using different approaches to enforce these conditions. In the majority of cases, a hybrid technique is used, whereby the elements are linked through an auxiliary conforming displacement frame which has the same form as in the conventional finite element method. This means that, in the Trefftz finite element approach, a conforming DEP field should be independently defined on the element boundary to enforce the field continuity between elements and also to link the coefficient **c** appearing in Eq. (151), with nodal DEP **d** (={*d*}). The frame is defined as:

$$\tilde{\mathbf{u}}(\mathbf{x}) = \begin{Bmatrix} \tilde{u}_1 \\ \tilde{u}_2 \\ \tilde{\phi} \end{Bmatrix} = \begin{Bmatrix} \tilde{\mathbf{N}}_1 \\ \tilde{\mathbf{N}}_2 \\ \tilde{\mathbf{N}}_3 \end{Bmatrix} \mathbf{d} = \tilde{\mathbf{N}}\mathbf{d}, \qquad (\mathbf{x} \in \Gamma_e) \tag{156}$$

for a two-dimensional problem, where the symbol "~" is used to specify that the field is defined on the element boundary only, **d**=**d**(**c**) stands for the vector of the nodal displacements which are the final unknowns of the problem, Γ_e represents the boundary of element *e*, and $\tilde{\mathbf{N}}$ is a matrix of the related shape functions which are similar to those in the conventional finite element formulation.

In the development of the present *p*-element, the following assumptions are adopted. First, the problem is assumed to be plane strain of a transversely isotropic piezoelectric solid referred to a Cartesian system (x_1, x_2). Secondly, the element may have a general quadrilateral shape or a triangular shape with three DOF (u_1, u_2, ϕ) at each corner node (see Fig. **4**). Thirdly, to achieve higher order variations, an optional number of extra hierarchic modes is introduced along with the hierarchic DOF, a_i for $\tilde{u}_1$, b_i for $\tilde{u}_2$, p_i for $\tilde{\phi}$, which are conveniently associated with mid-side node *C* (see Fig. **4**). Thus, along the side *A-C-B* of a particular element (see Fig. **4**), a simple interpolation of the frame DEP field can be given in the form:

$$\begin{Bmatrix} \tilde{u}_1 \\ \tilde{u}_2 \\ \tilde{\phi} \end{Bmatrix} = \begin{bmatrix} \tilde{N}_1 & 0 & 0 & \tilde{N}_2 & 0 & 0 \\ 0 & \tilde{N}_1 & 0 & 0 & \tilde{N}_2 & 0 \\ 0 & 0 & \tilde{N}_1 & 0 & 0 & \tilde{N}_2 \end{bmatrix} \begin{Bmatrix} u_{1A} \\ u_{2A} \\ \phi_A \\ u_{1B} \\ u_{2B} \\ \phi_B \end{Bmatrix} + \sum_{i=1}^{M} \gamma^{i-1} R_i \begin{Bmatrix} a_{Ci} \\ b_{Ci} \\ p_{Ci} \end{Bmatrix} \tag{157}$$

where *M* is the order of the hierarchic DOF.

$$\tilde{N}_1 = \frac{1-\xi}{2} \quad \tilde{N}_2 = \frac{1+\xi}{2} \quad R_i = \xi^{i-1}(1-\xi^2) \tag{158}$$

and where ξ is defined in Fig. (**4**).

The coefficient γ is equal to +1 or -1 according to the orientation of the side *A-C-B* (see Fig. **4**) in the global coordinate system (X,Y):

$$\gamma = \begin{cases} +1 \text{ if } X_B - X_A \le Y_B - Y_A \\ -1 \text{ if } X_B - X_A > Y_B - Y_A \end{cases} \tag{159}$$

The purpose of using the coefficient γ is to ensure a unequivocal definition of the frame functions $\tilde{\mathbf{u}}$ in terms of parameters a_i, b_i, and p_i, common to two elements sharing the mid-side node *Z*.

Using the definitions in (139), (143) and (144) the generalized boundary forces and electric displacements can be given as:

$$\mathbf{T} = \begin{Bmatrix} t_1 \\ t_2 \\ D_n \end{Bmatrix} = \begin{Bmatrix} \sigma_{1j} n_j \\ \sigma_{2j} n_j \\ D_j n_j \end{Bmatrix} = \begin{Bmatrix} \breve{t}_1 \\ \breve{t}_2 \\ \breve{D}_n \end{Bmatrix} + \begin{Bmatrix} \mathbf{Q}_1 \\ \mathbf{Q}_2 \\ \mathbf{Q}_3 \end{Bmatrix} \mathbf{c} = \breve{\mathbf{T}} + \mathbf{Qc} \tag{160a}$$

$$\tilde{\mathbf{T}} = \begin{Bmatrix} \tilde{t}_1 \\ \tilde{t}_2 \\ \tilde{D}_n \end{Bmatrix} = \begin{Bmatrix} \tilde{\sigma}_{1j} n_j \\ \tilde{\sigma}_{2j} n_j \\ \tilde{D}_j n_j \end{Bmatrix} = \begin{Bmatrix} \tilde{\mathbf{Q}}_1 \\ \tilde{\mathbf{Q}}_2 \\ \tilde{\mathbf{Q}}_3 \end{Bmatrix} \mathbf{c} = \tilde{\mathbf{Q}}\mathbf{c} \tag{160b}$$

where $\breve{t}_i$ and $\breve{D}_n$ are derived from $\breve{\mathbf{u}}$.

Particular Solutions

The particular solution of $\breve{\mathbf{u}}$ can be obtained by means of their Green's functions. The Green's functions of (138) are as follows [112]:

$$u_{ij}^*(p,q) = \frac{1}{4\pi} \left\{ a_{ij}^{(0)} \ln\left|r_{pq}\right| + \sum_{k=1}^{\infty} \frac{(-1)^{k+1}}{k} [a_{ij}^{(k)} \cos(2k\theta) + b_{ij}^{(k)} \sin(2k\theta)\right\} \tag{161}$$

With

$$r_{pq} = \sqrt{(x_q - x_p)^2 + (y_q - y_p)^2} \quad \textbf{(162)}$$

where $u^*_{ij}(p,q)$ designates the Green's function of the in-plane displacement (for i=1,2) or electric potential(for i=3) at field point p of an infinite plane when a unit point force is applied at the source point q in the j-direction. $a^{(k)}_{ij}$ and $b^{(k)}_{ij}$ are calculated as follows: $a^{(k)}_{ij} = 0$ if $i \neq j; k = 0,1,2,\ldots$, $b^{(k)}_{ij} = 0$ if $i = j; k = 1,2,\ldots$, and the non-zero values of $a^{(k)}_{ij}$ and $b^{(k)}_{ij}$ can be found in Table 1 of Ref. [112]

Thus, the particular solutions of (138) can be expressed as:

$$\breve{\mathbf{u}} = \begin{Bmatrix} \breve{u}_1 \\ \breve{u}_2 \\ \breve{\phi} \end{Bmatrix} = \iint_\Omega \bar{b}_j \begin{Bmatrix} \breve{u}^*_{1j} \\ \breve{u}^*_{2j} \\ \breve{u}^*_{3j} \end{Bmatrix} d\Omega \quad \textbf{(163)}$$

where $\bar{b}_3 = \bar{q}_b$. The area integration in (163) is performed by numerical quadrature using the Gauss-Legendre rule.

Modified Variational Principle

The Trefftz finite element equation for piezoelectric materials can be established by the variational approach [19]. Since the stationary conditions of the traditional potential and complementary variational functional cannot satisfy the inter-element continuity condition which is required in the Trefftz finite element analysis, some new variational functionals need to be developed. For this purpose, we present the following modified variational functional suitable for Trefftz finite element analysis:

$$\Pi_m = \sum_e \Pi_{me} = \sum_e \{\Pi_e + \int_{\Gamma_{\phi e}} (\bar{\phi} - \phi)\tilde{D}_n ds + \int_{\Gamma_{ue}} (\bar{u}_i - u_i)\tilde{t}_i ds - \int_{\Gamma_{Ie}} (\tilde{\phi} D_n + \tilde{u}_i t_i ds\} \quad \textbf{(164)}$$

where

$$\Pi_e = \iint_{\Omega_e} [H(\boldsymbol{\varepsilon}, \mathbf{E}) - \bar{b}_i u_i - \bar{q}_b \phi] d\Omega - \int_{\Gamma_{te}} \bar{t}_i u_i ds - \int_{\Gamma_{De}} \bar{D}_n \phi ds \quad \textbf{(165)}$$

in which Eq. (138) is assumed to be satisfied, a priori. The terminology "modified principle" refers here to the use of a conventional functional (Π_e here) and some modified terms for the construction of a special variational principle to account for additional requirements such as the condition defined in Eqs (146) and (147).

The boundary Γ_e of a particular element consists of the following parts:

$$\Gamma_e = \Gamma_{ue} \cup \Gamma_{te} \cup \Gamma_{Ie} = \Gamma_{\phi e} \cup \Gamma_{De} \cup \Gamma_{Ie} \tag{166}$$

Where

$$\Gamma_{ue} = \Gamma_u \cap \Gamma_e, \quad \Gamma_{te} = \Gamma_t \cap \Gamma_e, \quad \Gamma_{\phi e} = \Gamma_\phi \cap \Gamma_e, \quad \Gamma_{De} = \Gamma_D \cap \Gamma_e, \tag{167}$$

and Γ_{Ie} is the inter-element boundary of the element '*e*'.

Element Matrix Equation

The element matrix equation can be generated by setting $\delta\Pi_{me} = 0$. To simplify the derivation, we first transform all domain integrals in Eq. (164) into boundary ones. In fact, by reason of the solution properties of the intra-element trial functions, the functional Π_{me} can be simplified to:

$$\begin{aligned}\Pi_{me} = &\frac{1}{2}\int_{\Gamma_e}(t_i u_i + D_n\phi)ds - \frac{1}{2}\int_\Omega(\bar{b}_i u_i + \bar{q}_b\phi)d\Omega + \int_{\Gamma_{\phi e}}(\bar{\phi} - \phi)\tilde{D}_n ds \\ &+ \int_{\Gamma_{ue}}(\bar{u}_i - u_i)\tilde{t}_i ds - \int_{\Gamma_{Ie}}(D_n\tilde{\phi} + t_i\tilde{u}_i)ds - \int_{\Gamma_{te}}\bar{t}_i u_i ds - \int_{\Gamma_{De}}\bar{D}_n\phi ds\end{aligned} \tag{168}$$

Substituting the expressions given in Eqs (151), (157), and (160) into (132) produces:

$$\Pi_{me} = \frac{1}{2}\mathbf{c}^T\mathbf{H}\mathbf{c} + \mathbf{c}^T\mathbf{S}\mathbf{d} + \mathbf{c}^T\mathbf{r}_1 + \mathbf{d}^T\mathbf{r}_2 + \text{terms without } \mathbf{c} \text{ or } \mathbf{d} \tag{169}$$

in which the matrices **H**, **S** and the vectors $\mathbf{r}_1$, $\mathbf{r}_2$ are defined by:

$$\mathbf{H} = \int_{\Gamma_e}\mathbf{Q}^T\mathbf{N}ds \tag{170}$$

$$\mathbf{S} = -\int_{\Gamma_{\phi e}} \mathbf{N}_3^T \widetilde{\mathbf{Q}}_3 ds - \int_{\Gamma_{ue}} \begin{bmatrix} \mathbf{N}_1 \\ \mathbf{N}_2 \end{bmatrix}^T \begin{bmatrix} \widetilde{\mathbf{Q}}_1 \\ \widetilde{\mathbf{Q}}_2 \end{bmatrix} ds - \int_{\Gamma_{Ie}} \mathbf{Q}^T \widetilde{\mathbf{N}} ds \tag{171}$$

$$\mathbf{r}_1 = \frac{1}{2}\int_{\Gamma_e} (\mathbf{N}^T \breve{\mathbf{T}} + \mathbf{Q}^T \breve{\mathbf{u}}) ds - \frac{1}{2}\int_{\Omega} \mathbf{N}^T \overline{\mathbf{b}} d\Omega - \int_{\Gamma_{De}} \mathbf{N}_3^T \overline{D}_n ds - \int_{\Gamma_{te}} \begin{bmatrix} \mathbf{N}_1 \\ \mathbf{N}_2 \end{bmatrix}^T \begin{Bmatrix} \bar{t}_1 \\ \bar{t}_2 \end{Bmatrix} ds \tag{172}$$

$$\mathbf{r}_2 = \int_{\Gamma_{\phi e}} \widetilde{\mathbf{Q}}_3^T (\overline{\phi} - \breve{\phi}) ds + \int_{\Gamma_{\phi e}} \begin{bmatrix} \widetilde{\mathbf{Q}}_1 \\ \widetilde{\mathbf{Q}}_2 \end{bmatrix}^T \begin{Bmatrix} \overline{u}_1 - \breve{u}_1 \\ \overline{u}_2 - \breve{u}_2 \end{Bmatrix} ds - \int_{\Gamma_{Ie}} \begin{bmatrix} \widetilde{\mathbf{N}}_1 \\ \widetilde{\mathbf{N}}_2 \\ \widetilde{\mathbf{N}}_3 \end{bmatrix}^T \begin{Bmatrix} \breve{t}_1 \\ \breve{t}_2 \\ \breve{D}_n \end{Bmatrix} ds \tag{173}$$

To enforce inter-element continuity on the common element boundary, the unknown vector **c** should be expressed in terms of nodal DOF **d**. An optional relationship between **c** and **d** in the sense of variation can be obtained from:

$$\frac{\partial \Pi_{me}}{\partial \mathbf{c}^T} = \mathbf{Hc} + \mathbf{Sd} + \mathbf{r}_1 = 0 . \tag{174}$$

This leads to:

$$\mathbf{c} = \mathbf{Gd} + \mathbf{g} \quad , \tag{175}$$

where $\mathbf{G} = -\mathbf{H}^{-1}\mathbf{S}$ and $\mathbf{g} = -\mathbf{H}^{-1}\mathbf{r}_1$, directly expressing Π_{me} in terms of **d** and other known matrices only:

$$\Pi_{me} = \frac{1}{2}\mathbf{d}^T \mathbf{G}^T \mathbf{HGd} + \mathbf{d}^T (\mathbf{G}^T \mathbf{Hg} + \mathbf{r}_2) + \text{terms without } \mathbf{d} \tag{176}$$

Thus, the element displacement-force equation can be generated by taking the vanishing variation of the functional Π_{me} with respect to **d** as:

$$\frac{\partial \Pi_{me}}{\partial \mathbf{d}^T} = 0 \Rightarrow \quad \mathbf{Kd} = \mathbf{P} \tag{177}$$

where $\mathbf{K}=\mathbf{G}^T\mathbf{HG}$ and $\mathbf{P}=-\mathbf{G}^T\mathbf{Hg}-\mathbf{r}_2$ are, respectively, the element stiffness matrix and the equivalent nodal flow vector. The expression (177) is the elemental stiffness matrix equation for Trefftz finite element analysis.

Moreover, to ensure good numerical conditioning during the inversion of matrix **H**, the homogeneous solutions $\mathbf{N}_k$ in (151) must be expressed in terms of suitably scaled local coordinates (x,y) originating at the element centroid (see Fig. **5**):

$$x = \frac{X - X_c}{a}, \quad y = \frac{Y - Y_c}{a} \tag{178}$$

where X and Y are respectively global coordinates, X_c and Y_c stand for global coordinates at the centroid of the element under consideration, and a is the average distance between the centroid and node i of the element:

$$a = \sum_{i}^{N} \frac{(x_i^2 + y_i^2)^{1/2}}{N} \tag{179}$$

and where N is the nodal number of the element.

CONCLUSION AND FUTURE DEVELOPMENTS

On the basis of the preceding discussion, the following conclusions can be drawn. In contrast to conventional FE and boundary element methods, the advantages of the HT FE procedure are: (i) the element stiffness formulation involves integration along the element boundaries only. It enables convenient construction of arbitrary polygonal shaper- or even curve-sided elements. As a consequence, it may be considered a symmetric and boundary solution approach at element level, and has the advantages of the boundary element method. In contrast to traditional boundary element model, however, the hybrid Trefftz FE approach avoids dealing with singular integral equations and does not require the use of a fundamental solution, which might be very laborious to build; (ii) the hybrid FEM appears to represent the optimal expansion bases for hybrid-type FEs where inter-element continuity need not be satisfied, a priori, which is very important for developing a quasi-conforming plate bending element; (iii) the HT FE formulation providess the attractive possibility of constructing accurate corner, perforated or crack singular elements. This can be simply done by using appropriate known local solutions as the interpolation functions of the intra-element fields. Remarkable progress has been achieved in the fields of fracture mechanics, potential flow problems, plane elasticity, elastodynamics, thin and thick plate bending, and nonlinear problems of plate bending by the Trefftz FE model.

It is noted that the HT FEM has become increasingly popular as an efficient computational tool in mechanics, material science, and physics since its initiation in the late seventies. There are, however, still many possible extensions to various engineering problems as well as areas in need of further development. Among those problems and areas one could list the following:

1. Creation of efficient HT FE-BEM schemes for complex engineering materials and structures, with powerful capabilities of pre-processing and postprocessing..
2. Development of various special-purpose elements for handling efficiently singularities attributable to local load or geometrical effects. As discussed above, the special-purpose elements ensure that relatively accurate results are obtained without local mesh refinement and at minimal computational cost. Extensions of such element model could be applied to other cases such as the interaction between fluid and structure in fluid–structure problems, the boundary layer effect between two materials, and corner, circular hole, and load singularities.
3. Integration of Trefftz elements with machine learning approach and topology optimization scheme for efficient microstructure and metamaterial design.
4. Development of efficient adaptive procedures such as higher order p-capabilities, h-extension elements, error estimation, and convergence studies.
5. Extensions of HT FE to dynamic problems of fluid flow, soil mechanics, piezoelectric materials, deep shell structure, and rheology problems.

CONSENT FOR PUBLICATION

Not applicable.

CONFLICT OF INTEREST

The author declares no conflict of interest, financial or otherwise.

ACKNOWLEDGEMENTS

Declared none.

REFERENCES

[1] Q. H. Qin, *Green's function and boundary elements of multifield materials*. Oxford: Elsevier, 2007.

[2] Q. H. Qin and Q. S. Yang, *Macro-micro theory on multifield coupling behaivor of hetereogenous materials*. Beijing: Higher Education Press and Springer, 2008.

[3] Q. H. Qin and Y.-W. Mai, "A closed crack tip model for interface cracks in thermopiezoelectric materials," *International Journal of Solids and Structures,* vol. 36, no. 16, pp. 2463-2479, 1999.

[4] S. Yu and Q. H. Qin, "Damage analysis of thermopiezoelectric properties: Part II. Effective crack model," *Theoretical and Applied Fracture Mechanics,* vol. 25, no. 3, pp. 279-288, 1996.

[5] S. Yu and Q. H. Qin, "Damage analysis of thermopiezoelectric properties: Part I—crack tip singularities," *Theoretical and Applied Fracture Mechanics,* vol. 25, no. 3, pp. 263-277, 1996.

[6] Q. H. Qin and M. V. Swain, "A micro-mechanics model of dentin mechanical properties," *Biomaterials,* vol. 25, no. 20, pp. 5081-5090, 2004.

[7] Q. H. Qin, *Advanced mechanics of piezoelectricity*. Beijing: Higher Education Press and Springer, 2013.

[8] Q. H. Qin, Y.-W. Mai, and S.-W. Yu, "Effective moduli for thermopiezoelectric materials with microcracks," *International Journal of Fracture,* vol. 91, no. 4, pp. 359-371, 1998.

[9] Q. H. Qin, "Green's function for thermopiezoelectric plates with holes of various shapes," *Archive of Applied Mechanics,* vol. 69, no. 6, pp. 406-418, 1999.

[10] Q. H. Qin and Y.-W. Mai, "Crack branch in piezoelectric bimaterial system," *International Journal of Engineering Science,* vol. 38, no. 6, pp. 673-693, 2000.

[11] H. C. Martin and G. F. Carey, *Introduction to finite element analysis: Theory and application*. McGraw-Hill New York, 1973.

[12] J. Jirousek and N. Leon, "A powerful finite element for plate bending," *Computer Methods in Applied Mechanics and Engineering,* vol. 12, no. 1, pp. 77-96, 1977.

[13] Q. H. Qin, "Fundamental Solution Based Finite Element Method," *J Appl Mech Eng,* vol. 2, p. e118, 2013.

[14] Q. H. Qin and C. X. Mao, "Coupled torsional-flexural vibration of shaft systems in mechanical engineering—I. Finite element model," *Computers & Structures,* vol. 58, no. 4, pp. 835-843, 1996.

[15] J. Argyris, J. S. Doltsinis, and M. Kleiber, "Incremental formulation in nonlinear mechanics and large strain elasto-plasticity—Natural approach. Part II," *Computer Methods in Applied Mechanics and Engineering,* vol. 14, no. 2, pp. 259-294, 1978.

[16] J. Argyris, P. Dunne, M. Haase, and J. Orkisz, "Higher-order simplex elements for large strain analysis—natural approach," *Computer Methods in Applied Mechanics and Engineering,* vol. 16, no. 3, pp. 369-403, 1978.

[17] J. H. Argyris and M. Kleiber, "Incremental formulation in nonlinear mechanics and large strain elasto-plasticity—Natural approach. Part 1," *Computer Methods in Applied Mechanics and Engineering,* vol. 11, no. 2, pp. 215-247, 1977.

[18] H. Wang and Q. H. Qin, "FE approach with Green's function as internal trial function for simulating bioheat transfer in the human eye," *Archives of Mechanics,* vol. 62, no. 6, pp. 493-510, 2010.

[19] Q. H. Qin, *The Trefftz finite and boundary element method.* Southampton, WIT Press, 2000.

[20] J. Jirousek and Q. H. Qin, "Application of hybrid-Trefftz element approach to transient heat conduction analysis," *Computers & Structures,* vol. 58, no. 1, pp. 195-201, 1996.

[21] Q. H. Qin and H. Wang, *Matlab and C programming for Trefftz finite element methods.* New York: CRC Press, 2008.

[22] Q. H. Qin, "Transient plate bending analysis by hybrid Trefftz element approach," *Communications in Numerical Methods in Engineering,* vol. 12, no. 10, pp. 609-616, 1996.

[23] C. X. Mao and Q. H. Qin, "Coupled torsional-flexural vibration of shaft systems in mechanical engineering—II. FE-TM impedance coupling method," *Computers & structures,* vol. 58, no. 4, pp. 845-849, 1996.

[24] Q. H. Qin, "Nonlinear analysis of Reissner plates on an elastic foundation by the BEM," *International Journal of Solids and Structures,* vol. 30, no. 22, pp. 3101-3111, 1993.

[25] Q. H. Qin, "Micromechanics-BEM Analysis for Piezoelectric Composites," *Tsinghua Science & Technology,* vol. 10, no. 1, pp. 30-34, 2005.

[26] Y. Sun, X. He, and Q. H. Qin, "A new procedure for the nonlinear analysis of reissner plate by boundary element method," *Computers & Structures,* vol. 53, no. 3, pp. 649-652, 1994.

[27] Q. H. Qin, "Boundary Element Method," In: *Special Topics in the Theory of Piezoelectricity,* J. S. Yang, Ed.: Springer, Cambridge Massachusetts, 2009, pp. 137-168.

[28] I. Herrera, "General variational principles applicable to the hybrid element method," *Proceedings of the National Academy of Sciences,* vol. 74, no. 7, pp. 2595-2597, 1977.

[29] I. Herrera, "Boundary methods: a criterion for completeness," *Proceedings of the National Academy of Sciences,* vol. 77, no. 8, pp. 4395-4398, 1980.

[30] I. Herrera, R. E. Ewing, M. A. Celia, and T. F. Russell, "Eulerian-Lagrangian localized adjoint method: The theoretical framework," *Numerical Methods for Partial Differential Equations,* vol. 9, no. 4, pp. 431-457, 1993.

[31] I. Herrera and F. J. Sabina, "Connectivity as an alternative to boundary integral equations: Construction of bases," *Proceedings of the National Academy of Sciences,* vol. 75, no. 5, pp. 2059-2063, 1978.

[32] E. Trefftz, "Ein gegenstuck zum ritzschen verfahren," in *Proc. 2nd Int. Cong. Appl. Mech. Zurich*, 1926, pp. 131-137.

[33] H. Wang and Q. H. Qin, "Hybrid FEM with fundamental solutions as trial functions for heat conduction simulation," *Acta Mechanica Solida Sinica,* vol. 22, no. 5, pp. 487-498, 2009.

[34] J. T. De Freitas and Z. Y. Ji, "Hybrid-Trefftz equilibrium model for crack problems," *International Journal for Numerical Methods in Engineering,* vol. 39, no. 4, pp. 569-584, 1996.

[35] J. Jirousek and A. Venkatesh, "A simple stress error estimator for Hybrid-Trefftz p-version elements," *International journal for numerical methods in engineering,* vol. 28, no. 1, pp. 211-236, 1989.

[36] E. Kita, N. Kamiya, and Y. Ikeda, "A new boundary-type scheme for sensitivity analysis using Trefftz formulation," *Finite elements in analysis and design,* vol. 21, no. 4, pp. 301-317, 1996.

[37] S. Lifits, S. Reutskiy and B. Tirozzi "Trefftz spectral method for initial-boundary problems," *Comput. Assist. Mech. Eng. Sci.*, vol. 4: pp. 549–565, 1997.

[38] M. Markiewicz and O. Mahrenholtz, "Combined time-stepping and Trefftz approach for nonlinear wave-structure interaction," *Comput. Assist. Mech. Eng. Sci,* vol. 4, pp. 567-586, 1997.

[39] E. A. W. Maunder and J. Moitiniho de Almeida, "Hybrid-equilibrium elements with control of spurious kinematic modes," *Comput. Assist. Mech. Eng. Sci,* vol. 4, pp. 587-606, 1997.

[40] J. Melenk and I. Babuska, "Approximation with harmonic and generalized harmonic polynomials in the partition of unity method," *Comput. Assist. Mech. Eng. Sci,* vol. 4, pp. 607-632, 1997.

[41] S. Ohnimus, M. Rüter, and E. Stein, "General aspects of Trefftz method and relations to error estimation of finite element approximations," *Comput. Assist. Mech. Eng. Sci,* vol. 8, pp. 425-437, 2001.

[42] K. Peters, E. Stein, and W. Wagner, "A new boundary-type finite element for 2-D-and 3-D-elastic structures," *International journal for numerical methods in engineering,* vol. 37, no. 6, pp. 1009-1025, 1994.

[43] X. P. Zheng and Z. H. Yao, "Some applications of the Trefftz method in linear elliptic boundary-value problems," *Advances in Engineering Software,* vol. 24, no. 1-3, pp. 133-145, 1995.

[44] J. Jirousek and A. Venkatesh, "Hybrid trefftz plane elasticity elements with p-method capabilities," *International Journal for Numerical Methods in Engineering,* vol. 35, no. 7, pp. 1443-1472, 1992.

[45] R. Piltner, "Special finite elements with holes and internal cracks," *International journal for numerical methods in engineering,* vol. 21, no. 8, pp. 1471-1485, 1985.

[46] Q. H. Qin, "Dual variational formulation for Trefftz finite element method of elastic materials," *Mechanics Research Communications,* vol. 31, no. 3, pp. 321-330, 2004.

[47] H. Wang and Q.-H. Qin, "Numerical implementation of local effects due to two-dimensional discontinuous loads using special elements based on boundary integrals," *Engineering Analysis with Boundary Elements,* vol. 36, no. 12, pp. 1733-1745, 2012.

[48] Y. Cui and Q. H. Qin, "Fracture analysis of mode III problems by Trefftz finite element approach," in *WCCM VI in conjunction with APCOM,* vol. 4, pp. 380, 2004.

[49] Y. Cui, Q. H. Qin, and J. Wang, "Application of HT finite element method to multiple crack problems of Mode I, II and III," *Chinese Journal of Engineering Mechanics,* vol. 23, no. 3, pp. 104-110, 2006.

[50] Y. Cui, J. Wang, M. Dhanasek, and Q. H. Qin, "Mode III fracture analysis by Trefftz boundary element method," *Acta Mechanica Sinica,* vol. 23, no. 2, pp. 173-181, 2007.

[51] Q. H. Qin, "Solving anti-plane problems of piezoelectric materials by the Trefftz finite element approach," *Computational Mechanics,* vol. 31, no. 6, pp. 461-468, 2003.

[52] J. Jirouseka and L. Guex, "The hybrid-Trefftz finite element model and its application to plate bending," *International Journal for Numerical Methods in Engineering,* vol. 23, no. 4, pp. 651-693, 1986.

[53] Q. H. Qin, "Postbuckling analysis of thin plates by a hybrid Trefftz finite element method," *Computer Methods in Applied Mechanics and Engineering,* vol. 128, no. 1, pp. 123-136, 1995.

[54] Q. H. Qin, "Hybrid Trefftz finite-element approach for plate bending on an elastic foundation," *Applied Mathematical Modelling,* vol. 18, no. 6, pp. 334-339, 1994.

[55] F. Jin and Q.H. Qin, "A variational principle and hybrid Trefftz finite element for the analysis of Reissner plates," *Computers & structures,* vol. 56, no. 4, pp. 697-701, 1995.

[56] J. Jirousek, A. Wroblewski, Q.H. Qin, and X. He, "A family of quadrilateral hybrid-Trefftz p-elements for thick plate analysis," *Computer Methods in Applied Mechanics and Engineering,* vol. 127, no. 1-4, pp. 315-344, 1995.

[57] E. A. W. Maunder, "A Trefftz patch recovery method for smooth stress resultants and applications to Reissner-Mindlin equilibrium plate models," *Comput. Assist. Mech. Eng. Sci,* vol. 8, pp. 409-424, 2001.

[58] Q. H. Qin, "Hybrid-Trefftz finite element method for Reissner plates on an elastic foundation," *Computer Methods in Applied Mechanics and Engineering,* vol. 122, no. 3-4, pp. 379-392, 1995.

[59] J. Petrolito, "Hybrid-Trefftz quadrilateral elements for thick plate analysis," *Computer Methods in Applied Mechanics and Engineering,* vol. 78, no. 3, pp. 331-351, 1990.

[60] J. Petrolito, "Triangular thick plate elements based on a hybrid-Trefftz approach," *Computers & structures,* vol. 60, no. 6, pp. 883-894, 1996.

[61] R. Piltner, "A quadrilateral hybrid-Trefftz plate bending element for the inclusion of warping based on a three-dimensional plate formulation," *International journal for numerical methods in engineering,* vol. 33, no. 2, pp. 387-408, 1992.

[62] Q. H. Qin, "Nonlinear analysis of thick plates by HT FE approach," *Computers & Structures,* vol. 61, no. 2, pp. 271-281, Oct 1996.

[63] Q. H. Qin and S. Diao, "Nonlinear analysis of thick plates on an elastic foundation by HT FE with p-extension capabilities," (in English), *International Journal of Solids and Structures,* vol. 33, no. 30, pp. 4583-4604, Dec 1996.

[64] C.-Y. Lee, Q. H. Qin, and H. Wang, "Trefftz functions and application to 3D elasticity," *Computer Assisted Mechanics and Engineering Sciences,* vol. 15, pp. 251-263, 2008.

[65] J. A. T. de Freitas and M. Toma, "Hybrid-Trefftz stress and displacement elements for axisymmetric incompressible biphasic media," *Computer Methods in Applied Mechanics and Engineering,* vol. 198, no. 30-32, pp. 2368-2390, 2009.

[66] K. Y. Wang, L. Q. Zhang, and P. C. Li, "A four-node hybrid-Trefftz annular element for analysis of axisymmetric potential problems," *Finite Elements in Analysis and Design,* vol. 60, pp. 49-56, Nov 2012.

[67] A. Zielinski and O. C. Zienkiewicz, "Generalized finite element analysis with T-complete boundary solution functions," *International Journal for Numerical Methods in Engineering,* vol. 21, no. 3, pp. 509-528, 1985.

[68] L. Cao, Q.H. Qin, and N. Zhao, "A new RBF-Trefftz meshless method for partial differential equations," in *IOP conference series: materials science and engineering*, 2010, vol. 10, no. 1, p. 012217: IOP Publishing.

[69] W. Chen, Z.-J. Fu, and Q.H. Qin, "Boundary particle method with high-order Trefftz functions," *Computers, Materials & Continua (CMC),* vol. 13, no. 3, pp. 201-217, 2010.

[70] H. Wang, Q. H. Qin, and D. Arounsavat, "Application of hybrid Trefftz finite element method to non-linear problems of minimal surface," *International Journal for Numerical Methods in Engineering,* vol. 69, no. 6, pp. 1262-1277, 2007.

[71] H. Beckert, "On the realization of trefftz theory for thin shells," *Zeitschrift Fur Angewandte Mathematik Und Mechanik,* vol. 69, no. 11, pp. 401-408, 1989.

[72] P. H. C. Martins, F. L. S. Bussamra, and E. L. Neto, "Three dimensional hybrid-Trefftz stress finite elements for plates and shells," *International Journal for Numerical Methods in Engineering,* vol. 113, no. 11, pp. 1676-1696, Mar 2018.

[73] J. De Freitas, "Hybrid-Trefftz displacement and stress elements for elastodynamic analysis in the frequency domain," *Comput Assis Mech Eng Sci,* vol. 4, no. 3, pp. 345-368, 1997.

[74] C. Cao, Q. H. Qin, and A. Yu, "Micromechanical analysis of heterogeneous composites using hybrid trefftz FEM and hybrid fundamental solution based FEM," *Journal of Mechanics,* vol. 29, no. 4, pp. 661-674, 2013.

[75] C. Cao, A. Yu, and Q. H. Qin, "Evaluation of effective thermal conductivity of fiber-reinforced composites by boundary integral based finite element method," *International Journal of Architecture, Engineering and Construction,* vol. 1, no. 1, pp. 14-29, 2012.

[76] L. Cao, Q. H. Qin, and N. Zhao, "Application of DRM-Trefftz and DRM-MFS to Transient Heat Conduction Analysis," *Recent Patents on Space Technology (Open access),* vol. 2, pp. 41-50, 2010.

[77] N. Zhao, L.L. Cao, and Q.H. Qin, "Application of Trefftz Method to Heat Conduction Problem in Functionally Graded Materials," *Recent Patents on Space Technology,* vol. 1, no. 2, pp. 158-166, 2011.

[78] Z. J. Fu, Q. H. Qin, and W. Chen, "Hybrid-Trefftz finite element method for heat conduction in nonlinear functionally graded materials," *Engineering Computations,* vol. 28, no. 5, pp.578-599, 2011.

[79] Q. H. Qin, "Postbuckling analysis of thin plates on an elastic foundation by HT FE approach," *Applied Mathematical Modelling,* vol. 21, no. 9, pp. 547-556, 1997.

[80] J. T. De Freitas and Z. Wang, "Hybrid-Trefftz stress elements for elastoplasticity," *International Journal for Numerical Methods in Engineering,* vol. 43, no. 4, pp. 655-683, 1998.

[81] A. Zieliński, "Trefftz method: elastic and elastoplastic problems," *Computer methods in applied mechanics and engineering,* vol. 69, no. 2, pp. 185-204, 1988.

[82] Q. H. Qin, "Formulation of hybrid Trefftz finite element method for elastoplasticity," *Applied Mathematical Modelling,* vol. 29, no. 3, pp. 235-252, 2005.

[83] Q.H. Qin, "Trefftz plane elements of elastoplasticity with p-extension capabilities," *Journal of Mechanical Engineering,* vol. 56, no. 1, pp. 40-59, 2005.

[84] Q. H. Qin and K.-Y. Wang, "Application of hybrid-Trefftz finite element method fractional contact problems," *Computer Assisted Mechanics and Engineering Sciences,* vol. 15, pp. 319-336, 2008.

[85] K. Y. Wang, Q. H. Qin, Y. L. Kang, J. S. Wang, and C. Y. Qu, "A direct constraint-Trefftz FEM for analysing elastic contact problems," *International Journal for Numerical Methods in Engineering,* vol. 63, no. 12, pp. 1694-1718, 2005.

[86] M. Dhanasekar, K. Y. Wang, Q. H. Qin, and Y. L. Kang, "Contact analysis using Trefftz and interface finite elements," *Computer Assisted Mechanics and Engineering Sciences,* vol. 13, no. 3, pp. 457-471, 2006.

[87] Q. H. Qin, "Variational formulations for TFEM of piezoelectricity," *International Journal of Solids and Structures,* vol. 40, no. 23, pp. 6335-6346, 2003.

[88] Q. H. Qin, "Mode III fracture analysis of piezoelectric materials by Trefftz BEM," *Structural Engineering and Mechanics,* vol. 20, no. 2, pp. 225-240, 2005.

[89] Q. H. Qin, "Fracture Analysis of Piezoelectric Materials by Boundary and Trefftz Finite Element Methods," *WCCM VI in conjunction with APCOM'04, Sept. 5-10, 2004, Beijing, China,* 2004.

[90] Q. H. Qin, "Trefftz Plane Element of Piezoelectric Plate with p-Extension Capabilities," in *IUTAM Symposium on Mechanics and Reliability of Actuating Materials*, W. Yang, Ed. Dordrecht: Springer, 2006, pp. 144-153.

[91] M. Dhanasekar, J. Han, and Q. H. Qin, "A hybrid-Trefftz element containing an elliptic hole," *Finite Elements in Analysis and Design,* vol. 42, no. 14, pp. 1314-1323, 2006.

[92] Q. H. Qin and X.-Q. He, "Special elliptic hole elements of Trefftz FEM in stress concentration analysis," *Journal of Mechanics and MEMS,* vol. 1, no. 2, pp. 335-348, 2009.

[93] H.-T. Huang and Z.-C. Li, "Global superconvergence of Adini's elements coupled with the Trefftz method for singular problems," *Engineering analysis with boundary elements,* vol. 27, no. 3, pp. 227-240, 2003.

[94] I. Herrera, "Trefftz-Herrera Method," *Computer Assisted Mechanics and Engineering Sciences,* vol. 4, pp. 369-382, 1997.

[95] J. Jirousek and A. Zieliński, "Survey of Trefftz-type element formulations," *Computers & structures,* vol. 63, no. 2, pp. 225-242, 1997.

[96] E. Kita and N. Kamiya, "Trefftz method: an overview," *Advances in Engineering Software,* vol. 24, no. 1-3, pp. 3-12, 1995.

[97] Q. H. Qin, "Trefftz finite element method and its applications," *Applied Mechanics Reviews,* vol. 58, no. 5, pp. 316-337, 2005.

[98] D. L. Colton, *Partial differential equations in the complex domain* (no. 4). Pitman Publishing, 1976.

[99] I. Herrera, *Boundary methods: an algebraic theory*. Pitman Advanced Pub. Program, 1984.

[100] Y.-c. Fung, *Foundations of solid mechanics*. Englewood Cliffs, NJ: Prentice-Hall, 1965.

[101] L. E. Malvern, *Introduction to the Mechanics of a Continuous Medium*. Englewood Cliffs, NJ: Prentice-Hall, 1969.

[102] N. I. Muskhelishvili, *Some basic problems of the mathematical theory of elasticity*. Groningen, Holland: Noordhoff, 1963.

[103] E. Sternberg, "On some recent developments in the linear theory of elasticity," presented at the 1st Symp. On Naval Structural Mechanics, New York, 1960.

[104] M. Williams, "Stress singularities resulting from various boundary conditions in angular corners of plates in extension," *Journal of applied mechanics,* vol. 19, no. 4, pp. 526-528, 1952.

[105] P. K. Banerjee, "Two-and Threedimensional Problems of Elasto-plasticity," In: *Developments in boundary element methods-1*, P. K. Banejee and B. R., Eds. London: Applied Science Publishers, 1979, pp. 65-95.

[106] Q. H. Qin and Y. Huang, "BEM of postbuckling analysis of thin plates," *Applied Mathematical Modelling,* vol. 14, no. 10, pp. 544-548, 1990.

[107] G. Nayak and O. Zienkiewicz, "Elasto-plastic stress analysis. A generalization for various contitutive relations including strain softening," *International Journal for Numerical Methods in Engineering,* vol. 5, no. 1, pp. 113-135, 1972.

[108] S. Y. Ma, "The solution of two-dimensional frictionless elastic contact problems using boundary element method," *Journal of Hebei Institute of Technology,* vol. 2, pp. 11-21, 1984.

[109] E. L. Wilson, "The static condensation algorithm," *International Journal for Numerical Methods in Engineering,* vol. 8, no. 1, pp. 198-203, 1974.

[110] Y. H. Zhang, M. C. Xi, and C. S. Cheng, *Computational Methods and Their Algorithms*. Beijing: Science Press, 2000.

[111] Q. H. Qin, *Fracture mechanics of piezoelectric materials*. WIT Press, Southampton, 2001.

[112] N. Khutoryansky, H. Sosa, and W. Zu, "Approximate Green's functions and a boundary element method for electro-elastic analyses of active materials," *Computers & structures,* vol. 66, no. 2-3, pp. 289-299, 1998.

CHAPTER 2

Hybrid-Trefftz Finite Elements for Non-homogeneous Parabolic Problems

Ionut D. Moldovan[1*], Ana Coutinho[2] and Ildi Cismasiu[2]

[1]*CERIS, Instituto Superior Técnico, Universidade de 1649004 Lisboa, Portugal*

[2]*Faculdade de Ciências e Tecnologia, Universidade Nova de 1649004 Lisboa, Portugal*

Abstract: The solution of transient parabolic problems using a novel hybrid-Trefftz finite element is presented in this chapter. The governing equations are first discretised in time and reduced to a series of non-homogeneous elliptic problems in space variables only. The complementary and particular solutions of each elliptic problem are approximated independently. The complementary solution is expanded in Trefftz bases, designed to satisfy the homogeneous form of the problem. Trefftz bases include regular functions of arbitrary orders and are defined independently for each finite element. A novel dual reciprocity method is used for the approximation of the particular solution to avoid domain integration. The same regular basis is used for the expansions of the source function and particular solution, avoiding the cumbersome expressions of the latter that typify conventional dual reciprocity techniques. Moreover, the bases of the complementary and particular solutions are defined by the same expressions, with different wave numbers. The finite element formulation is obtained by enforcing weakly the domain equations and boundary conditions. The formulation is implemented in the computational platform FreeHyTE, where it takes advantage of the pre-programmed numerical procedures and graphical user interfaces. The resulting software is open-source, user-friendly and freely distributed to the scientific community.

Keywords: Dual reciprocity method, Hybrid-Trefftz finite element, Particular solution, Non-homogeneous parabolic problem.

INTRODUCTION

Hybrid-Trefftz finite elements combine favourable traits of the finite and boundary element methods. Like the former, hybrid-Trefftz elements are based on the subdivision of the domain into finite elements, where complex solutions can be more easily approximated, regular approximation functions, and Hermitian, sparse and localized solving systems. On the other hand, similar to the boundary elements, they feature physically meaningful approximations bases built on the exact (free-

***Corresponding author Ionut D. Moldovan:** CERIS, Instituto Superior Técnico, Universidade de 1649004 Lisboa, Portugal; E-mail: dragos.moldovan@tecnico.ulisboa.pt field) solutions of the differential equation governing the problem. This option reduces the formulation to the boundaries of the elements, considerably enhances their convergence under basis (p-) refinement, and increases their robustness to mesh distortion, near-incompressibility of the constituents, and high frequency content of the excitation (*e.g.* [1, 2]).

However, exact particular solutions may not exist when the boundary value problem is non-homogeneous, especially if the source function (*i.e.* the non-homogeneous term) includes the effect of the initial conditions, as typical of transient problems (*e.g.* [3]). In such cases, the particular solution can be calculated using the fundamental solution of the differential operator, or it can be approximated using the Dual Reciprocity Method (DRM). While the former approach should provide an exact particular solution, it requires the domain integration of the (singular) fundamental solution, which is numerically demanding and a drawback from the perspective of the boundary methods. The DRM was proposed by Nardini and Brebbia in 1982 [4]. As opposed to the fundamental solution approach, the DRM originates formulations free of domain integrals at the cost of abandoning the quest for an exact particular solution and settling for an approximation instead. The method involves two (dependent) approximations: one for the source function and one for the particular solution. They are obtained in two steps. In the first step, the source function is approximated using some trial basis, whose weights are obtained by domain collocation. In the second step, the particular solution approximation is obtained by solving analytically the governing differential equation with the approximation of the source function as the non-homogeneous term. The method was applied successfully to different kinds of boundary value problems, including the Poisson's and biharmonic equations [5], parabolic equations [6], elliptic problems [7] and hyperbolic problems [8]. Reviews of the DRM can be found in references [9, 10].

Radial basis functions are typically used in the DRM to approximate the source function [8], but polynomial and trigonometric functions have also been tried [11]. While for simple (*e.g.*, Laplace) differential operators, the expressions of the particular solution trial functions are relatively simple [9], cumbersome (and, occasionally, singular) expressions are obtained for parabolic and hyperbolic boundary value problems [12]. Moreover, the solving system is not symmetric and frequently ill-conditioned, especially if high-order interpolation functions are collocated in a high number of points. Such situations are not easy to avoid if the domain is not partitioned into subdomains.

Two alternative strategies were proposed by the authors to avoid both the domain integration of the fundamental solution and the cumbersome expression of the particular solution basis in parabolic and hyperbolic boundary value problems. In references [3, 13], the particular solution basis is built using Trefftz-compliant functions of the corresponding static problem, that is, the problem obtained by eliminating all time-dependent terms from the original differential equation. This ensures that the particular solution basis has a very simple expression. Moreover, the particular and complementary solution trial functions are not uncoupled, as happens in the DRM, but merged in the same basis and used to enforce the same equations in a single analysis step. This option enables the complementary solution basis to compensate for deficiencies of the particular solution basis, meaning that accurate total solutions can be obtained with relatively poor particular solution approximations. The price to be paid, however, is the emergence of a small number of terms in the solving system that are defined by domain integrals, although the integrands are quite simple and do not present points of singularity.

Conversely, in references [14, 15], a new DRM technique for time-dependent boundary value problems is presented. The same (Trefftz-type) trial functions are used for both source term and particular solution approximations, except for a constant multiplier. Moreover, the trial functions are simple and involve no points of singularity, except when physically justified. The finite elements constructed in this way are highly convergent under p-refinement and free of domain integrals. In the context of the Method of Fundamental Solutions, and still, for hyperbolic problems, a similar approach was independently developed by Alves [16].

This technique is described in this chapter and applied to the solution of non-homogeneous parabolic problems. The original problem defined in time and space is first discretized in time using transform-based techniques, finite difference (time-stepping) algorithms, or weighted residual methods. Regardless of the technique adopted for the time discretization, the original problem is reduced to a series of non-homogeneous, elliptic problems in space, which are solved using hybrid-Trefftz finite elements. The elements are based on the independent approximations of the state field in their domains and of the normal flux field on their boundaries. The domain approximation is built on two function spaces, one for the complementary solution, and the other for the particular solution. These spaces are formally similar, but feature distinct wave numbers, to avoid linear dependence. As opposed to other hybrid-Trefftz elements (*e.g.* [17]), all approximations are hierarchical and independent of the element's nodes. Therefore, different approximations can be used for different boundaries and finite elements, and localized p-refinement is possible.

To establish a physically meaningful nomenclature, the formulation is presented using a well-known parabolic boundary value problem, the heat diffusion equation. To preserve generality, however, the source term (*i.e.* internal heat generation) can assume any distribution in time and space and may or may not have analytical expression. The presentation of the formulation is followed by a description of its implementation in the FreeHyTE framework. FreeHyTE is an open-source and user-friendly computational platform that uses hybrid-Trefftz finite elements [18] and boundary elements [19] for the solution of boundary value problems. Besides the solvers themselves, FreeHyTE provides a collection of pre-programmed numerical procedures and graphical user interfaces that can be included in new modules with minimal coding effort. All FreeHyTE modules are freely distributed to the scientific community through the FreeHyTE web page [20].

DEFINITION OF THE PROBLEM

Consider the domain Ω presented in Fig. (**1**). The transient heat diffusion problem defined in Ω is described by the heat diffusion equation (1) and the Fourier law (2),

$$k \cdot \nabla^2 T(\mathbf{x},t) + Q(\mathbf{x},t) = \rho c \cdot \dot{T}(\mathbf{x},t) \tag{1}$$

$$\mathbf{q}(\mathbf{x},t) = -k \cdot \nabla T(\mathbf{x},t) \tag{2}$$

where $T(\mathbf{x},t)$ and $\mathbf{q}(\mathbf{x},t)$ are the (scalar) temperature and (vectorial) heat flux fields, $Q(\mathbf{x},t)$ is the internal heat source, and k, ρ and c are thermal conductivity, density and specific heat capacity of the material, respectively. Bold notations are used for vectorial quantities. The differential operators ∇ and ∇^2 designate the gradient and Laplace operators, respectively, and the dot over a field variable denotes its differentiation with respect to time.

Differential equations (1) and (2) are complemented by the initial condition,

$$T(\mathbf{x},0) = T_0(\mathbf{x}) \tag{3}$$

and boundary conditions,

$$T(\mathbf{x},t) = T_\Gamma(\mathbf{x},t), \text{ on } \Gamma_d \tag{4}$$

$$\mathbf{N} \cdot \mathbf{q}(\mathbf{x},t) = q_\Gamma\left(\mathbf{x},t\right), \text{ on } \Gamma_n \quad \mathbf{(5)}$$

$$\mathbf{N} \cdot \mathbf{q}(\mathbf{x},t) = h \cdot \left[T\left(\mathbf{x},t\right) - T_f(t)\right], \text{ on } \Gamma_r \quad \mathbf{(6)}$$

Boundary conditions (4) to (6) are used to enforce the temperature $T_\Gamma\left(\mathbf{x},t\right)$ on the Dirichlet $\left(\Gamma_d\right)$ boundary, the normal heat flux $q_\Gamma\left(\mathbf{x},t\right)$ on the Neumann $\left(\Gamma_n\right)$ boundary, and a convection-type relation between the temperature and the normal heat flux on the Robin $\left(\Gamma_r\right)$ boundary, respectively. It is assumed that Dirichlet, Neumann and Robin boundaries are complementary and disjoint: $\Gamma = \Gamma_d \cup \Gamma_n \cup \Gamma_r$ and, $\phi = \Gamma_d \cap \Gamma_n = \Gamma_d \cap \Gamma_r = \Gamma_n \cap \Gamma_r$. In expressions (5) and (6), vector $\mathbf{N}$ collects the components of the outward normal to the boundary. In the convection law (6), $T_f(t)$ the temperature of the fluid in contact with boundary Γ_r, is assumed (piecewise) constant in space, but variable in time, and h is the heat transfer coeiffcient.

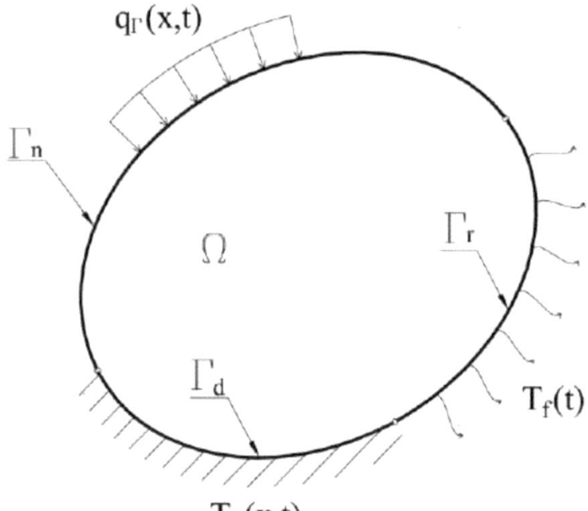

Fig. (1). Problem's domain, Neumann, Dirichlet and Robin boundaries.

DISCRETIZATION IN TIME

The problem defined in the previous section is first discretized in time, to remove the time dependence of the involved variables. This can be achieved using transform-based approaches [21], finite difference techniques [22], and weighted residual methods [23]. Regardless of the technique chosen for the time discretization, the original problem in time and space is transformed into a series of N elliptic problems in space, defined by the following equations,

$$\nabla^2 T_n(\mathbf{x}) + \omega_n^2 T_n(\mathbf{x}) = f_{0_n}(\mathbf{x}) \tag{7}$$

$$\mathbf{q}_n(\mathbf{x}) = -k \cdot \nabla T_n(\mathbf{x}) \tag{8}$$

complemented by boundary conditions,

$$T_n(\mathbf{x}) = T_{\Gamma_n}(\mathbf{x}), \text{ on } \quad \Gamma_d \tag{9}$$

$$\mathbf{N} \cdot \mathbf{q}_n(\mathbf{x}) = q_{\Gamma_n}(\mathbf{x}), \text{ on } \quad \Gamma_n \tag{10}$$

$$\mathbf{N} \cdot \mathbf{q}_n(\mathbf{x}) = h \cdot \left[T_n(\mathbf{x}) - T_{f_n} \right], \text{ on } \Gamma_r \tag{11}$$

Taking advantage of the formal similarity of the elliptic problems defined by equations (7)-(11) the same finite element formulation can be used for all time discretization techniques [24]. The explicit expressions of the terms in expressions (7)-(11) are derived next for some techniques belonging to each class of time discretization methods.

Discrete Fourier Transform

The Discrete Fourier Transform is a transform-based time discretization technique most adequate for periodic time-dependent problems. All time-dependent fields present in the definition of the original problem are decomposed into harmonic components by taking their Fourier expansion over the period P, using a finite number N of spectral frequencies, $\xi_n = n \cdot 2\pi / P$,

$$\Diamond(\mathbf{x},t)=\frac{1}{P}\sum_{n=0}^{N-1}\Diamond_n(\mathbf{x})\exp(i\,\xi_n t) \qquad \textbf{(12)}$$

where $\Diamond(\mathbf{x},t)$ represents a generic time-dependent field and i is the imaginary unit. Expansions of type (12) are enacted for the temperature $T(\mathbf{x},t)$ and heat flux $\mathbf{q}(\mathbf{x},t)$ fields, the heat source $Q(\mathbf{x},t)$ and the boundary fields $T_\Gamma(\mathbf{x},t)$, $q_\Gamma(\mathbf{x},t)$ and $T_f(t)$. Substituting expansions (12) into the diffusion equation (1) and equating the coefficients corresponding to each of the spectral components, one immediately gets an equation of type (7), where,

$$f_{0_n}(\mathbf{x})=-\frac{1}{k}Q_n(\mathbf{x}) \qquad \textbf{(13)}$$

$$\omega_n^2=-i\frac{\xi_n}{\alpha} \qquad \textbf{(14)}$$

and $\alpha = {}^{k}\!/_{\rho c}$ is the thermal diffusivity.

Equation (8) and the boundary conditions (9)-(11) are trivially recovered using the same technique.

It is noted that the initial conditions are not included in the definition of the source function $f_{0_n}(\mathbf{x})$ as the problem is assumed periodic, and is defined and solved over the duration of one period. This assumption is typically an issue when this approach is used to model transient problems. This restriction is lifted when the time discretization is performed using finite difference and weighted residual techniques, as shown in the next sections.

Finite Difference Techniques

Finite difference techniques are based on the division of the total duration of the analysis in (relatively small) time steps, followed by the collocation of the governing equations (1)-(6) at some 'critical' point in each time step, typically at its end. A generalized mid-point algorithm is exemplified next.

Let the total time of the analysis be split into time intervals of duration Δt, and let $T_0(\mathbf{x})$ designate the temperature field at the beginning of the current time step. $T_0(\mathbf{x})$ is known from the previous time step, or from the user's input if the current time step is the first. The time derivative of the solution $\dot{T}_{\Delta t}(\mathbf{x})$ at the end of the current time step is approximated as,

$$\dot{T}_{\Delta t}(\mathbf{x}) = \frac{1}{\bar{\theta}\,\Delta t}\left(T_{\Delta t}(\mathbf{x}) - T_0(\mathbf{x})\right) - \frac{1-\bar{\theta}}{\bar{\theta}}\dot{T}_0(\mathbf{x}) \tag{15}$$

where the time derivative of the temperature field at the beginning of the time step $\dot{T}_0(\mathbf{x})$ is obtained by collocating the heat diffusion equation (1) at $t=0$ and susbstituting the initial condition (3) in the resulting expression. In expression (7), $\bar{\theta}$ is a parameter that controls the algorithmic properties of the time stepping scheme. Common options are $\bar{\theta}=1/2$ (the Crank-Nicolson scheme), $\bar{\theta}=2/3$ (the Galerkin scheme) and $\bar{\theta}=1$ (the implicit Euler scheme).

The time-discretized form of the heat diffusion equation is obtained by collocating expression (1) at the end of the current time step and inserting approximation (15), to yield,

$$k\cdot\nabla^2 T_{\Delta t}(\mathbf{x}) + Q_{\Delta t}(\mathbf{x}) = \rho c\cdot\left[\frac{1}{\bar{\theta}\,\Delta t}\left(T_{\Delta t}(\mathbf{x}) - T_0(\mathbf{x})\right) - \frac{1-\bar{\theta}}{\bar{\theta}}\dot{T}_0(\mathbf{x})\right] \tag{16}$$

where $Q_{\Delta t}(\mathbf{x})$ is the heat source $Q(\mathbf{x},t)$ at $t=\Delta t$.

Further algebraic processing of equation (16) leads to the spectral form of the time-discretized heat diffusion equation,

$$\nabla^2 T_{\Delta t}(\mathbf{x}) + \omega^2 T_{\Delta t}(\mathbf{x}) = f_0(\mathbf{x}) \tag{17}$$

where the source function f_0 is defined as,

$$f_0(\mathbf{x}) = -\frac{1}{k}Q_{\Delta t}(\mathbf{x}) - \frac{1}{\alpha}\left(\frac{1}{\bar{\theta}\,\Delta t}T_0(\mathbf{x}) + \frac{1-\bar{\theta}}{\bar{\theta}}\dot{T}_0(\mathbf{x})\right) \tag{18}$$

Expression (17) is consistent with the general elliptic form (7) where,

$$\omega^2 = -\frac{1}{\alpha \bar{\theta} \Delta t} \tag{19}$$

It is noted that in expression (17) the source function $f_0(\mathbf{x})$ accounts for the influence of both heat source and initial conditions.

The generalized mid-point technique presented above is an implicit time-stepping algorithm. All values of $\bar{\theta}$ larger than 0.5 render it unconditionally stable, meaning that the time steps do not need to be smaller than a critical value in order to keep the error bounded. Explicit time-stepping algorithms could, in principle, also be used, but they require very small time steps to maintain stability. This may result in large values for the algorithmic frequency (19) which is detrimental for the stability of the solution in space, as the (frequency-dependent) Trefftz approximation functions may become hard to integrate numerically. Therefore, the use of implicit time-stepping algorithms is recommended.

An equation of form (8) is obtained by directly collocating equation (2) at the end of the time step. Boundary conditions (4)-(6) are also collocated in the same way to yield expressions of type (9)-(11), respectively, where the right-hand side terms are simply the values of the respective fields at the end of the time interval.

Weighted Residual Techniques

Assume that the temperature field $T(\mathbf{x},t)$, its 'velocity' field $V(\mathbf{x},t)$, and heat flux field $\mathbf{q}(\mathbf{x},t)$ are independently approximated using the time basis $\mathbf{W}(t)$,

$$T(\mathbf{x},t) = \sum_{n=0}^{N-1} W_n(t) T_n(\mathbf{x}) \tag{20}$$

$$V(\mathbf{x},t) = \sum_{n=0}^{N-1} W_n(t) V_n(\mathbf{x}) \tag{21}$$

$$\mathbf{q}(\mathbf{x},t) = \sum_{n=0}^{N-1} W_n(t) \mathbf{q}_n(\mathbf{x}) \tag{22}$$

The independent approximations of the temperature field and of its 'velocity' derivative mean that the definition of the latter must be explicitly enforced. This is done using the Galerkin weighted residual principle,

$$\int_0^{\Delta t} \mathbf{W}^* \left(V - \dot{T} \right) dt = \mathbf{0} \tag{23}$$

where $\mathbf{W}^*$ is the conjugate transposed of the time basis $\mathbf{W}$ and Δt is the size of the time step (typically, much larger than in finite difference methods). The second term in expression (23) is integrated by parts and the initial condition (3) is inserted in the resulting expression. After the mathematical manipulations described in reference **[23]**, one gets,

$$\Delta t\, \mathbf{H}\, V_n = \mathbf{G}\, T_n - \mathbf{W}^*(0)\, T_0 \tag{24}$$

where matrices $\mathbf{H}$ and $\mathbf{G}$ depend only on the time basis $\mathbf{W}$,

$$\mathbf{H} = \frac{1}{\Delta t} \int_0^{\Delta t} \mathbf{W}^* \mathbf{W} dt \tag{25}$$

$$\mathbf{G} = \mathbf{W}^* (\Delta t) \mathbf{W}(\Delta t) - \int_0^{\Delta t} \left(\dot{\mathbf{W}} \right)^* \mathbf{W} dt \tag{26}$$

Equation (24) can be uncoupled if the time basis is constructed such that matrices $\mathbf{H}$ and $\mathbf{G}$ are related by a diagonal matrix of constants,

$$\mathbf{H} \cdot \mathbf{\Psi} = \mathbf{G} \tag{27}$$

The mathematical procedure to secure this is presented in detail in reference [23] and can be applied to any function basis. Under condition (27), the algebraic system (24) is uncoupled into,

$$\Delta t\, V_n = \Psi_n\, T_n - \psi_n\, T_0 \tag{28}$$

where,

$$\psi_n = \sum_{m=0}^{N-1} \bar{H}_{nm} \hat{W}_m(0) \tag{29}$$

where $\hat{W}_m$ is the complex conjugate of function W_m, Ψ_n is the n-th diagonal term of matrix $\boldsymbol{\Psi}$ and $\bar{H}_{nm}$ is the general term of the inverse of matrix $\mathbf{H}$.

The weak enforcement of the governing equation (1), using the same time basis for weighting yields,

$$\int_0^{\Delta t} \mathbf{W}^* \left[k\nabla^2 T - \rho c \cdot \dot{T} \right] dt = -\int_0^{\Delta t} \mathbf{W}^* Q\, dt \tag{30}$$

Approximations (20) and (21) are inserted into equation (30), along with condition (28), to yield,

$$\mathbf{H} \cdot \left[k\nabla^2 T_n - \rho c \cdot \frac{\Psi_n}{\Delta t} T_n + \rho c \cdot \frac{\psi_n}{\Delta t} T_0 \right] = -\frac{1}{\Delta t} \int_0^{\Delta t} \mathbf{W}^* Q\, dt \tag{31}$$

which uncouples to,

$$\nabla^2 T_n - \frac{\Psi_n}{\alpha\, \Delta t} T_n = -\frac{1}{\alpha\, \Delta t} Q_n - \frac{\psi_n}{\alpha\, \Delta t} T_0 \tag{32}$$

$$Q_n = \sum_{m=0}^{N-1} \bar{H}_{nm} \int_0^{\Delta t} \hat{W}_m Q\, dt \tag{33}$$

Equation (32) corresponds to the elliptic form (7) making the following definitions:

$$\omega_n^2 = -\frac{\Psi_n}{\alpha\, \Delta t} \tag{34}$$

$$f_{0_n}(\mathbf{x}) = -\frac{1}{\alpha\, \Delta t} Q_n(\mathbf{x}) - \frac{\psi_n}{\alpha\, \Delta t} T_0(\mathbf{x}) \tag{35}$$

As was the case of the finite difference techniques, the source function $f_{0_n}(\mathbf{x})$ collects both the effect of the internal heat source and that of the initial conditions.

Boundary conditions (9) to (11) are recovered by enforcing weakly equations (4) to (6) using the time basis $\mathbf{W}$ for weighting and substituting approximations (20) and (22) into the resulting expressions. The right-hand side terms in the boundary conditions are defined as,

$$T_{\Gamma_n} = \frac{1}{\Delta t}\sum_{m=0}^{N-1} \bar{H}_{nm} \int_0^{\Delta t} \hat{W}_m T_\Gamma \, dt \tag{36}$$

$$q_{\Gamma_n} = \frac{1}{\Delta t}\sum_{m=0}^{N-1} \bar{H}_{nm} \int_0^{\Delta t} \hat{W}_m q_\Gamma \, dt \tag{37}$$

$$T_{f_n} = \frac{1}{\Delta t}\sum_{m=0}^{N-1} \bar{H}_{nm} \int_0^{\Delta t} \hat{W}_m T_f \, dt \tag{38}$$

The finite difference technique presented above is adopted for the reminder of this presentation. All subscripts n present in equations (7) to (11) are dropped from this point, to keep the notations simple.

SOLUTION IN SPACE

The temperature model of the hybrid-Trefftz finite elements is used to solve the time-discretized problem defined *via* the generalized mid-point algorithm presented above, in every time step. The trademark feature of the hybrid-Trefftz elements is that their approximation bases collect functions that satisfy locally the domain equations (7) (or (17)) and (8).

The main difficulty in solving non-homogeneous problems is that closed form particular solutions of the domain equations may not be available, especially since the source function $f_0(\mathbf{x})$ depends on the initial conditions of the current time step. In order to preserve the boundary formulation typical of the hybrid-Trefftz finite elements, the option here is to approximate the particular solution using a novel variant of the DRM.

The formulation of the hybrid-Trefftz finite elements for the solution of the elliptic boundary value problem (7)-(11) is presented below.

Discretization in Space

Let the domain Ω of the problem be discretized in an arbitrary number of finite elements, and let Ω^e, Γ_d^e, Γ_n^e, and Γ_r^e designate the domain, and the Dirichlet, Neumann and Robin boundaries of a generic element (Fig. **2**). No geometric restrictions are enforced on the finite elements, which may not be convex or simply-connected, and may have arbitrary numbers of edges.

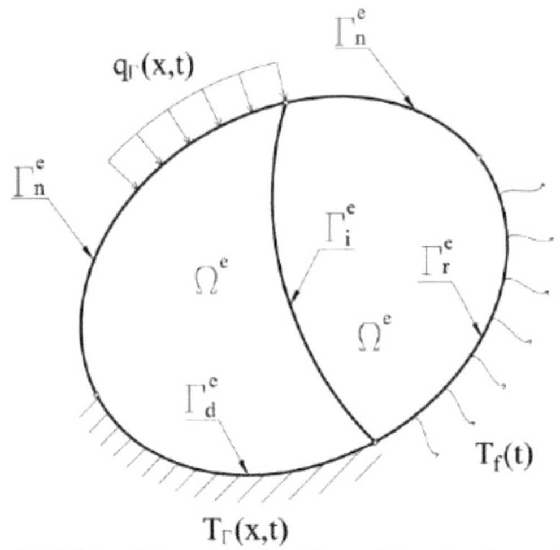

Fig. (2). Finite elements, Dirichlet, Neumann, Robin and interior boundaries.

Governing equations (7) and (8) must now be satisfied within each finite element and on the external boundaries. Due to the meshing process, however, additional equations must be formulated on the interior $\left(\Gamma_i^e\right)$ boundaries of the mesh, to enforce the temperature and normal flux continuity,

$$T^j\left(\mathbf{x}\right) = T^k\left(\mathbf{x}\right), \text{ on } \Gamma_i^e \quad \textbf{(39)}$$

$$\mathbf{N}^j \cdot \mathbf{q}^j(\mathbf{x}) + \mathbf{N}^k \cdot \mathbf{q}^k(\mathbf{x}) = 0, \text{ on } \Gamma_i^e \quad \textbf{(40)}$$

where j and k designate the elements that share the interior boundary Γ_i^e.

Approximations

The temperature model of the hybrid-Trefftz finite element is constructed on the independent approximations of the temperature field in the domain of each element and of the normal heat flux on its Dirichlet, Robin and interior boundaries.

The domain temperature approximation reads,

$$T(\mathbf{x}) = \mathbf{U}_c(\mathbf{x})\cdot\mathbf{X}_c + \mathbf{U}_p(\mathbf{x})\cdot\mathbf{X}_p, \text{ on } \Omega^e \tag{41}$$

In definition (41), $\mathbf{U}_c$ and $\mathbf{U}_p$ designate the approximation bases of the complementary and particular solutions, and $\mathbf{X}_c$ and $\mathbf{X}_p$ are their respective weights. The complementary solution basis $\mathbf{U}_c$ is required to satisfy locally the homogeneous form of equation (7),

$$\nabla^2\mathbf{U}_c(\mathbf{x}) + \omega^2\,\mathbf{U}_c(\mathbf{x}) = \mathbf{0} \tag{42}$$

The free-field solutions of the homogeneous equation (42) are well-known in both $\mathbb{R}^2$ and $\mathbb{R}^3$ [9]. For the computational implementation of this formulation in the FreeHyTE environment, the solutions in polar coordinates (r,θ) are of interest. They are given by,

$$U_{c_n}(r,\theta) = J_n(\omega r)\cdot\exp(in\theta) \tag{43}$$

where U_{c_n} is the generic term of basis $\mathbf{U}_c$, $J_n(z)$ is the Bessel function of the first kind, order n and argument z, and i is the imaginary unit. As opposed to the complementary solution basis $\mathbf{U}_c$, closed form expressions for the particular solution basis $\mathbf{U}_p$ are not available, in general. The construction of this basis is discussed in the next section.

Dependent approximations of the heat flux fields in the finite elements are derived from the approximation (43) by enforcing the Fourier law (8),

$$\mathbf{q}(\mathbf{x}) = \mathbf{S}_c(\mathbf{x})\cdot\mathbf{X}_c + \mathbf{S}_p(\mathbf{x})\cdot\mathbf{X}_p, \text{ on } \Omega^e \tag{44}$$

$$\mathbf{S_i}(\mathbf{x}) = -k \cdot \nabla \mathbf{U_i}(\mathbf{x}),\ i = \{c, p\} \tag{45}$$

Substitution of the complementary solution (43) into equation (45) yields,

$$S_{c_n}(r,\theta) = -\frac{1}{2}k\omega \cdot \begin{bmatrix} J_{n-1}(\omega r) - J_{n+1}(\omega r) \\ i \cdot \left(J_{n-1}(\omega r) + J_{n+1}(\omega r)\right) \end{bmatrix} \cdot \exp(in\theta) \tag{46}$$

Moreover, independent normal flux approximations are enacted on the Robin, Dirichlet and interior boundaries of the mesh,

$$\mathbf{N} \cdot \mathbf{q}(\mathbf{x}) = \mathbf{Z}(\mathbf{x}) \cdot \mathbf{Y_r},\ \text{on } \Gamma_\mathrm{r}^\mathrm{e} \tag{47}$$

$$\mathbf{N} \cdot \mathbf{q}(\mathbf{x}) = \mathbf{Z}(\mathbf{x}) \cdot \mathbf{Y_e},\ \text{on } \Gamma_\mathrm{e}^\mathrm{e} \tag{48}$$

where it is convenient to define the essential boundary of the element as the union of its Dirichlet and interior boundaries, $\Gamma_e^e = \Gamma_d^e \cup \Gamma_i^e$.

No restrictions are enforced on the choice of the boundary bases $\mathbf{Z}$, nor on the dimension of their weight vector $\mathbf{Y}$. In the FreeHyTE implementation, Chebyshev polynomial bases are used, with their degrees freely chosen by the user.

Particular Solution

The particular solution basis $\mathbf{U_p}(\mathbf{x})$ present in the temperature approximation (41) is constructed using the DRM. This option avoids completely the domain integration, preserving most of the key advantages of the Trefftz methods. The source function $f_0(\mathbf{x})$ is approximated first, using domain collocation, and the particular solution basis is subsequently found such as to satisfy exactly equation (7) having the *approximation* of the source function as the non-homogeneous term. The main advantage of the DRM formulation presented here as compared to the conventional approaches [9] is that the same basis is used for both source function and particular solution approximations, thus eliminating the issue of the cumbersome and, sometimes, singular expressions of the latter.

Approximation of the Source Function

Let the source function $f_0(\mathbf{x})$ be approximated in the domain of the element as,

$$f_0(\mathbf{x}) = \mathbf{F}_\mathbf{p}(\mathbf{x}) \cdot \mathbf{X}_\mathbf{p}, \text{ on } \Omega^e \tag{49}$$

and let the source function basis $\mathbf{F}_\mathbf{p}(\mathbf{x})$ satisfy the homogeneous Helmholtz equation,

$$\nabla^2 \mathbf{F}_\mathbf{p}(\mathbf{x}) + \lambda^2 \mathbf{F}_\mathbf{p}(\mathbf{x}) = \mathbf{0} \tag{50}$$

where λ is one or some arbitrary number of algorithmic parameters, and $\lambda \neq \omega$. The solution of equation (50) in a polar referential yields,

$$F_{p_n}(r,\theta) = J_n(\lambda r) \cdot \exp(in\theta) \tag{51}$$

for the nth term of the basis $\mathbf{F}_\mathbf{p}(\mathbf{x})$. However, the DRM approach suggested here holds for any choice of referential, in any dimension. Typically, only regular solutions of equation (51) are included in the basis, but fundamental solutions may also be added if the source function is expected to present points of singularity.

The weights $(\mathbf{X}_\mathbf{p})$ of the functions collected in basis $\mathbf{F}_\mathbf{p}$ are now obtained by collocation in a set of points $\bar{\mathbf{x}} \in \Omega^e$:

$$\mathbf{F}_\mathbf{p}(\bar{\mathbf{x}}) \cdot \mathbf{X}_\mathbf{p} = f_0(\bar{\mathbf{x}}) \tag{52}$$

The solution of system (52) defines completely the source function approximation.

It is noted that equation (50) is a Trefftz-type expression, similar to equation (42). Therefore, no additional coding effort is required for its implementation and limited computational effort is needed for its calculation. The convergence and completeness properties of the Trefftz-compliant basis (51) were investigated in reference [14]. It was shown that it corresponds to a source function expansion in a spherical harmonic series in the angular direction, and in a Dini series in the radial direction. Further convergence and density results are reported in reference [16].

Approximation of the Particular Solution

The particular solution basis $\mathbf{U}_p(\mathbf{x})$ is found by solving exactly equation (7) having the approximation of the source function as the non-homogeneous term,

$$\nabla^2 \mathbf{U}_p(\mathbf{x}) + \omega^2 \mathbf{U}_p(\mathbf{x}) = \mathbf{F}_p(\mathbf{x}) \quad \textbf{(53)}$$

We search for solutions of type,

$$\mathbf{U}_p(\mathbf{x}) = \zeta \, \mathbf{F}_p(\mathbf{x}) \quad \textbf{(54)}$$

where ζ represents some constant to be determined. Substituting definition (54) into equation (53) yields,

$$\zeta \nabla^2 \mathbf{F}_p(\mathbf{x}) + \left(\zeta \omega^2 - 1\right) \mathbf{F}_p(\mathbf{x}) = \mathbf{0} \quad \textbf{(55)}$$

Taking into account property (50), it follows that,

$$\zeta = \frac{1}{\omega^2 - \lambda^2} \quad \textbf{(56)}$$

and therefore, the particular solution and source function bases are related through a scalar multiplier:

$$\mathbf{U}_p(\mathbf{x}) = \frac{1}{\omega^2 - \lambda^2} \mathbf{F}_p(\mathbf{x}) \quad \textbf{(57)}$$

Property (57) is the key feature of the dual reciprocity variant presented in this work, as it preserves the boundary integral formulation of the Trefftz finite elements, while avoiding the difficulties associated with the cumbersome expressions of the particular solution bases reported by other authors [9].

Complementary Solution

Once particular solution approximations are obtained for all finite elements, the generalized temperatures $\mathbf{X}_c$ corresponding to the complementary solution basis

$\mathbf{U}_c$ and the weights $\mathbf{Y}_r$ and $\mathbf{Y}_e$ corresponding to the normal flux approximations (47) and (48) must be computed. This is done by enforcing weakly equation (7) in the domain of the element, the convection law (11) on its Robin boundary, and the temperature compatibility conditions (9) and (39) on the essential (Dirichlet and interior) boundaries of the mesh.

Domain Equations

Basis $\mathbf{U}_c$ is used to enforce weakly equation (7) in the domain of the element,

$$\int \mathbf{U}_c^* \left(\nabla^2 T + \omega^2 T\right) d\Omega^e = \int \mathbf{U}_c^* f_0 \, d\Omega^e \tag{58}$$

where $\mathbf{U}_c^*$ denotes the complex conjugate transposed of vector $\mathbf{U}_c$.

The first term in expression (58) is integrated by parts and the Neumann boundary condition (10) and heat flux approximations (47) and (48) are inserted in the resulting boundary integrals, to yield,

$$\begin{aligned} &-\frac{1}{k}\int \mathbf{U}_c^* q_\Gamma \, d\Gamma_n^e - \frac{1}{k}\int \mathbf{U}_c^* \mathbf{Z} d\Gamma_r^e \mathbf{Y}_r - \frac{1}{k}\int \mathbf{U}_c^* \mathbf{Z} d\Gamma_e^e \mathbf{Y}_e - \\ &\int (\nabla \mathbf{U}_c)^* (\nabla T) d\Omega^e + \omega^2 \int \mathbf{U}_c^* T d\Omega^e = \int \mathbf{U}_c^* f_0 \, d\Omega^e \end{aligned} \tag{59}$$

Substitution of approximations (41) and (49) into expression (59) reads,

$$\begin{aligned} &-\frac{1}{k}\bar{\mathbf{q}}_\Gamma - \frac{1}{k}\mathbf{B}_c \mathbf{Y}_r - \frac{1}{k}\mathbf{B}_c \mathbf{Y}_e - \int (\nabla \mathbf{U}_c)^* (\nabla \mathbf{U}_c) d\Omega^e \mathbf{X}_c - \\ &\int (\nabla \mathbf{U}_c)^* (\nabla \mathbf{U}_p) d\Omega^e \mathbf{X}_p + \omega^2 \int \mathbf{U}_c^* \mathbf{U}_c d\Omega^e \mathbf{X}_c + \\ &\omega^2 \int \mathbf{U}_c^* \mathbf{U}_p d\Omega^e \mathbf{X}_p = \int \mathbf{U}_c^* \mathbf{F}_p d\Omega^e \mathbf{X}_p \end{aligned} \tag{60}$$

$$\bar{\mathbf{q}}_\Gamma = \int \mathbf{U}_c^* q_\Gamma \, d\Gamma_n^e \tag{61}$$

$$\mathbf{B}_c = \int \mathbf{U}_c^* \mathbf{Z} d\Gamma_{e(r)}^e \tag{62}$$

Integrating by parts the terms of type $\int (\nabla \mathbf{U_i})^* (\nabla \mathbf{U_j}) d\Omega^e$ in expression (60) and taking into account properties (42) and (53), all domain integrals terms vanish,

$$-\frac{1}{k}\bar{\mathbf{q}}_\Gamma - \frac{1}{k}\mathbf{B_c Y_r} - \frac{1}{k}\mathbf{B_c Y_e} - \int \mathbf{U_c}^* \mathbf{N}(\nabla \mathbf{U_c})\, d\Gamma^e \mathbf{X_c} - \int \mathbf{U_c}^* \mathbf{N}(\nabla \mathbf{U_p})\, d\Gamma^e \mathbf{X_p} = \mathbf{0} \qquad \textbf{(63)}$$

Finally, using definition (45), the domain statement of the hybrid-Trefftz temperature element becomes,

$$\mathbf{K_c X_c} - \mathbf{B_c Y_r} - \mathbf{B_c Y_e} = \bar{\mathbf{q}}_\Gamma - \mathbf{K_p X_p} \qquad \textbf{(64)}$$

$$\mathbf{K_i} = \int \mathbf{U_c}^* (\mathbf{N S_i})\, d\Gamma^e,\ i = \{c, p\} \qquad \textbf{(65)}$$

Convection Equations

The convection law (11) is enforced weakly on the Robin boundaries, using the heat flux bases $\mathbf{Z}$ for testing,

$$\int \mathbf{Z}^* \left(h^{-1}\mathbf{Nq} - (T - T_f)\right) d\Gamma_r^e = \mathbf{0} \qquad \textbf{(66)}$$

Substitution of the temperature (41) and normal heat flux (47) approximations into the above expression reads,

$$\int \mathbf{Z}^* h^{-1} \mathbf{Z} d\Gamma_r^e \mathbf{Y_r} - \int \mathbf{Z}^* \mathbf{U_p} d\Gamma_r^e \mathbf{X_p} - \int \mathbf{Z}^* \mathbf{U_c} d\Gamma_r^e \mathbf{X_c} = -\int \mathbf{Z}^* T_f\, d\Gamma_r^e \qquad \textbf{(67)}$$

Making notations,

$$\mathbf{H} = \int \mathbf{Z}^* h^{-1} \mathbf{Z} d\Gamma_r^e \qquad \textbf{(68)}$$

$$\bar{\mathbf{T}}_\mathbf{f} = \int \mathbf{Z}^* T_f\, d\Gamma_r^e \qquad \textbf{(69)}$$

$$\mathbf{B_p} = \int \mathbf{U_p}^* \mathbf{Z} d\Gamma_{e(r)}^e \qquad \textbf{(70)}$$

and taking into account definition (62) yields,

$$-\mathbf{B}_{\mathbf{c}}^{*}\mathbf{X}_{\mathbf{c}}+\mathbf{H}\mathbf{Y}_{\mathbf{r}}=-\bar{\mathbf{T}}_{\mathbf{f}}+\mathbf{B}_{\mathbf{p}}^{*}\mathbf{X}_{\mathbf{p}} \tag{71}$$

Temperature Compatibility Equations

On the exterior Dirichlet boundaries, the heat flux bases $\mathbf{Z}$ are used to enforce weakly the temperature compatibility equation (9), reading,

$$\int \mathbf{Z}^{*}(T-T_{\Gamma})\,d\Gamma_{d}^{e}=\mathbf{0} \tag{72}$$

Substituting approximation (41) into the above expression reads,

$$\mathbf{B}_{\mathbf{c}}^{*}\mathbf{X}_{\mathbf{c}}=\bar{\mathbf{T}}_{\Gamma}-\mathbf{B}_{\mathbf{p}}^{*}\mathbf{X}_{\mathbf{p}} \tag{73}$$

$$\bar{\mathbf{T}}_{\Gamma}=\int \mathbf{Z}^{*}T_{\Gamma}\,d\Gamma_{d}^{e} \tag{74}$$

On the interior boundaries Γ_{i}^{e}, temperature continuity (39) is enforced in the same way,

$$\int \mathbf{Z}^{*}\left(T^{j}-T^{k}\right)d\Gamma_{i}^{e}=\mathbf{0} \tag{75}$$

The same procedure as above immediately yields,

$$\mathbf{B}_{\mathbf{c}}^{\mathbf{j}*}\mathbf{X}_{\mathbf{c}}^{\mathbf{j}}-\mathbf{B}_{\mathbf{c}}^{\mathbf{i}*}\mathbf{X}_{\mathbf{c}}^{\mathbf{i}}=-\mathbf{B}_{\mathbf{p}}^{\mathbf{j}*}\mathbf{X}_{\mathbf{p}}^{\mathbf{j}}+\mathbf{B}_{\mathbf{p}}^{\mathbf{i}*}\mathbf{X}_{\mathbf{p}}^{\mathbf{i}} \tag{76}$$

Solving System

The solving system of the hybrid-Trefftz temperature element is constructed on the domain, convection, and temperature compatibility statements (64), (71) and (73) (or (76) for the case of internal boundaries),

$$\left[\begin{array}{c|cc}\mathbf{K}_{\mathbf{c}} & -\mathbf{B}_{\mathbf{c}} & -\mathbf{B}_{\mathbf{c}}\\ \hline -\mathbf{B}_{\mathbf{c}}^{*} & \mathbf{H} & \mathbf{0}\\ -\mathbf{B}_{\mathbf{c}}^{*} & \mathbf{0} & \mathbf{0}\end{array}\right]\left(\begin{array}{c}\mathbf{X}_{\mathbf{c}}\\ \hline \mathbf{Y}_{\mathbf{r}}\\ \mathbf{Y}_{\mathbf{e}}\end{array}\right)=\left(\begin{array}{c}\bar{\mathbf{q}}_{\Gamma}-\mathbf{K}_{\mathbf{p}}\mathbf{X}_{\mathbf{p}}\\ \hline -\bar{\mathbf{T}}_{\mathbf{f}}+\mathbf{B}_{\mathbf{p}}^{*}\mathbf{X}_{\mathbf{p}}\\ -\bar{\mathbf{T}}_{\Gamma}+\mathbf{B}_{\mathbf{p}}^{*}\mathbf{X}_{\mathbf{p}}\end{array}\right) \tag{77}$$

System (77) is sparse, Hermitian, and strongly localized, since the temperature variables are element-specific and the heat flux variables can be shared by, at most, two neighbouring elements. Also, all coefficients of the solving system are defined by boundary integral expression. Upon the solution of system (77), the temperature and heat flux fields in the domain of the elements are recovered using approximations (41) and (44).

IMPLEMENTATION ON THE FREEHYTE PLATFORM

For full reproducibility, the solution technique described above is implemented in the FreeHyTE environment. FreeHyTE is an open-source hybrid-Trefftz finite element platform which aims at taking the advantages of this class of formulations to the fingertips of scientists worldwide. FreeHyTE offers user-friendly interfaces to get new users started without needing to master the subtleties of the formulations, and features pre-programmed routines to help developers implement and test new formulations with minimal coding effort. FreeHyTE is based on a unifying approach to the solution of elliptic, parabolic and hyperbolic boundary value problems, endorsing their coding into similar frameworks. Its general description is available in reference [18]. Here, the presentation is limited to the the specific features of the non-homogeneous parabolic solver.

Workflow of the Implementation

The workflow of the computational implementation is presented in Fig. (**3**). The acquisition of the user input using the Graphical User Interface (GUI) is followed by the computation of the initial fields and the construction of the coefficient matrix of system (77). This matrix is independent of the time step and computed only once.

The two calculation steps that typify the DRM are performed in every time step. The particular solution is computed first, using the collocation procedure presented in the *Particular solution* section. System (52) is solved element by element. The right-hand side vector of system (77) is obtained next, and the system is solved to obtain the complementary solution. Finally, the temperature, its time derivative and the heat fluxes in the Cartesian directions are obtained using approximations (41), (15) and (44). The post-processing involves plotting the instantaneous solution and writing of the solution files.

Some key features of these steps are described below:

Input

The definitions of the structure, finite element mesh, heat source, and initial and boundary conditions are made using a series of GUIs. Their layout is fully described in the User's Manual of the module [20].

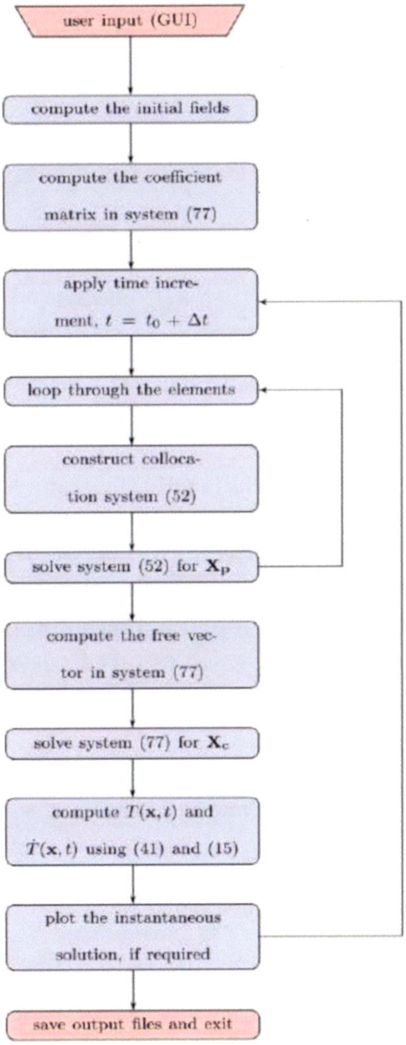

Fig. (3). Workflow of the implementation.

Geometry and Meshing

Two geometry and mesh generators are currently implemented in FreeHyTE, namely a regular mesh generator and a non-regular mesh generator.

The regular mesh generator (Fig. **4**) can only be used for rectangular domains, which are meshed using a regular grid of finite elements. It therefore suffices to specify the dimensions of the domain and the number of elements in the Cartesian directions. All elements are rectangular and identical in size.

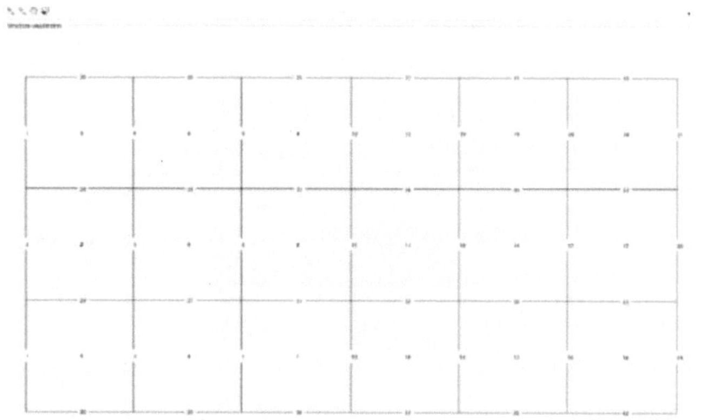

Fig. (4). Sample of a regular mesh in the FreeHyTEvisualization interface.

The non-regular mesh generator (Fig. **5**) can be used for any structural geometry. The domain is meshed into triangular finite elements, whose size can be controlled by specifying the largest admissible boundary and the boundary growth rate. Since all boundaries are linear, the representation of curved geometries is imperfect. It is noted that this is not a limitation of the formulation, but of its FreeHyTE implementation.

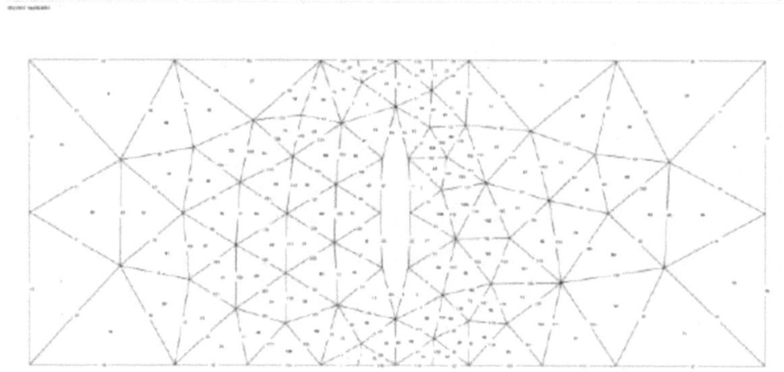

Fig. (5). Sample of a non-regular mesh in the FreeHyTEvisualization interface.

Definition of the Problem

Any valid mathematical expression is acceptable for the definition of the heat source term $Q(\mathbf{x},t)$, present in the heat diffusion equation (1). In Matlab, it is straightforward to define fields involving branch cuts or discontinuities using a single expression, so the definition is not limited to continuous fields. The same is true for the definition of the initial temperature field $T_0(\mathbf{x})$ and its time derivative $\dot{T}_0(\mathbf{x})$. Although the latter could be calculated analytically by collocating equation (1) at $t=0$, its analytic expression is generally easy to obtain, and requested from the user to avoid including the Symbolic Math Toolbox in the module's requirements. The compliance with equation (1) is not explicitly checked, meaning that the user may input physically inconsistent initial fields.

The boundary conditions must be defined by expressions of the type,

$$T_\Gamma\left(\mathbf{x},t\right)=T_x(\mathbf{x})\cdot T_t(t), \text{ on } \Gamma_\mathrm{d} \tag{78}$$

$$q_\Gamma\left(\mathbf{x},t\right)=q_x(\mathbf{x})\cdot q_t(t), \text{ on } \Gamma_\mathrm{n} \tag{79}$$

The space distribution of the enforced temperature and normal heat flux can be defined by any polynomial expression, input by specifying the field's values in as many equally spaced points along the boundary as needed to interpolate the desired polynomial. The time variations of the enforced temperature $T_t(t)$, normal heat flux $q_t(t)$, and surrounding fluid temperature $T_f(t)$ can be defined by any valid mathematical expression (and may include points of discontinuity), or may be specified by time-value pairs loaded from an input file, to account for irregular time variations as caused, for instance, by the action of a fire. Space and time boundary conditions are defined separately and can be different for different boundaries.

Algorithmic Definitions

The algorithmic definitions specified by the user include the time discretization parameters Δt and $\bar{\theta}$, the generalized wave numbers λ present in the definition of the source function basis (50), the orders of the domain and boundary bases, and the number of integration points per finite element side.

Regarding the time discretization parameters, while their choices are free, it is recalled that the definition of $\bar{\theta} < 0.5$ leads to conditionally stable time stepping algorithms, which may lead to numerical instability if the time step is chosen too large.

Considerable flexibility is granted to the user regarding the construction of the domain and boundary bases present in definitions (41), (47) and (48). The domain bases $\mathbf{U}_c$ and $\mathbf{U}_p$ can be constructed independently for each finite element and their orders need not match. Three generalized wave numbers λ are used to construct the particular solution basis $\mathbf{U}_p$. As shown in the numerical example presented in section *Thermal shock problem*, the source function approximation is fairly robust to sub-optimal choices of λ, as long as they are not large enough to cause numerical instability. The predefined values, $\lambda = \{1,2,3\}$ typically yield stable approximations, but lower values may be used if the solution is unstable. The boundary flux bases $\mathbf{Z}$ are constructed using Chebyshev polynomials. Their orders are also free to choose on each Robin and essential boundary, but should be small enough to ensure that the problem is kinematically indeterminate (see next section for further details on the kinematic indeterminacy). All approximation bases are hierarchical, meaning that the addition of a new approximation function does not require the redefinition of those previously present in the bases.

Finally, the user is also free to define the number of (Gauss-Legendre) integration points on the boundaries of the elements. This number is fixed for all boundaries and integrals, so it should be large enough to ensure that more steep or oscillating integrands are adequately handled.

Processing

FreeHyTE uses a uniform processing framework regardless of the boundary value problem that needs to be solved. This framework is described in detail in reference [24] and is not repeated here. Therefore, the following presentation is limited to the features specific to the solver of parabolic equations.

Construction of the Particular Solution

The particular solution approximation is obtained by solving the collocation system (52) to obtain the weights $\mathbf{X}_p$ of the functions collected in the particular solution basis (57). This is done element by element.

Let n_{p_i} be the order of the source function basis (51) constructed for the generalized wave number λ_i. As mentioned above, the choices of n_{p_i} and λ_i are free and input by the user through the GUI, but the number of generalized wave numbers is always equal to three, that is, $i = \{1, 2, 3\}$. The layout of the source function basis (51) is,

$$\mathbf{F}_{\mathbf{p}}(r,\theta) = \underbrace{\left[J_{-n_{p_i}}(\lambda_i r)\cdot\exp\left(-\mathrm{i}\cdot n_{p_i}\theta\right)\cdots J_0(\lambda_i r)\cdots J_{n_{p_i}}(\lambda_i r)\cdot\exp\left(\mathrm{i}\cdot n_{p_i}\theta\right)\right]}_{\text{for } i=\{1,2,3\}} \quad \textbf{(80)}$$

meaning that the total number of approximation functions per element is,

$$N_p = \sum_{i=1}^{3}\left(2\cdot n_{p_i} + 1\right) \quad \textbf{(81)}$$

N_p also denotes the number of columns in the collocation system (52), for the respective finite element. Its number of lines is equal to the number of collocation points in the domain of the element.

The collocation points are not chosen by the user. They are the Gauss-Legendre quadrature points of the element. The number of quadrature points is chosen to be the square number which is closest to, but higher than the total number of functions $\left(N_p\right)$ in the particular solution basis. This condition ensures that the collocation system (52) is not underdetermined. Since system (52) is generally overdetermined, it is solved using the Moore-Penrose pseudoinverse.

Construction of the Complementary Solution

Let n_{c_i} be the order of the complementary solution basis $\mathbf{U}_{\mathbf{c}}$ of element i, defined by equation (43). The wave number ω present in definition (43) is not arbitrary (as was the case of the particular solution basis), but dependent on the material properties and algorithmic parameters of the time integration method, as defined by expression (19). The total number of approximation functions collected in the complementary solution bases of all finite elements is,

$$N_c = \sum_{i=1}^{NEL}\left(2\cdot n_{c_i} + 1\right) \quad \textbf{(82)}$$

where NEL is the total number of finite elements in the mesh.

Let $n_{\Gamma_{rj}}$ be the degree of the Chebyshev polynomial basis $\mathbf{Z}$ on the Robin boundary $\Gamma^e_{r_j}$. The total number of approximation functions collected in the bases of all Robin boundaries is,

$$N_{\Gamma_r} = \sum_{j=1}^{NR} \left(n_{\Gamma_{rj}} + 1 \right) \tag{83}$$

where NR is the total number of Robin boundaries in the mesh.

Finally, let $n_{\Gamma_{ej}}$ be the degree of the Chebyshev polynomial basis $\mathbf{Z}$ on the essential (*i.e.* Dirichlet and interior) boundary $\Gamma^e_{e_j}$. The total number of approximation functions collected in the bases of all essential boundaries is,

$$N_{\Gamma_e} = \sum_{j=1}^{NE} \left(n_{\Gamma_{ej}} + 1 \right) \tag{84}$$

where NE is the total number of essential boundaries in the mesh.

Under the assumption that all equations in system (77) are linearly independent, a necessary condition to ensure that the relative dimensions of the domain and boundary bases do not induce spurious modes is to require that the structure is both *statically* and *kinematically* indeterminate[1]. System (77) is said to be statically indeterminate if the generalized heat fluxes collected in vectors $\mathbf{Y_r}$ and $\mathbf{Y_e}$ cannot be computed strictly from the heat diffusion equation (7). This is equivalent to require that the first set of equations in system (77) is underdetermined. The static indeterminacy condition is thus,

$$\beta_s = \underbrace{\left(N_c + N_{\Gamma_r} + N_{\Gamma_e} \right)}_{\text{no of unknowns}} - \underbrace{N_c}_{\text{no of eqns}} = \left(N_{\Gamma_r} + N_{\Gamma_e} \right) > 0 \tag{85}$$

---

[1] Because of the lack of terminology adapted to heat diffusion problems, the nomenclature used in Computational Mechanics is adopted here.

The static indeterminacy number β_s is always positive if at least one Dirichlet or Robin boundary exists. System (77) is said to be kinematically indeterminate if the generalized temperatures collected in vector $\mathbf{X}_c$ cannot be computed strictly from its second and third sets of equations. The kinematic indeterminacy number β_k is thus defined as,

$$\beta_k = \underbrace{\left(N_c + N_{\Gamma_r}\right)}_{\text{no of unknowns}} - \underbrace{\left(N_{\Gamma_r} + N_{\Gamma_e}\right)}_{\text{no of eqns}} = \left(N_c - N_{\Gamma_e}\right) > 0 \tag{86}$$

Unlike the static indeterminacy condition (85), the kinematic indeterminacy condition is not trivial. It is satisfied if the total number of approximation functions collected in the heat flux bases on the Dirichlet boundaries is inferior to the total number of approximation functions collected in the domain bases. Robin bases are neutral to the kinematic indeterminacy condition.

The closer the kinematic indeterminacy number is to zero, the more constrained is the temperature solution by the enforcement of the Dirichlet boundary conditions. To avoid overconstrainment, it is recommended that condition (86) is satisfied at each element's level, instead of at the structure's level.

System (77) is pre-conditioned before it is solved using a technique designed to reduce the diagonal terms of the system to unity, while preserving its symmetry properties. This approach is described in detail in reference [18] and has been shown to be effective in improving the condition number [25].

The matrix of coefficients is computed only once, since only the free vector changes between successive iterations. If the inverse of its condition number is larger than the machine's precision, the matrix of coefficients is stored in its LU-factored form, and only back-substitutions are performed at every time step. If the inverse of the condition number is inferior to the machine's precision, the matrix of coefficients is considered ill-conditioned and its Moore-Penrose pseudo-inverse is calculated and stored for the solution of the system at every time step. However, the results may contain errors.

It is noted that ill-conditioning of the solving systems in Trefftz methods is typically caused by excessive orders of p-refinement. The condition number can therefore be reduced by decreasing the order of p-refinement and refining the finite element mesh instead.

Post-Processing

Post-processing consists of the calculation of the temperature, its time derivative, and heat flux fields at the end of the time interval, and of the output of the results.

The temperature and heat flux estimates at the end of the time interval are computed from approximations (41) and (44), respectively. The time derivative of the temperature is obtained from approximation (15).

The values of the temperature field and its time derivative are stored in the collocation points $(\bar{\mathbf{x}})$ of each element. They are required for the calculation of the source function f_0 in the next time step, according to definition f_0.

The output in FreeHyTE consists of solution files containing the temperature and heat flux fields at the end of each time step and colour plots of the same quantities with a periodicity defined by the user.

The solution files contain a list of temperature and heat flux values in a grid of points within each finite element. The points where the solutions are listed belong, by default, to a Gauss-Legendre quadrature constructed with as many points (in each Cartesian direction) as defined by the user in the GUI. Header rows are added to the solution files to render them directly compatible with the visualization softwares Tecplot and Paraview, but their contents are, of course, compatible with most mainstream visualization packages. Separate solution files are produced at every time step.

Colour plots of the solutions are not produced at each time step, since the process is notoriously time consuming, particularly in Matlab. Instead, the user is required to provide the periodicity of these plots in the GUI (entering zero produces no plots at all). Further information on the output produced by FreeHyTE can be found in reference [18].

NUMERICAL EXAMPLES

Three numerical tests of the formulation proposed in this paper are presented in this section.

The first two testing sequences focus well-known benchmark problems with analytic solutions. One problem involves a thermal shock applied to a uniformly heated domain (section *Thermal shock problem*). The sudden variation of the temperature puts considerable difficulties to the time discretization scheme and is used to assess the influence of the time step and choice of the parameter $\bar{\theta}$ on the quality of the results. Also, the effect of sub-optimal choices of the algorithmic parameters λ used to construct the particular solution basis in the quality of the solution is assessed. The other problem involves a domain with a non-smooth boundary and is used to illustrate the convergence of the solution under mesh and basis refinements (section *Non-smooth boundary problem*).

The last example is a real world application involving a steel I beam subjected to a fire near one of its flanges (Section *I-beam subjected to fire*).

Thermal Shock Problem

The thermal shock problem solved in this section is a well-known benchmark for testing transient heat conduction solvers. It was first proposed in 1974 by Bruch [26], and solved using the boundary element method by Brebbia [27], and Trefftz finite elements by Jirousek [17]. The 'conventional' variant of the Dual Reciprocity Method, involving radial basis functions for the expansion of the source function, was also applied to the solution of this problem by Partridge [28] and Cao [29].

The problem involves the square domain presented in Fig. (**6**), with an $L=3$ side and constant initial temperature $T_0(x,y)=30$, cooled by the sudden application of a temperature field $T_\Gamma(x,y,t)=0$ on all its exterior boundaries, which remains constant for the whole duration of the analysis, $t_{max}=1.2$. The material properties are $\rho=c=1$, and $k=1.25$. No internal heat generation occurs, that is, $Q(x,y,t)=0$.

The analytic solution of this problem is,

$$T(x,y,t)=\sum_{i=1}^{\infty}\sum_{j=1}^{\infty}A_{ij}\sin\left(\frac{i\pi x}{3}\right)\sin\left(\frac{j\pi y}{3}\right)\exp\left(\frac{-k\pi^2\left(i^2+j^2\right)t}{3^2}\right) \tag{87}$$

where,

$$A_{ij}=120\cdot\frac{\left[(-1)^i-1\right]\cdot\left[(-1)^j-1\right]}{ij\pi^2} \tag{88}$$

The main difficulty posed by this problem is the C^0 time discontinuity of the boundary fields. The sudden drop in temperature causes a Gibbs-type effect in the vicinity of the exterior boundaries, which can be controlled by the choice of time discretization parameters, Δt and $\overline{\theta}$.

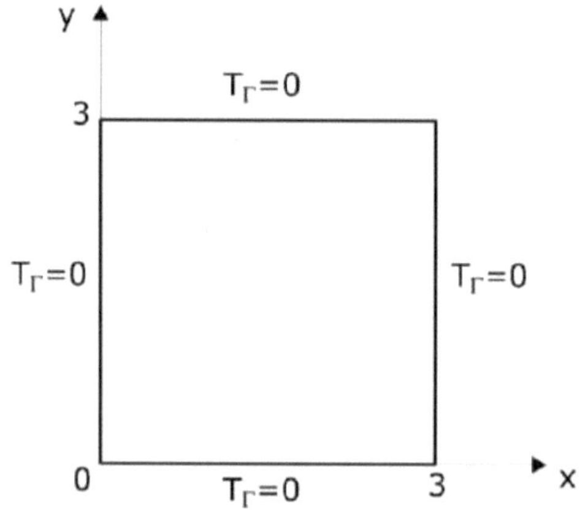

Fig. (6). Description of the thermal shock problem.

The problem is solved using a mesh of four square hybrid-Trefftz elements of 1.5 side. The source function is approximated using bases of type (80), of order $n_{p_i} = 3$. The wave numbers used for the construction of the bases are $\lambda_1 = 1$, $\lambda_2 = 2$, and $\lambda_3 = 3$. The particular solution approximation basis is obtained from the source function (80) by applying definition (57). The total number of approximation functions for the particular solution is $N_p = 84$, in all elements.

The complementary solution approximation is constructed on functions of type (43). Their order is $n_c = 8$ in all elements. The total number of kinematic degrees of freedom is $N_c = 68$. Chebyshev polynomials of degree $n_{\Gamma_e} = 3$ are used for the normal heat flux approximation on all essential boundaries. The total number of static degrees of freedom is $N_{\Gamma_e} = 48$. The dimension of the solving system (77) is 116.

For the time discretization of the problem, the total time $t_{max} = 1.2$ is split into time steps of size $\Delta t = \{0.1, 0.05, 0.03, 0.02, 0.01, \text{and}\, 0.005\}$. The number of time steps varies from 12, for $\Delta t = 0.1$, to 600, for $\Delta t = 0.005$. For each time refinement, the calibration parameter $\bar{\theta}$ is set to 0.5 (Crank-Nicolson), 0.67 (Galerkin), and 1.0 (backward Euler).

The peak values of the temperature obtained with each time refinement at the end of the analysis, that is, at $t_{max} = 1.2$, are listed in Table 1. The exact solution is obtained from equation (87), $T(1.5,1.5,1.2) = 1.812$.

Table 1. Maximum temperature values at the end of the analysis (exact solution is 1.812).

Description	$\bar{\theta}$	$\Delta t = 0.1$	$\Delta t = 0.05$	$\Delta t = 0.03$	$\Delta t = 0.02$	$\Delta t = 0.01$	$\Delta t = 0.005$
Hybrid-Trefftz elements in FreeHyTE, $N_{\Gamma_e} = 48$	$\frac{1}{2}$	2.099	1.913	1.857	1.828	1.802	1.815
	$\frac{2}{3}$	2.248	2.014	1.919	1.870	1.820	1.808
	1	2.644	2.221	2.039	1.951	1.860	1.818
Hybrid-Trefftz elements in [17], $N_{\Gamma_e} = 121$	$\frac{1}{2}$	1.929	1.894	-	-	1.852	-
Hybrid-Trefftz elements in [17], $N_{\Gamma_e} = 961$	$\frac{1}{2}$	1.874	1.829	-	-	1.816	-
DRM-Trefftz and DRM-MFS in [29]	1	-	1.776	-	-	-	-

The results obtained with the present formulation are also compared with some of those reported in the literature, namely in references [17] and [29]. In reference [17], Jirousek uses hybrid-Trefftz finite elements with linear frame functions for the solution of the thermal shock problem. Although cubic 'frame' functions are used in this work, Jirousek uses much more finite elements (from 25 to 225 for a quarter of the domain, as compared to a single one), so the total number of frame degrees of freedom is much larger (from 121 to 961, as compared to 48). The particular solution is computed using Green's functions.

In reference [29], Cao solves the thermal shock problem with two meshless methods, involving regular (Trefftz) and fundamental solutions for the domain temperature approximation. The particular solution is computed using the DRM, with radial basis functions for the approximation of the source term.

It is clear that the maximum temperature converges to the exact solution as the time step is decreased, in all cases. The precision for smaller time steps is quite high, and similar for the four models. For larger time steps, the numerical damping induced to the solution by values of $\overline{\theta}$ superior to 0.5 slightly over-damps the solution, causing the (decreasing) temperature amplitude to be higher than the target value. For smaller time steps, all choices of $\overline{\theta}$ yield accurate solutions.

In order to assess the convergence of the solutions under time step refinement over the whole domain, a global error norm is defined as,

$$\epsilon = \frac{\| T_{FE} - T \|}{\| T \|} \tag{89}$$

where $\| T \|$ is the L^2-norm of the exact temperature field (87) computed over the domain, and T_{FE} is the temperature field approximated using hybrid-Trefftz finite elements.

The variation of the error norm (89) is plotted as a function of the time step Δt in Fig. (**7**). The solid, dashed, and dash-dotted lines correspond to the error norms obtained with calibration parameters $\left(\overline{\theta}\right)$ of 0.5, 0.67 and 1.0, respectively. Both time step and error axes are logarithmic.

For all values of $\overline{\theta}$, monotonical convergence is obtained over the tested range of time steps. As for the peak temperature values, the lowest error norms are generally recovered for $\overline{\theta} = 0.5$, but the differences become smaller for the smallest time step.

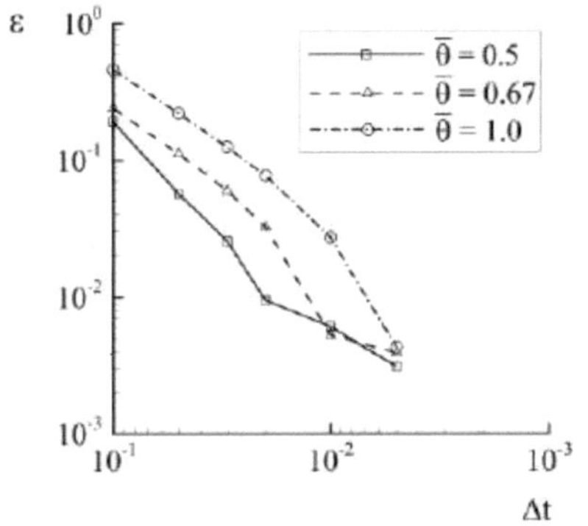

Fig. (7). Convergence of the temperature solution under time step refinement.

It is, however, misleading to conclude that the calibration parameter $\bar{\theta} = 0.5$ simply works better for the whole range of time steps. In fact, higher values of $\bar{\theta}$ are effective in damping the spurious oscillations caused by the sudden temperature drop in the vicinity of the boundary, and this may be more important for the analyst than the overall error. To illustrate this, plots of the temperature field at $t_{max} = 1.2$ are given in Fig. (**8**) for $\Delta t = 0.1$ and $\bar{\theta} = \{0.5, 0.67, 1.0\}$.

Although closer to the analytic solution both in terms of amplitude and average values, the solution obtained with $\bar{\theta} = 0.5$ is strongly affected by the spurious oscillations in the vicinity of the boundary. Conversely, the spurious oscillations are damped by the alternative choices of $\bar{\theta}$, causing the boundary conditions to be well recovered over the whole boundary. However, the higher numerical damping also damps the cooling effect itself, causing the temperature to be overestimated in the domain.

A source of uncertainty in the solution process is the choice of the algorithmic wave numbers λ_i, $i = \{1, 2, 3\}$, present in the approximation (51) of the source function. While in the solutions presented above, the (trivial) choice $\lambda_i = i$ seems to yield good results, it is important to assess the robustness of the solutions to a wider range of wave numbers. To achieve this, the value of the first wave number is fixed at $\lambda_1 = 1$ and ten thousand random combinations of λ_2 and λ_3 in the interval $0 < \lambda_{2,3} \le 9$ are used to compute the temperature error norm (89). The time step is set to $\Delta t = 0.01$ and the calibration parameter $\bar{\theta}$ is set to $2/3$.

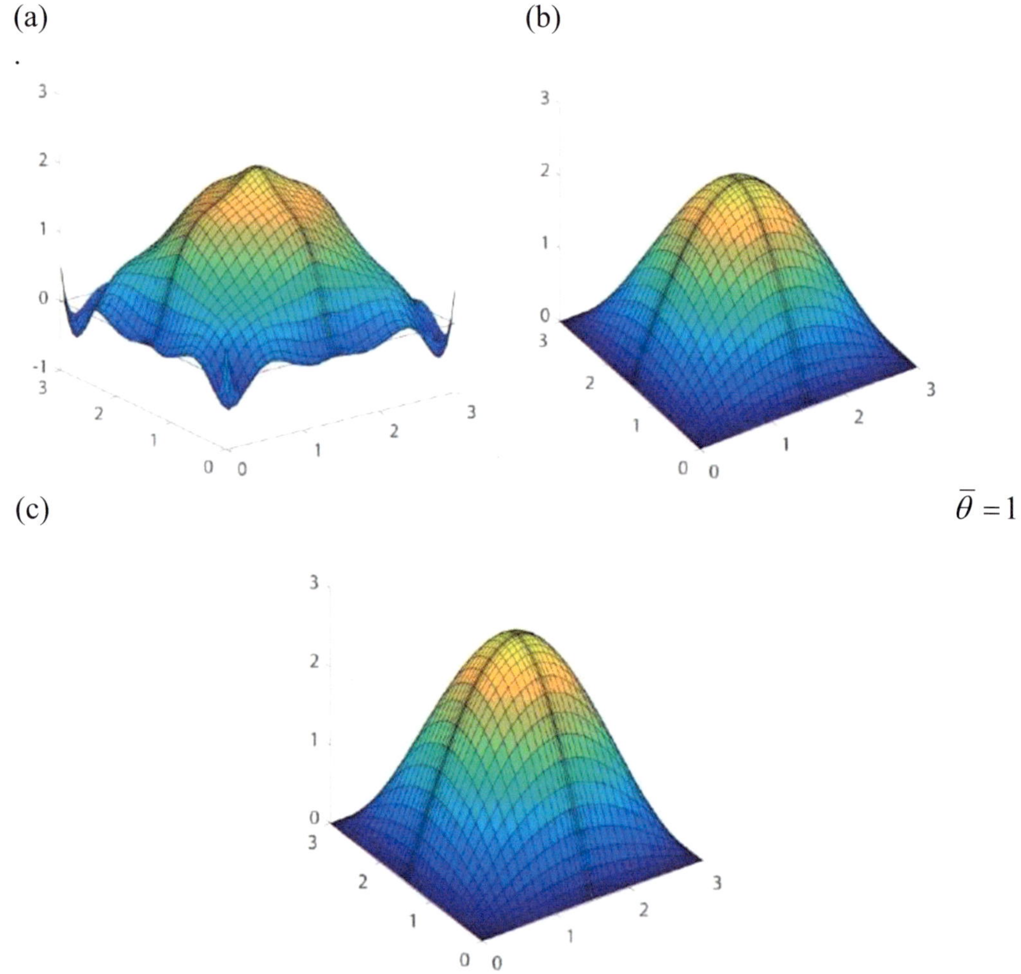

Fig. (8). Temperature field at $t_{max} = 1.2$ for various values of the calibration parameter.

The error norms obtained for these combinations are plotted in Fig. (**9**). The plot presents a large and continuous small error region located in the area where the wave number values are relatively small and balanced. The error norms remain below roughly 1% for all wave number combinations below $\lambda_{2,3} = 4$. The errors remain small even for larger wave numbers (roughly below 6) as long as the values of λ_2 and λ_3 remain balanced. They may, however, not be equal, since that would cause dependencies in the particular solution solving system (52). Conversely, when the values of λ_2 and λ_3 become unbalanced, localized escalations of the error are observed. Somewhat surprinsingly, the absolute smallest solution error (0.18%) was actually observed for quite unbalanced cobinations of wave numbers, namely for $\{\lambda_2, \lambda_3\} = \{2.292, 8.572\}$, a point that corresponds to a depression between two local maxima. However, for a robust solution, insensitive to sub-optimal choices of the wave numbers, it is recommended that the wave numbers are kept relatively small and balanced. Finally, it is observed that large wave numbers tend to render the source function basis highly oscillatory, causing serious numerical issues and dramatically increasing the error.

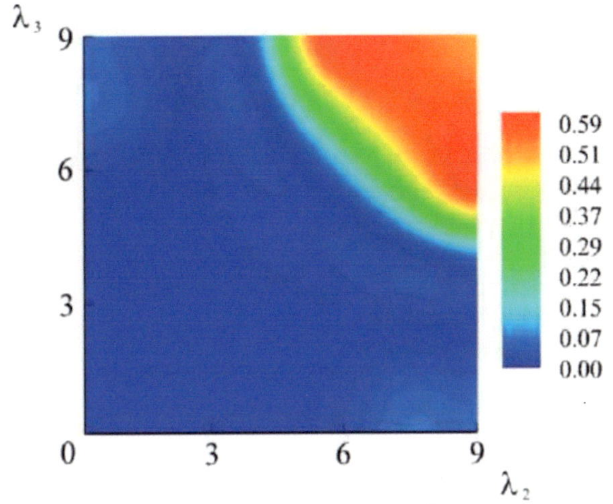

Fig. (9). Error norm variation with two random wave numbers.

Non-Smooth Boundary Problem

The second benchmark problem is solved by Cao [29], using two meshless approaches where the complementary solution basis is constructed using regular Trefftz functions and the Method of Fundamental Solution. In both cases, the DRM with radial basis functions is used for the source function approximation.

The problem is defined on the L-shaped domain presented in Fig. (**10**). The material properties are $\rho = c = k = 1$, and the internal heat generation is null $Q(x, y, t) = 0$.

The analytic solution of the problem is defined by,

$$T(x, y, t) = \cos(3x) \cdot \sinh(2y) \cdot \exp(-5t) \quad \textbf{(90)}$$

On all exterior boundaries, temperature fields consistent with the analytic solution (90) are prescribed.

The problem is solved using the two meshes of hybrid-Trefftz finite elements illustrated in Fig. (**11**), and the levels of basis (p-)refinement listed in Table **2**. For each mesh, five p-refinement levels are used. The first four levels correspond to a low (and constant) order of p-refinement of the particular solution basis, and increasing orders of p-refinement of the complementary solution and boundary bases. For the fifth level of p-refinement, the complementary and boundary bases are fixed, and the particular solution basis is refined. The orders listed in Table **2** are constant for all finite elements and essential boundaries.

The discretization in time is performed by dividing the total duration $t_{max} = 1.0$ of the problem into 50 time steps of $\Delta t = 0.02$. The calibration parameter $\overline{\theta}$ is fixed at 0.67 (Galerkin time stepping scheme).

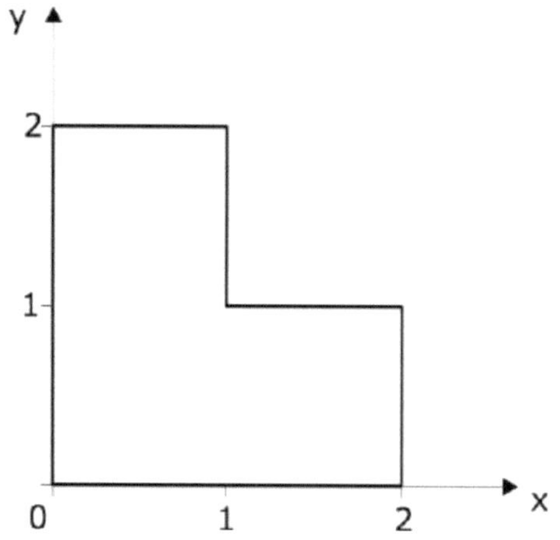

Fig. (10). Description of the non-smooth boundary problem.

The global error norms (89) at $t = t_{max}$, obtained with refinements A and F (see Table **2**), are listed in Table **3** and compared with those reported in reference [29]. The latter results were obtained with the Method of Fundamental Solutions using 20 and 80 source and collocation points, 50 time steps, and $\overline{\theta} = 1.0$.

To assess the influence of the mesh and basis refinements on the finite element solution, the error norms ε computed using all refinements listed in Table **2** are plotted against the total number of degrees of freedom of the model (the total dimension of the domain and boundary bases, NDoF) in Fig. (**12**). Both axes are logarithmic.

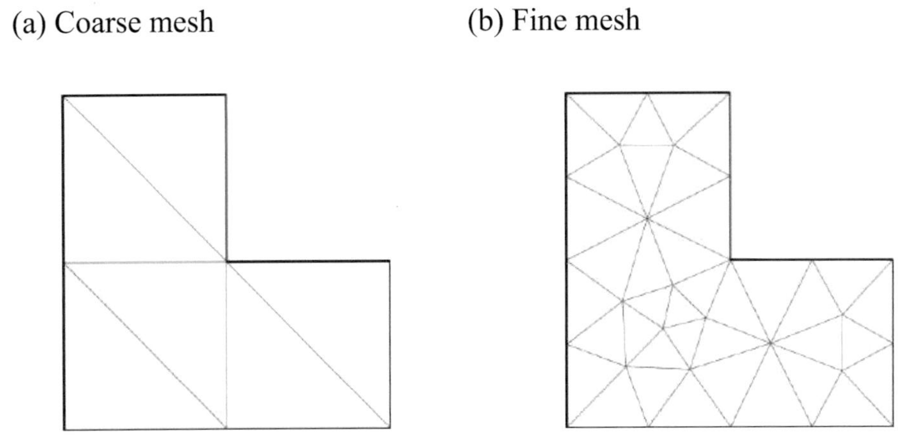

Fig. (11). Meshes of hybrid-Trefftz elements.

Table 2. Orders of p-refinement.

	Coarse mesh					**Fine mesh**				
n_c	5	7	9	11	11	4	5	7	9	9
$n_{\boxtimes}$	2	3	4	6	6	1	2	3	4	4
n_p	4	4	4	4	6	3	3	3	3	5
Labels	A	B	C	D	E	F	G	H	I	J

Table 3. Temperature error norms.

	Hybrid-Trefftz finite elements		**MFS in [29]**	
	refinements		*source points*	
	A	F	20	80
Error Norm	0.0255	0.0124	0.218	0.118

It is clear that the refinement of the approximation bases is more effective in reducing the average error than the refinement of the mesh. Although the coarsest levels of p-refinement produce less accurate solutions with the coarse mesh (point A) than with the fine mesh (point F), the errors obtained with the coarse mesh decrease fast with the p-refinement, reaching 0.37% in point E. For the finer mesh, the steep decrease of the error only occurs for the coarsest p-refinements (point F to point G). The same level of error as in point E is reached in point H, but at the expense of a total number of degrees of freedom roughly 3.5 times larger. From points H to J, the error does not decrease further, because it is bounded by the discretization in time, and not in space. Therefore, it can only be further decreased by reducing the time step and optimizing the calibration parameter $\overline{\theta}$.

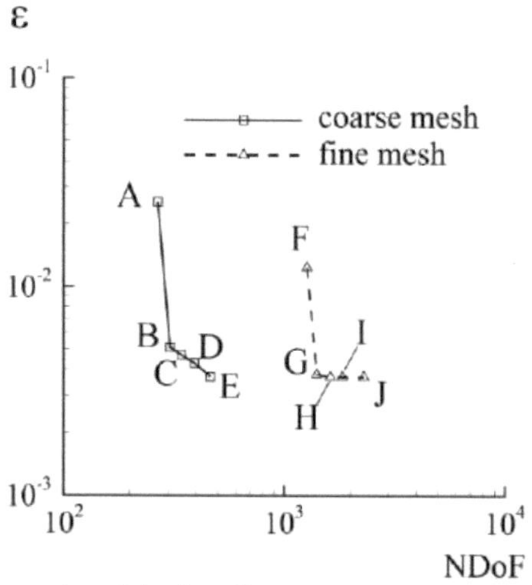

Fig. (12). Convergence under mesh and basis refinements.

However, the difference between the solutions obtained with the coarsest and finest refinement may not justify the time investment into exceedingly refining the model. Indeed, little difference is visually perceivable between the temperature fields recovered at $t = t_{max}$ with refinement A (Fig. **13a**) and with refinement J (Fig. **13b**). Both solutions recover accurately the enforced boundary conditions and the inter-element temperature continuity. However, the computational time to obtain the latter result (18s) was roughly 4.5 times higher than for the former (4s).

I-Beam Subjected To Fire

The last numerical example involves a steel I-beam subjected to a fire. The beam is a European profile IPE300, its cross section presented in Fig. (**14**). The material is a standard European structural steel, with conductivity $k = 40W/(mK)$ mass density $\rho = 7860 kg/m^3$, and specific heat $c = 490 J/(kg K)$.

The I-beam is insulated along its web and on the interior side of its flanges $\left(q_{\Gamma} = 0W/m^2\right)$, and exposed to the surrounding air on the exterior side of its flanges (Fig. **14a**). Accordingly, the insulated sides are modelled as Neumann boundaries, and the sides exposed to air are modelled as Robin boundaries. The initial conditions of the problem correspond to a constant temperature $T_0(x,y) = 32.5^{\circ}C$ throughout the cross section and the surrounding air.

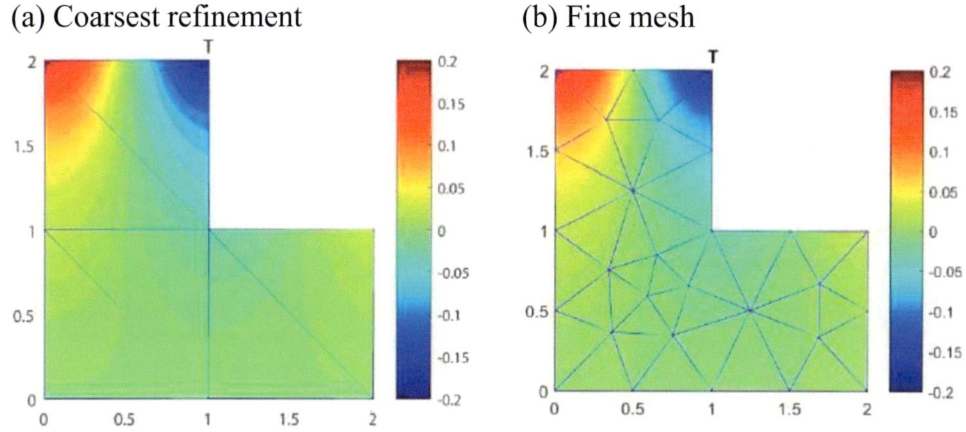

Fig. (13). Temperature fields at $t = t_{max}$ obtained with the coarsest and finest refinements.

The I-beam is subjected to a fire which causes the heating of the air underneath its bottom flange, while the air around its top flange boundary remains at the initial temperature (Fig. **14b**).

The fire event is described by the time-temperature curve presented in Fig. (**15**). It is a customized fire curve for tunnel design, which takes into account the activation of the sprinklers and the response of the fire squad after roughly 25-30 minutes [30]. The effect of the fire is applied by means of Robin (convection) boundary conditions of type (6). The fluid temperature $T_f(t)$ is defined by the time-temperature curve in Fig. (**15**), on the lower boundary, and maintained constant at $T_f = 32.5^{\circ}C$ on the upper boundary. The heat transfer coefficient is set to $h = 28\,W/m^2K$ on the bottom flange and $h = 22.5\,W/m^2K$ on the top flange, to account for the higher air speed near the bottom flange, as compared to the top flange.

The time discretization is performed by dividing the total duration $t_{max} = 2100$ s of the problem into 210 time steps of $\Delta t = 10$ s. The calibration parameter $\bar{\theta}$ is fixed at 0.5 (Crank-Nicolson time stepping scheme).

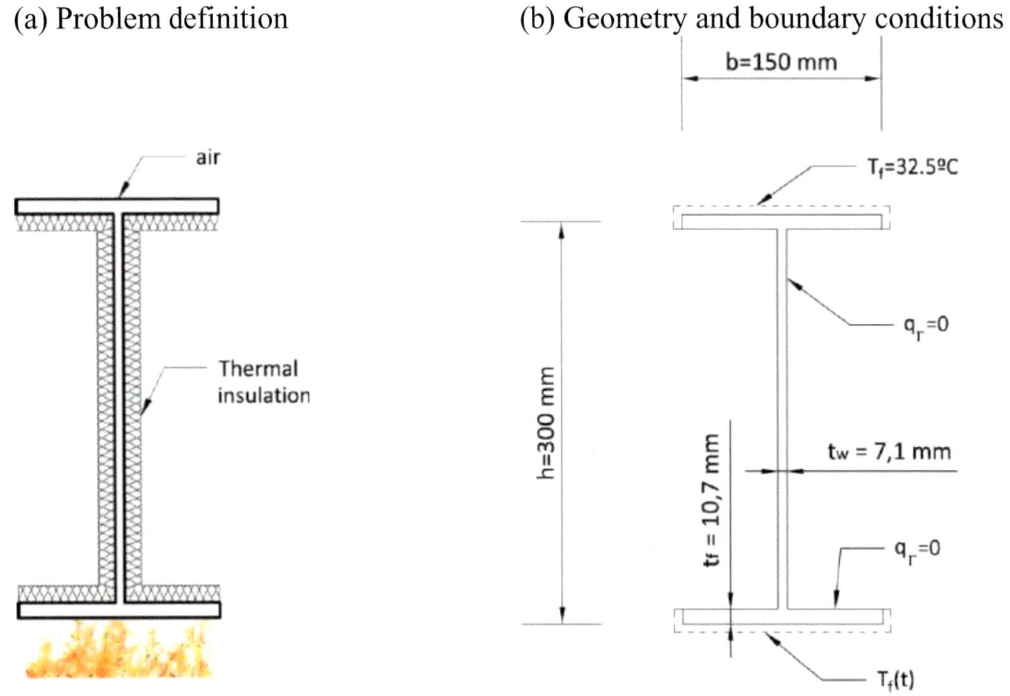

Fig. (14). I-beam subjected to fire: problem definition and boundary conditions.

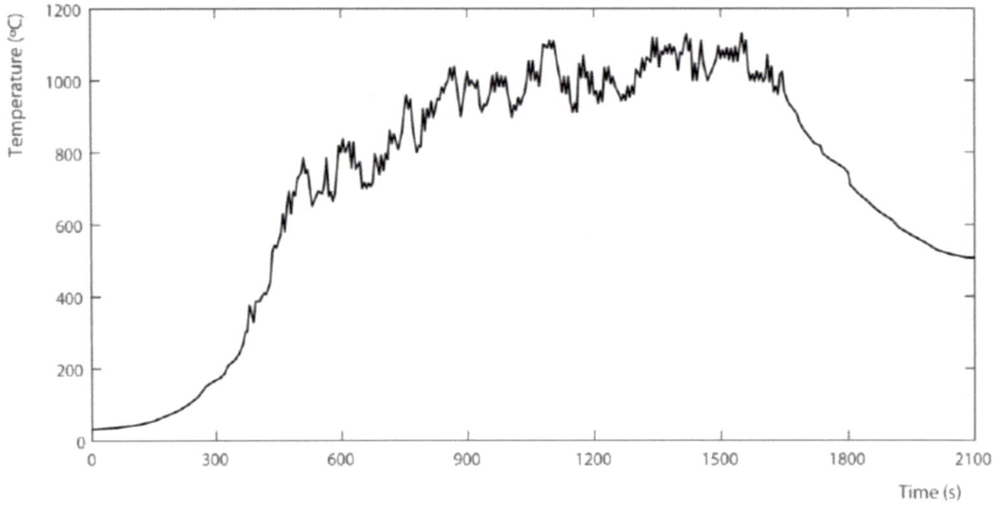

Fig. (15). Time-temperature plot of the fire event.

The source function is approximated in each element using bases of type (80), and of order $n_{p_i} = 3$. The wave numbers used for the construction of the bases are $\lambda_1 = 1$, $\lambda_2 = 2$, and $\lambda_3 = 3$. The particular solution basis is obtained from the source function (80) by applying definition (57). The total number of approximation functions for the particular solution is $N_p = 2352$, in all elements.

The complementary solution approximation is constructed using functions of type (43). Their order is $n_c = 6$ in all elements. The total number of kinematic degrees of freedom is $N_c = 1456$. Chebyshev polynomials of degree $n_{\Gamma_e} = 2$ are used for the normal heat flux approximation on all essential boundaries. The total number of static degrees of freedom is $N_{\Gamma_e} = 453$. The dimension of the solving system (77) is 1909.

The I-beam is meshed in 110 finite elements, as presented in Fig. (**16**), with superior refinement at the web-flange joints, where large heat flux concentrations (and discontinuities) are expected. The total computational time on an Intel i7 CPU, at 2.60GHz, is roughly 127s.

The temperature graphs throughout the I-beam starting at $t = 6.0\,\text{min}$ and plotted every four minutes are presented in Fig. (**17**). The same scale is used in all plots, to enable the clear observation of the heating front, that is, the high temperature gradient area where the material is heating.

In all plots, the applied boundary conditions and the temperature continuity between adjacent elements are well recovered. The convective heat transmitted by the fire sharply increases the temperature of the bottom flange, particularly near its lateral edges, where the upward heat flux is prevented by the insulation (Fig. **14a**). Conversely, the temperature is slightly lower near the centre of the flange (Fig. **17b** and Fig. **17c**), where a large vertical heat flux occurs. The temperature values drop swiftly at the heating front and stabilize close to the initial temperature at the upper part of the web and throughout the top flange. The heating front propagates upwards through the flange at all times, as the convective heat flux continues to heat the bottom flange, despite the air temperature drop after roughly 27 minutes (1700s, see Fig. (**15**)).

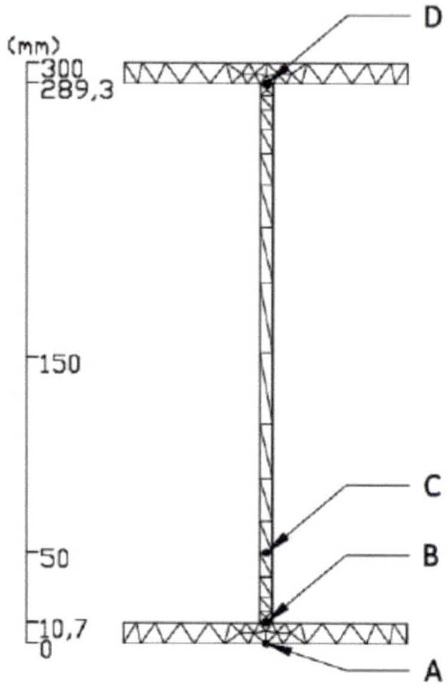

Fig. (16). Finite element mesh and solution points.

To clarify the temperature evolution through the bottom flange and web of the I-beam, time-history plots of the air temperatures in points A to D (identified in Fig. (**17**) are presented in Fig. (**18**).

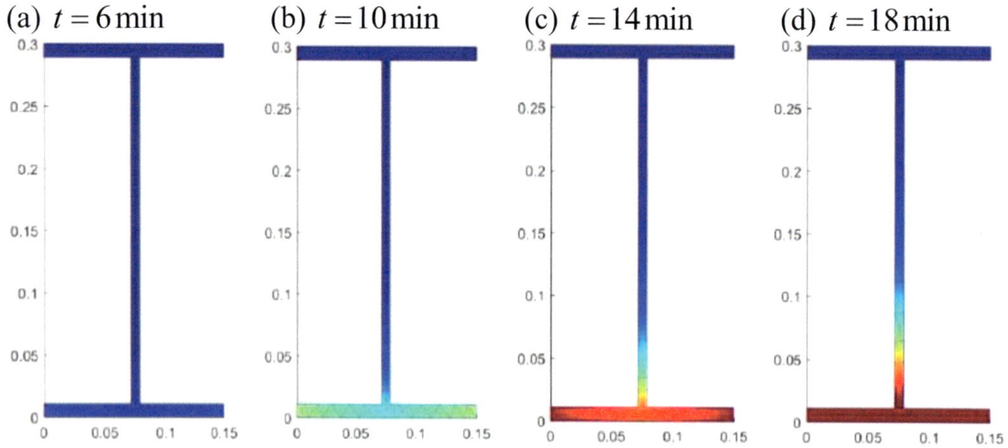

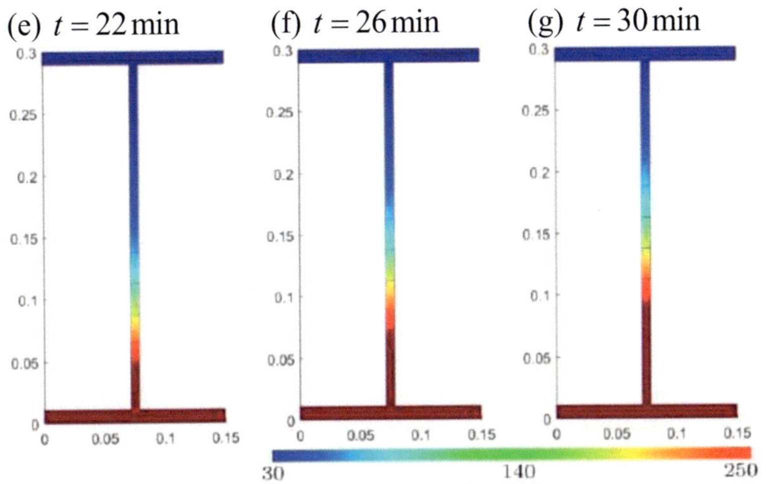

Fig. (17). Plots of the temperature field at selected instants.

Points A and B, belonging to the bottom flange, have a similar heating pattern, although the temperature is larger in point A, which is directly exposed to the fire. The temperature remains idle during more time in point C, situated 5cm up the web from point A. However, once the heat front reaches it, its heating pattern is similar to those of points A and B. The drop in the air temperature caused by the fire squad intervention after roughly 27 minutes does not cause the cooling of the I-beam, because the air temperature always remains larger than the temperature of the beam. However, the temperature increase rate is drastically reduced in points A, B and C as a consequence of the cooling of the surrounding air. The fire event seems to have little impact on the temperature recorded at the base of the top flange (point D). Only at the end of the analysis, some slight temperature gradient occurs as the heating front moves closer to the top of the web.

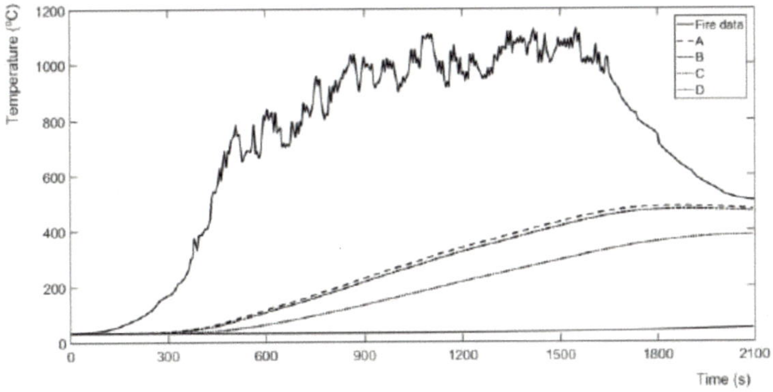

Fig. (18). Time-histories of the air temperature and the temperatures at points A to D.

CONCLUSIONS

A Trefftz-based approach to the solution of parabolic non-homogeneous boundary value problems is presented in this paper. The original problem, defined in time and space, is discretized using a generalized time stepping technique, which leads to a series of non-homogeneous Helmholtz equations in space variables only. The particular solution of each Helmholtz equation is obtained using a novel dual reciprocity approach, where the source function and the particular solution are approximated using the same basis (except for a constant multiplier). This feature avoids the cumbersome (and sometimes singular) expressions of the approximation functions for the particular solution that result from the mainstream dual reciprocity variants. Here, the particular solution basis is regular, as no fundamental solutions are required, and straightforward to compute, since it is defined by the same type of functions as the (Trefftz) basis used for the approximation of the complementary solution.

The complementary solution of the Helmholtz equation is obtained using the temperature formulation of the hybrid-Trefftz finite elements. As typical of Trefftz methods, physically meaningful approximation functions endorse the use of very large elements, as compared to conventional finite element models, and the convergence under basis refinement is very fast. Moreover, no domain integration is required to compute the coefficients of the solving system.

The formulation is validated using benchmark tests reported in the literature, and its convergence under time, mesh, and basis refinements is studied. A thermal shock problem shows that the results are convergent under the refinement of the time step independently of the calibration parameter $\overline{\theta}$ of the time stepping scheme. For larger time steps, values of the calibration parameter closer to 0.5 yield smaller average errors, but are less effective in damping the spurious oscillations caused by the application of the thermal shock. The approximation of the particular solution is robust to sub-optimal choices of the algorithmic wave numbers λ, but it is recommended that their values are rather small and balanced, to avoid highly oscillatory functions in the source function basis. The convergence of the solutions under mesh and basis refinements is studied on a benchmark problem involving a non-smooth boundary. It is concluded that the basis refinement is much more effective in reducing the average errors than the mesh refinement, but excessively high orders of the approximation bases may yield the solving system ill-conditioned. For both benchmark problems, the accuracy of the solutions is comparable with alternative Trefftz-type methods, and better than obtained with conventional finite elements. Finally, a more complex problem, involving a steel I-beam subjected to a fire is solved. The propagation of the high temperature gradients caused by the sudden heating of the air in the vicinity of the I-beam is correctly recovered at all times, with no need for remeshing.

The formulation is implemented in the FreeHyTE environment and deployed as an open-source and user-friendly application, fostering the reproducibility and dissemination of this work.

CONSENT FOR PUBLICATION

Not applicable.

CONFLICT OF INTEREST

The author declares no conflict of interest, financial or otherwise.

ACKNOWLEDGEMENTS

This research was supported by Fundação para a Ciência e a Tecnologia through grant PTDC/EAM- GTC/29923/2017.

REFERENCES

[1] I. Moldovan, D. Cao and J. Freitas, "Hybrid-Trefftz elements for biphasic elastostatics," *Finite Elem. Anal. Des.*, vol. 66, pp. 68-82, 2013.

[2] I. Moldovan, D. Cao and J. Freitas, "Elastic wave propagation in unsaturated porous media using hybrid-Trefftz stress elements," *Int. J. Numer. Meth. Eng.*, vol. 97, pp. 32-67, 2014.

[3] I. Moldovan, "A new particular solution strategy for hyperbolic boundary value problems using hybrid-Trefftz displacement elements," *Int. J. Numer. Meth. Eng.*, vol. 102, pp. 1293-1315, 2015.

[4] D. Nardini and C. Brebbia, "A new approach to free vibration analysis using boundary elements," In : *Boundary Element Methods in Engineering*, Berlin, Heidelberg, Springer-Verlag, 1982, pp. 312-326.

[5] A. Mayo, "The fast solution of Poisson's and the biharmonic equations on irregular regions," *SIAM J. Numer. Anal.*, vol. 21, pp. 285-299, 1984.

[6] L. Wrobel and C. Brebbia, "Boundary elements for nonlinear heat conduction problems," *Commun. Appl. Numer. M.*, vol. 4, pp. 617-622, 1988.

[7] G. Fairweather and A. Karageorghis, "The method of fundamental solutions for elliptic boundary value problems," *Adv. Comput. Math.*, vol. 9, pp. 69-95, 1998.

[8] M. Golberg and C. Chen, The method of fundamental solutions for potential, Helmholtz and diffusion problems, CMP/WIT Press, Boston, Southampton: Boundary Integral Methods: Numerical and Mathematical Aspects, 1998.

[9] H. Cho, M. Golberg, A. Muleshkov and X. Li, "Trefftz methods for time dependent partial differential equations," *CMC-Comput. Mater. Con.*, vol. 1 (1), pp. 1-37, 2004.

[10] K. Yu, A. Kadarman and H. Djojodihardjo, "Development and implementation of some BEM variants. A critical review.," *Eng. Anal. Bound. Elem.*, vol. 34, pp. 884-899, 2010.

[11] X. Li and C. Chen, "A mesh-free method using hyperinterpolation and fast fourier transform for solving differential equations," *Eng. Anal. Bound. Elem.*, vol. 28, pp. 1253-1260, 2004.

[12] J. Agnantiaris, D. Polyzos and D. Beskos, "Some studies on dual reciprocity BEM for elastodynamic analysis," *Comput. Mech.*, vol. 17, pp. 270-277, 1996.

[13] I. Moldovan, "A new approach to non-homogeneous hyperbolic boundary value problems using hybrid-Trefftz stress finite elements," *Eng. Anal. Bound. Elem.*, vol. 69, pp. 57-71, 2016.

[14] I. Moldovan and L. Radu, "Trefftz-based dual reciprocity method for hyperbolic boundary value problems," *Int. J. Numer. Meth. Eng.*, vol. 106 (13), pp. 1043-1070, 2016.

[15] I. Moldovan, A. Coutinho and I. Cismasiu, "Hybrid-Trefftz finite elements for non-homogeneous parabolic problems using a novel dual reciprocity variant," *Eng. Anal. Bound. Elem.*, vol. 106, pp. 228-242, 2019.

[16] C. Alves, N. Martins and S. Valtchev, "Extending the method of fundamental solutions to non-homogeneous elastic wave problems," *Appl. Numer. Math.*, vol. 115, pp. 299-313, 2017.

[17] J. Jirousek and Q.H. Qin, "Application of hybrid-Trefftz element approach to transient heat conduction analysis," *Comput. Struct.*, vol. 58, pp. 195-201, 1996.

[18] I. Moldovan and I. Cismasiu, "FreeHyTE: a hybrid-Trefftz finite element platform," *Adv. Eng. Softw.*, vol. 121, pp. 98-119, 2018.

[19] M. Borkowski and I. Moldovan, "Direct Boundary Method Toolbox for some elliptic problems in FreeHyTE framework," *Eng. Anal Bound. Elem.*, 2020.

[20] FreeHyTE Release Page, www.sites.google.com/site/ionutdmoldovan/freehyte, accessed 01.01.2017, 2016.

[21] C. Chen, Y. Rashed and M. Golberg, "A mesh free method for linear diffusion equations," *Numer. Heat Transfer*, vol. Part B 33, pp. 469-486, 1998.

[22] H. Huang and A. Usmani, Finite Element Analysis for Heat Transfer, London: Springer-Verlag, 1994.

[23] J. Freitas, "Time integration and the trefftz method. Part I - Firstorder and parabolic problems," *Comput. Assisted Mech. Eng. Sci.*, vol. 10, pp. 453-463, 2003.

[24] I. Moldovan, I. Cismasiu and J. Freitas, "Unified Hybrid-Trefftz Finite Element Formulation for Dynamic Problems," In : *Advances in Trefftz Methods and Their Applications*, Springer, 2020, pp. 157-188.

[25] I. Moldovan and J. Freitas, "Hybrid-Trefftz displacement and stress elements for bounded poroelasticity problems," *Comput. Geotech.*, vol. 42, pp. 129-144, 2012.

[26] J. Bruch and G. Zyvoloski, "Transient two-dimensional heat conduction problems solved by the finite element method," *Int. J. Numer. Meth. Eng.*, vol. 8, pp. 481-494, 1974.

[27] C. Brebbia, J. Telles and L. Wrobel, Boundary element techniques, Berlin, Heidelberg: Springer Verlag, 1984.

[28] P. Partridge, C. Brebbia and L. Wrobel, The dual reciprocity boundary element method, London, Southampton: *Comput. Mech. Publications, Els. Appl. Sci.*, 1992.

[29] L. Cao, Q.H. Qin and N. Zhao, "Application of DRM-Trefftz and DRM-MFS to transient heat conduction analysis," *Recent Patents on Space Technology*, vol. 2, pp. 41-50, 2010.

[30] S. Zlatanic, Emerging best practices in tunnel liner design, Online; accessed March, 15, 2007: Tunnel Business Magazine 10, https://tunnelingonline.com/tunnel-design-fire-loads/.

CHAPTER 3

Thermal Analysis of Composites with Trefftz Finite Element Methods

Keyong Wang[1*] and **Ze She**[1]

[1] *School of Mechanical and Automotive Engineering, Shanghai University of Engineering Science, Shanghai, China*

Abstract: Hybrid Trefftz polygonal finite elements are proposed for the thermal analysis of composites reinforced by dispersed fibers. In addition to a homogenous matrix, three types of heterogeneities are considered in each polygon-shaped element: a circular inclusion, an elliptical inclusion or a coated inclusion. Based on T-complete functions satisfying the heat conduction governing equation, an interior temperature field is assumed in inclusion, matrix as well as interphase if any. Whereas an auxiliary frame temperature field is independently defined along the element outer-boundary. The piecewise T-complete functions satisfy not only the governing equations but also guarantees the temperature continuity on the interfaces by means of conformal mapping technology. By using the divergence theorem, all the integrals involved in the single hybrid functional are finally performed along the element outer-boundary only. This facilitates the finite element modelling of heterogeneous materials. Several examples are presented to demonstrate the accuracy and efficiency of the proposed method. It is also concluded that there exists a linear relationship between the maximum number of T-complete functions and the number of Gauss points sampled on each element side.

Keywords: Circular inclusion, Coating, Elliptical inclusion, Fiber-reinforced composite, Heat conduction, Piecewise T-complete function, Polygonal element, Single hybrid functional.

INTRODUCTION

Heterogeneous materials have wide applications in automotive, aerospace and other engineering fields. In the past decades, plenty of attention has been paid to composites reinforced by nature fibers thanks to environmentally sound characteristics [1]. Generally, these materials possess better physical properties in some ways [2]. Although the gross property has been extensively investigated through the volume-average methods, the influence of randomly dispersed fibers

*__Corresponding author Keyong Wang:__ School of Mechanical and Automotive Engineering, Shanghai University of Engineering Science, Shanghai, China; E-mail: keyong.wang@hotmail.com

(or inclusions) on the local behavior has not been completely understood. Much effort has been endeavored to conventional finite element methods for modelling heterogeneities with built-in local effects. As the second phase, the inclusion often couples with shape in geometry and randomness in distribution, which leads to great difficulty in mesh generation. Lots of studies reported the property of circular fiber-reinforced composites on the microscopic or macroscopic scale. In engineering applications, however, the external loads often make circular fibers creep to elliptical ones [3]. On the other hand, non-circular fibers like the elliptical, square or irregular fibers are widely used in engineering [4, 5]. Due to the chemical inactivity or smooth surface of artificial fibers such as p-phenylene benzobisoxazole (PBO) fiber [6], and carbon fiber [7], the poor adhesion between fiber and matrix resin greatly limits its application. In addition, there are fibers covered by one or more coatings. Despite the thickness of each coating, it is extremely thin; the coatings act as the interphase which controls the local material behavior or even the gross property of composites [8-12]. In the conventional finite element method, a refined mesh is usually needed around fibers and coatings to capture the local effects. As a consequence, a great number of degrees of freedom (DOFs) are involved, which leads to a poor computational efficiency. To address this, many approaches such as higher-order structural theory [13] and scaled boundary finite element method [14] have sprouted.

A polygon mesh provides greater flexibility and high efficiency for modelling the problems with singularities or local effects [15]. Ghosh and Mukhopadhyay [16] presented Voronoi cell finite elements for modelling two-dimensional (2D) heterogeneous media. These polygonal elements can generate a coarse mesh to describe the complex mechanical field near the inclusion. But domain integrals appear in the finite element formulation [17]. Alternatively, the hybrid Trefftz finite element method (HT-FEM) avoids this shortcoming. HT-FEM was first proposed by Jirousek and Leon [18] and has been successfully applied to many engineering areas [19, 20]. One of the greatest advantages of HT-FEM is the easy construction of polygonal elements. Owning to the boundary integral-type finite element formulation, HT-FEM may capture most of the singularities and local effects [21, 22]. For heterogeneous materials, a tessellation based mesh generation technique is usually employed to produce Voronoi polygonal elements, each of which consists of a heterogeneity surrounded by the neighborhood matrix [16]. Based on the Hellinger-Reissner principle, Zhang and Katsube [23-25] constructed special polygonal elements with a circular or elliptical elastic/rigid inclusion. Dong and Atluri [26] proposed T-Trefftz Voronoi cell finite elements (VCFEM-TTs) for micromechanical response of composite or porous materials. Subsequently, they extended the work of VCFEM-TTs to the three-dimensional (3D) case with spherical and ellipsoidal inclusions/voids [27,28]. Using the T-complete functions and Lekhnitskii's formulation, Bishay and coworkers developed Trefftz-Lekhnitsakii grains (TLGs) [29] and multi-region Trefftz collocation grains (MTCGs) [30] for piezoelectric, composite and porous materials. Later, Bishay *et al.* [31] dealt with multi-physics computational grains (MPCGs) to simulate the anisotropic micromechanics of piezoelectric composites. Chen and Ping [32] studied the inplane singularity deduced by inclusion corners in elastic materials based on the hybrid-stress FEM. The rhombic or rectangular inclusions can be easily simulated by combining two inclusion corner elements. It is noted that the calculations in the previous studies available have to be conducted along with both the matrix-inclusion interface and element outer-boundary. This causes more difficulty in the derivation of formulas and more complexity in numerical calculation. Although many contributions have been done to mechanical and thermomechanical analyses, only a few studies are related to heat conduction problems. Recently, Wang *et al.* [33-35] proposed special *n*-sided interphase/fiber and fiber/matrix elements based on special fundamental solution kernels. In their model, a single hybrid functional is used for inclusion region and surrounding matrix while the interface continuity is introduced into fundamental solution based trial functions. The resultant finite element formulation requires integrations along the element outer-boundary only. This greatly reduces the computational effort.

Inspired by the single hybrid functional [33, 35], the 2D heat conduction problems with circular, elliptical or coated inclusions are investigated in the framework of HT-FEM. T-complete functions are divided into separate parts for the inclusion, matrix and interphases. Interface continuity is considered by the method of piecewise function. Numerical results demonstrate that the proposed inclusion elements are very satisfactory for microthermal analyses, especially for characterizing the thermal transportation in heterogeneous composite structures due to the flexible meshing strategy.

POLYGONAL ELEMENT WITH A CIRCULAR INCLUSION

Consider a Trefftz polygonal element with a circular inclusion as shown in Fig. (**1**) [36]. The inclusion and surrounding matrix are denoted by Ω_1 and Ω_2, respectively. Both phases are assumed to be isotropic, homogeneous and perfectly bonded with each other. The 2D heat conduction governing equation (Laplace equation) for each phase Ω_i is written as:

$$\frac{\partial^2 T_i}{\partial x_1^2}+\frac{\partial^2 T_i}{\partial x_2^2}=0 \qquad \text{in } \Omega_i \tag{1}$$

where T_i is the temperature, $i=1, 2$, (x_1, x_2) and (r, θ) stand for the local Cartesian and polar coordinate systems originating at the centroid of inclusion, (X_1, X_2) is the global Cartesian coordinate system. The continuity conditions at the matrix-inclusion interface are given by:

$$T_1 = T_2 \quad \text{on } \Omega_1 \cap \Omega_2 \tag{2}$$

$$k_1 \frac{\partial T_1}{\partial \mathbf{n}} = k_2 \frac{\partial T_2}{\partial \mathbf{n}} \quad \text{on } \Omega_1 \cap \Omega_2 \tag{3}$$

where k_1 and k_2 are the thermal conductivities of the matrix and inclusion, respectively, $\mathbf{n}$ denotes the unit normal on the interface pointing towards the surrounding matrix, $\tilde{\mathbf{n}}$ denotes the outward unit normal to the element outer-boundary.

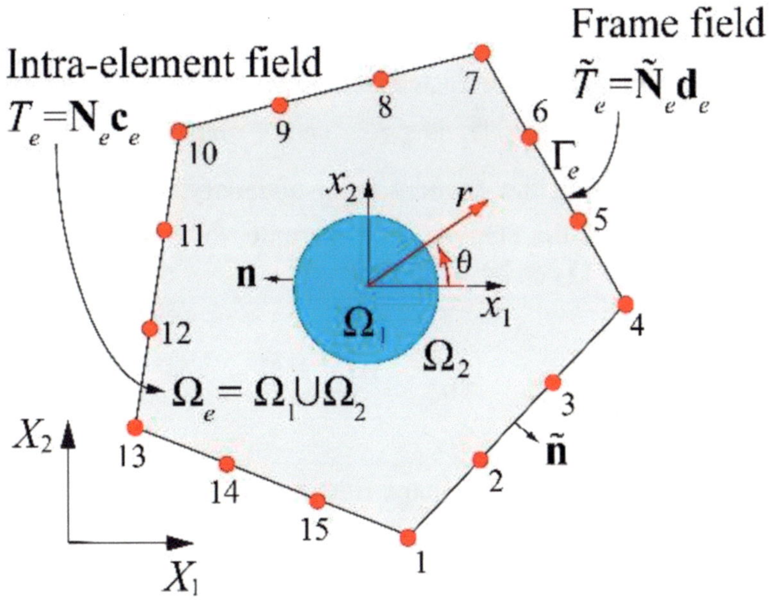

Fig. (1). A Trefftz polygonal element containing a circular inclusion and two assumed fields.

Trefftz Finite Element Formulation

Assumed Temperature Fields

For a typical polygonal element *e* as shown in Fig. (**1**), two temperature fields are independently assumed inside the element domain Ω_e as well as on the element outer-boundary Γ_e:

(a) Non-conforming temperature intra-element field

$$T_e = \sum_{i=1}^{V} N_i c_i = \mathbf{N}_e \mathbf{c}_e \qquad \text{on } \Omega_e \tag{4}$$

(b) Auxiliary conforming temperature frame field

$$\tilde{T}_e = \tilde{\mathbf{N}}_e \mathbf{d}_e \qquad \text{on } \Gamma_e \tag{5}$$

where $\mathbf{N}_e$ is a row vector collecting truncated homogeneous solutions N_i to Eq. (1), $\mathbf{c}_e$ is a column vector collecting undetermined coefficients c_i, V is the number of truncated homogeneous solutions (also called T-complete functions). It is noted that T-complete functions are preselected from Eqs. (25) and (26) addressed below. This makes V an even number. $\tilde{\mathbf{N}}_e$ is a row vector obtained by one-dimensional (1D) shape functions along the element outer-boundary, $\mathbf{d}_e$ is a column vector collecting nodal DOFs of the element. For example, the temperature distribution along Side 1-2-3-4 (Fig. **1**) can be written as:

$$\tilde{T}_{1-2-3-4} = \sum_{i=1}^{4} \tilde{N}_i T_i \tag{6}$$

where $\tilde{N}_i (i = 1, 2, 3, 4)$ denotes 1D shape functions in terms of natural coordinate ξ [36] such that $\tilde{N}_1 = -(\xi - 1)(9\xi^2 - 1)/16$, $\tilde{N}_2 = 9(3\xi - 1)(\xi^2 - 1)/16$, $\tilde{N}_3 = -9(3\xi + 1)(\xi^2 - 1)/16$ and $\tilde{N}_4 = (\xi + 1)(9\xi^2 - 1)/16$. According to individual contribution to the entire element outer-boundary, 1D shape functions for each side are assembled into the vector $\tilde{\mathbf{N}}_e$.

Based on Fourier's law, the heat flux components can be expressed in terms of temperature gradient as:

$$q_i = -k_i \frac{\partial T}{\partial x_i}; \quad i = 1, 2 \tag{7}$$

Then, substituting Eq. (4) into Eq. (7) leads to

$$q_{ei} = -k_i \frac{\partial \mathbf{N}_e \mathbf{c}_e}{\partial x_i} = \mathbf{\Theta}_{ei} \mathbf{c}_e; \quad i = 1, 2 \tag{8}$$

where

$$\mathbf{\Theta}_{ei} = -k_i \frac{\partial \mathbf{N}_e}{\partial x_i} \tag{9}$$

Based on the heat flux components q_{ei} and direction cosines n_i, it is possible to express the outward normal heat flux q_n as

$$q_{en} = q_{e1} n_1 + q_{e2} n_2 = \mathbf{Q}_{ei} \mathbf{c}_e \tag{10}$$

where

$$\mathbf{Q}_{ei} = \mathbf{\Theta}_{e1} n_1 + \mathbf{\Theta}_{e2} n_2 \tag{11}$$

Element Stiffness Equation

To link the two aforementioned temperature fields, the single hybrid functional [37] coming from the Hellinger-Reissner principle is used:

$$\Pi_{me} = -\frac{1}{2} \int_{\Omega_e} k_i \left[\left(\frac{\partial T_e}{\partial x_1} \right)^2 + \left(\frac{\partial T_e}{\partial x_2} \right)^2 \right] d$$
$$\Omega - \int_{\Gamma_{eq}} \bar{q}_e \tilde{T}_e \, d\Gamma + \int_{\Gamma_e} q_{en} \left(\tilde{T}_e - T_e \right) d\Gamma \tag{12}$$

where $\bar{q}_e$ is the prescribed normal heat flux along the boundary Γ_{eq}.

Applying the divergence theorem to the functional (12) yields

$$\Pi_{me} = -\frac{1}{2}\int_{\Gamma_e} q_{en} T_e \, d\Gamma - \int_{\Gamma_{eq}} \bar{q}_e \tilde{T}_e \, d\Gamma + \int_{\Gamma_e} q_{en} \tilde{T}_e \, d\Gamma \tag{13}$$

Then, substituting Eqs. (4), (5) and (10) into the functional (13) produces

$$\Pi_{me} = -\frac{1}{2}\mathbf{c}_e^{\mathrm{T}}\mathbf{H}_e\mathbf{c}_e - \mathbf{d}_e^{\mathrm{T}}\mathbf{P}_e + \mathbf{c}_c^{\mathrm{T}}\mathbf{G}_e\mathbf{d}_e + \text{terms without } \mathbf{c}_e \text{ or } \mathbf{d}_e \tag{14}$$

where

$$\mathbf{H}_e = \int_{\Gamma_e} \mathbf{Q}_e^{\mathrm{T}}\mathbf{N}_e d\Gamma, \; \mathbf{G}_e = \int_{\Gamma_e} \mathbf{Q}_e^{\mathrm{T}}\tilde{\mathbf{N}}_e d\Gamma, \; \mathbf{P}_e = \int_{\Gamma_{eq}} \tilde{\mathbf{N}}_e^{\mathrm{T}}\bar{q}_e \, d\Gamma$$

To enforce the temperature continuity across the element outer-boundaries, it is necessary to find the relationship between $\mathbf{c}_e$ and $\mathbf{d}_e$ vectors. Minimizing Π_{me} with regard to $\mathbf{c}_e$ and $\mathbf{d}_e$, respectively, will lead to

$$\frac{\partial \Pi_{me}}{\partial \mathbf{c}_e} = -\mathbf{H}_e\mathbf{c}_e + \mathbf{G}_e\mathbf{d}_e = \mathbf{0} \tag{15}$$

$$\frac{\partial \Pi_{me}}{\partial \mathbf{d}_e} = \mathbf{G}_e^{\mathrm{T}}\mathbf{c}_e - \mathbf{P}_e = \mathbf{0} \tag{16}$$

Combining Eqs. (15) and (16) gives rise to

$$\mathbf{c}_e = \mathbf{H}_e^{-1}\mathbf{G}_e\mathbf{d}_e \tag{17}$$

and

$$\mathbf{K}_e\mathbf{d}_e = \mathbf{P}_e \tag{18}$$

where $\mathbf{K}_e = \mathbf{G}_e^{\mathrm{T}} \mathbf{H}_e^{-1} \mathbf{G}_e$ and $\mathbf{P}_e$ are the element stiffness matrix and the equivalent nodal force vector, respectively. It is worth noting that the matrices $\mathbf{G}_e$, $\mathbf{H}_e$ and $\mathbf{P}_e$ involve boundary integrals only, which can be easily evaluated by means of 1D Gaussian quadrature along the element outer-boundary [37]. As is known, Eq. (18) is the element stiffness equation.

Construction of Piecewise T-Complete Functions

According to the methodology of HT-FEM, we establish T-complete functions by truncating homogeneous solutions to the heat transfer equation. It is noted that the rigid body motion is excluded to avoid numerical problem in matrix inversion of $\mathbf{H}_e$. A general solution of 2D Laplace equation in a polar coordinate system (r, θ) gives [38]

$$T(r,\theta) = a_0 + b_0 \ln r + \sum_{l=1}^{\infty} \left[\left(a_l r^l + b_l r^{-l} \right) \cos l\theta + \left(c_l r^l + d_l r^{-l} \right) \sin l\theta \right] \quad \textbf{(19)}$$

where a_0, b_0, a_l, b_l, c_l and d_l are the constants.

Because of bi-material property of the element, T-complete functions of the inclusion and surrounding matrix should be provided separately. To make sense in the inclusion region, b_0, b_l and d_l must be zero. Thus, Eq. (19) is rewritten as

$$T_1(r,\theta) = c_0 + \sum_{j=1}^{\infty} \left[c_{2j-1} A_j r^j \cos j\theta + c_{2j} B_j r^j \sin j\theta \right] \quad \textbf{(20)}$$

whereas in the surrounding matrix region, the solution may be sought by rewriting Eq. (19) as

$$T_2(r,\theta) = c_0 + b_0 \ln r + \sum_{j=1}^{\infty} \left[c_{2j-1} \left(E_j r^j + F_j r^{-j} \right) \cos j\theta + c_{2j} \left(P_j r^j + Q_j r^{-j} \right) \sin j\theta \right] \quad \textbf{(21)}$$

where A_j, B_j E_j, F_j, P_j and Q_j are undetermined coefficients.

Performing partial derivatives of Eqs. (20) and (21) with respect to the radial coordinate r gives rise to

$$\frac{\partial T_1}{\partial r} = \sum_{j=1}^{\infty} \left(j c_{2j-1} A_j r^{j-1} \cos j\theta + j c_{2j} B_j r^{j-1} \sin j\theta \right) \tag{22}$$

$$\frac{\partial T_2}{\partial r} = \frac{b_0}{r} + \sum_{j=1}^{\infty} \left[c_{2j-1} \left(jE_j r^{j-1} - jF_j r^{-j-1} \right) \cos j\theta + c_{2j} \left(jP_j r^{j-1} - jQ_j r^{-j-1} \right) \sin j\theta \right] \tag{23}$$

To derive the hybrid Trefftz finite element formulation, the interface continuity between phases should be taken into account. When $r = r_0$ (the inclusion radius), $b_0 = 0$ holds true by considering Eqs. (2) and (3). Based on Eqs. (20) through (23), the relationships between the coefficients appeared in Eq. (21) are expressed as follows

$$\frac{A_j}{E_j} = \frac{2}{\varepsilon + 1}, \quad \frac{B_j}{P_j} = \frac{2}{\varepsilon + 1}, \quad \frac{F_j}{E_j} = \frac{1-\varepsilon}{1+\varepsilon} r_0^{2j}, \quad \frac{Q_j}{P_j} = \frac{1-\varepsilon}{1+\varepsilon} r_0^{2j} \tag{24}$$

where $\varepsilon = k_1 / k_2$ denotes the ratio of thermal conductivity of the inclusion to surrounding matrix. Thanks to the proportional relationship, let us assume E_j and P_j to be unity. In doing so, the vector of T-complete functions is expressed in the following manner

$$\mathbf{N}_e = \left[\left(r^j + \frac{1-\varepsilon}{1+\varepsilon} r_0^{2j} r^{-j} \right) \cos j\theta \quad \left(r^j + \frac{1-\varepsilon}{1+\varepsilon} r_0^{2j} r^{-j} \right) \sin j\theta \quad \cdots \right]; \quad j = 1, 2, 3, \cdots, \frac{\nu}{2} \tag{25}$$

for $(r, \theta) \in \Omega_2$ and

$$\mathbf{N}_e = \left[\frac{2}{\varepsilon + 1} r^j \cos j\theta \quad \frac{2}{\varepsilon + 1} r^j \sin j\theta \quad \cdots \right] \quad j = 1, 2, 3, \cdots, \frac{\nu}{2} \tag{26}$$

for $(r,\theta)\in\Omega_1$. In a particular case, if the thermal conductivity of the inclusion is identical to that of the matrix, $k_1=k_2=k$, Eqs. (25) and (26) are combined into a united vector

$$\mathbf{N}_e=\left[r^j\cos j\theta \quad r^j\sin j\theta \quad \cdots\right];\quad j=1,2,3,\cdots,\frac{\nu}{2} \tag{27}$$

which corresponds to the heat conduction in an infinite isotropic homogeneous material.

As is known, a Trefftz-type element largely depends on the selection of trial functions for the intra-element field. It is worth noting that the element stiffness matrix $\mathbf{K}_e$ involves the inverse of $\mathbf{H}_e$ matrix. To make $\mathbf{K}_e$ be full rank, a necessary but not sufficient condition $\nu_{min}\geq n_{dof}-1$ should be considered. Here, n_{dof} is the number of nodal DOFs of the element. As addressed by many researchers [37,39], the minimal value ν_{min} of T-complete functions does not always guarantee the resulting element stiffness matrix with full rank. In numerical implementation, therefore, full rank can be achieved by selecting more terms of T-complete functions in Eq. (4). However, too many terms may cause ill-conditioned coefficient matrix or numerical overflow. Generally, an optimal value of ν for a considered element needs to be determined by numerical experiments [37,39]. Moreover, to make $\mathbf{H}_e$ be full rank, all rigid-body mode motions are discarded from T-complete functions in advance [39,40].

POLYGONAL ELEMENT WITH AN ELLIPTICAL INCLUSION

In this section, we consider a special polygonal element embedded with an inclined elliptical inclusion [40]. As shown in Fig. (**2**), φ is the inclination angle and (x_1,x_2) the local Cartesian coordinate system originating at the centroid of elliptical inclusion.

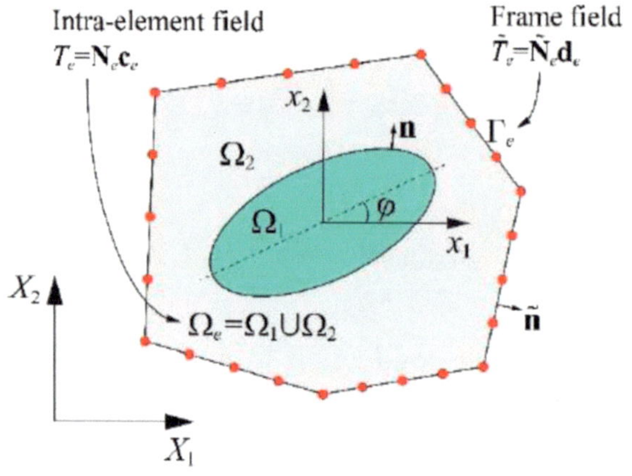

Fig. (2). A Trefftz polygonal element containing an elliptical inclusion and two assumed fields.

Obviously, the Trefftz finite element formualtions are in essence the same with the circular inclusion problem. The key step is to seek proper complete solutions reflecting local thermal effects caused by the elliptical inclusion. For this purpose, we introduce a conformal mapping function as folows [41]:

$$z = e^{i\varphi} R\left(\varsigma + \frac{m}{\varsigma}\right) \tag{28}$$

where $z = x_1 + ix_2$, $\varsigma = \xi_1 + i\xi_2$, $i = \sqrt{-1}$, $R = (a+b)/2, m = (a-b)/(a+b)$. By virtue of such a mapping function, an annulus in the ς -plane is converted to an ellipse in the *z*-plane (Fig. **3**). Readily, the inverse mapping function can be written as

$$\varsigma = \frac{z + \sqrt{z^2 - 4e^{2i\varphi} R^2 m}}{2e^{i\varphi} R} \quad \text{with } |\varsigma| \geq \sqrt{m} \tag{29}$$

which maps the ellipse back onto the annulus bounded by a unit circle and an inner circle of radius $\sqrt{m}$. In the end, the complex function defined in the original domain (the z-plane) is converted into a new function in terms of ς :

$$f(z) = f(z(\varsigma)) = \phi(\varsigma) \tag{30}$$

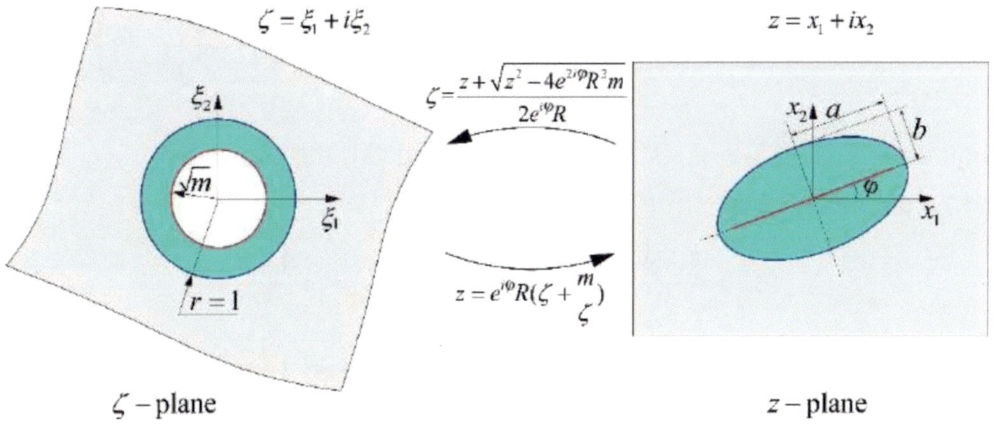

Fig. (3). Conformal mapping of an elliptical domain.

Now, let us focus on the ζ-plane. As mentioned above, the converted inclusion (fiber) contains a circular hole of radius $\sqrt{m}$. As shown in Fig. (**3**), T-complete functions for the special polygonal element including a hollow inclusion are given by [37]:

$$T_1 = c_0 + \sum_{j=1}^{\infty}\left[c_{2j-1}\operatorname{Re}\left(\zeta^j + m^j\zeta^{-j}\right) + c_{2j}\operatorname{Im}\left(\zeta^j - m^j\zeta^{-j}\right)\right] \tag{31}$$

It is worth noting that the innermost circle in the ζ-plane (denoted by red circle) is mapped onto the line segment with ellipse's bifocus as endpoints in the *z*-plane (denoted by red line). We obtain

$$\begin{cases} \operatorname{Re}\left(e^{-i\varphi}z\right) \le 2R\sqrt{m} \\ \operatorname{Im}\left(e^{-i\varphi}z\right) = 0 \end{cases} \tag{32}$$

It is obvious that the temperature symmetry of the innermost circle with respect to the ξ_1 axis should be introduced into T-complete functions for the matrix/elliptical inclusion element. Therefore, Eq. (31) needs to be rewritten as:

$$T_1 = c_0 + \sum_{j=1}^{\infty}\left[c_{2j-1}\operatorname{Re}\left(\zeta^j + m^j\zeta^{-j}\right) + c_{2j}\operatorname{Im}\left(\zeta^j + m^j\zeta^{-j}\right)\right] \tag{33}$$

For the converted matrix region, the temperature field can be written as [37, 39]:

$$T_2 = c_0 + \sum_{j=1}^{\infty}\left[c_{2j-1}\,\mathrm{Re}\left(E_j\zeta^j + F_j\zeta^{-j}\right) + c_{2j}\,\mathrm{Im}\left(P_j\zeta^j - Q_j\zeta^{-j}\right)\right] \tag{34}$$

Combining Eqs. (33) and (34) with Eqs. (2) and (3), the coefficients E_j, F_j, P_j and Q_j are obtained as:

$$\begin{aligned} E_j &= \frac{1-\varepsilon}{2}m^j + \frac{1+\varepsilon}{2}, \quad F_j = \frac{1+\varepsilon}{2}m^j + \frac{1-\varepsilon}{2} \\ P_j &= \frac{\varepsilon-1}{2}m^j + \frac{\varepsilon+1}{2}, \quad Q_j = -\frac{1+\varepsilon}{2}m^j + \frac{1-\varepsilon}{2} \end{aligned} \tag{35}$$

where the definition of ε is the same as done for the circular inclusion element, $\mathbf{N}_e$ composes of the fiber's and matrix's T-complete functions $\mathbf{N}_{e1}$ and $\mathbf{N}_{e2}$ which are truncated homogeneous solutions from Eqs. (31) and (33):

$$\mathbf{N}_{e1} = \left[\mathrm{Re}\left(\zeta^j + m^j\zeta^{-j}\right) \quad \mathrm{Im}\left(\zeta^j + m^j\zeta^{-j}\right) \quad \cdots\right]; \quad j = 1, 2, 3, \cdots, \frac{\nu}{2} \tag{36}$$

$$\mathbf{N}_{e2} = \left[\mathrm{Re}\left(E_j\zeta^j + F_j\zeta^{-j}\right) \quad \mathrm{Im}\left(P_j\zeta^j - Q_j\zeta^{-j}\right)\cdots\right]; \quad j = 1, 2, 3, \cdots, \frac{\nu}{2} \tag{37}$$

POLYGONAL ELEMENT WITH A MULTI-COATED CIRCULAR INCLUSION

Fig. (**4**) shows a Trefftz polygonal element including a fiber (inclusion) bonded by *n*-layer coating (interphase) [42]. The fiber, interphase and surrounding matrix are denoted by Ω_0, Ω_j $(j = 1, 2, \cdots, n)$, Ω_{n+1} respectively. All phases are assumed to be isotropic, homogeneous and perfectly-connected. It is noted that Ω_j denotes the *j*th-layer interphase.

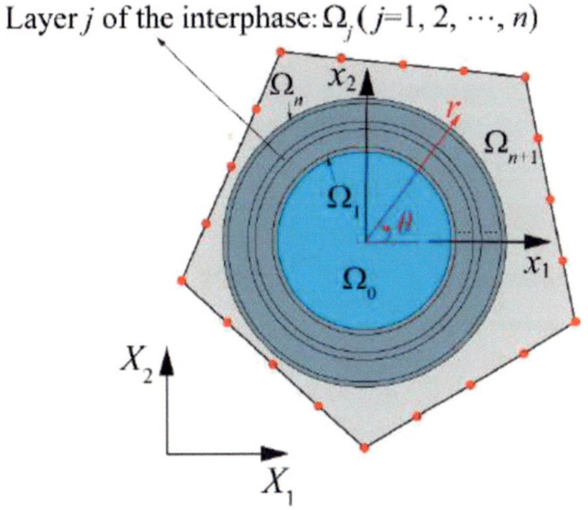

Fig. (4). A Trefftz polygonal element containing a multi-coated circular inclusion.

In the fiber, n-layer interphase and matrix, the heat conduction governing equations (Laplace equations) are given by [33,34].

$$\frac{\partial^2 T_i}{\partial x_1^2}+\frac{\partial^2 T_i}{\partial x_2^2}=0 \quad \text{in } \Omega_i \tag{38}$$

where T_i denote the temperature in the region Ω_i $(i=0,1,2,\cdots,n,n+1)$. It is worth noting that the interface conditions read

$$T_i = T_{i+1} \quad \text{on } \left(\Omega_j \cap \Omega_{j+1}\right) \tag{39}$$

$$k_i \frac{\partial T_i}{\partial \mathbf{n}} = k_{i+1} \frac{\partial T_{i+1}}{\partial \mathbf{n}} \quad \text{on } \left(\Omega_i \cap \Omega_{i+1}\right) \tag{40}$$

where $i=1,2,\cdots,n$, k_0 and k_{n+1} are the thermal conductivities of the fiber and surrounding matrix, respectively, k_i is the thermal conductivity of the ith-layer interphase and $\mathbf{n}$ represents the outward unit normal to the ith interface.

Construction of Piecewise T-Complete Functions

In HT-FEM, T-complete functions are truncated from homogeneous solutions to the heat conduction equation. A general solution of 2D Laplace equation can be given in polar coordinates [38]

$$T(r,\theta) = a_0 + b_0 \ln r + \sum_{n=1}^{\infty} \left[\left(a_n r^n + b_n r^{-n} \right) \cos n\theta + \left(c_n r^n + d_n r^{-n} \right) \sin n\theta \right] \tag{41}$$

with a_0, b_0, a_n, b_n, c_n and d_n are the constants. Considering the heterogeneous characteristics within a element of interest, T-complete functions for the fiber, each layer of coatings and matrix should be investigated separately. For each region of the element, therefore, Eq. (41) is rewritten in the following form

$$T_j(r,\theta) = c_0 + b_j \ln r + \sum_{i=1}^{\infty} \left[c_{2i-1} \left(E_{ji} r^i + F_{ji} r^{-i} \right) \cos i\theta + c_{2i} \left(P_{ji} r^i + Q_{ji} r^{-i} \right) \sin i\theta \right]; \quad j = 0, 1, 2, \cdots, n, n+1 \tag{42}$$

where c_0, c_{2i-1}, c_{2i}, b_j, E_{ji}, F_{ji}, P_{ji} and Q_{ji} are the constants. When $j = 0$, Eq. (42) represents the temperature field of the fiber. The three constants b_0, F_{0i} and Q_{0i} must be zero to make sense.
We have

$$T_0(r,\theta) = c_0 + \sum_{i=1}^{\infty} \left[c_{2i-1} E_{0i} r^i \cos i\theta + c_{2i} P_{0i} r^i \sin i\theta \right] \tag{43}$$

Replacing j with $(j+1)$ in Eq. (42) leads to the temperature field in the region Ω_{j+1}:

$$T_{j+1}(r,\theta) = c_0 + b_{(j+1)} \ln r + \sum_{i=1}^{\infty} \left[c_{2i-1} \left(E_{(j+1)i} r^i + F_{(j+1)i} r^{-i} \right) \cos i\theta + c_{2i} \left(P_{(j+1)i} r^i + Q_{(j+1)i} r^{-i} \right) \sin i\theta \right], \quad (j = 0, 1, 2, \cdots, n) \tag{44}$$

The partial derivatives of Eqs. (42) and (44) with respect to the radial coordinate r are thus given by

$$\frac{\partial T_j}{\partial r} = \frac{b_j}{r} + \sum_{i=1}^{n} \left[c_{2i-1} \left(iE_{ji} r^{i-1} - iF_{ji} r^{-i-1} \right) \cos i\theta + c_{2i} \left(iP_{ji} r^{i-1} - iQ_{ji} r^{-i-1} \right) \sin i\theta \right] \tag{45}$$

$$\begin{aligned} \frac{\partial T_{j+1}}{\partial r} = \frac{b_{(j+1)}}{r} + \sum_{i=1}^{n} \Big[& c_{2i-1} \left(iE_{(j+1)i} r^{i-1} - iF_{(j+1)i} r^{-i-1} \right) \cos i\theta \\ & + c_{2i} \left(iP_{(j+1)i} r^{i-1} - iQ_{(j+1)i} r^{-i-1} \right) \sin i\theta \Big] \end{aligned} \tag{46}$$

With Eqs. (42) and (43), we can obtain

$$b_j = b_{(j+1)} \tag{47}$$

$$\begin{aligned} E_{(j+1)i} &= \frac{1-\varepsilon_j}{2} F_{ji} r_j^{-2i} + \frac{1+\varepsilon_j}{2} E_{ji}, \quad F_{(j+1)i} = \frac{1+\varepsilon_j}{2} F_{ji} + \frac{1-\varepsilon_j}{2} E_{ji} r_j^{2i} \\ P_{(j+1)i} &= \frac{1-\varepsilon_j}{2} Q_{ji} r_j^{-2i} + \frac{1+\varepsilon_j}{2} P_{ji}, \quad Q_{(j+1)i} = \frac{1+\varepsilon_j}{2} Q_{ji} + \frac{1-\varepsilon_j}{2} P_{ji} r_j^{2i} \end{aligned} \tag{48}$$

where r_j is the radius of the jth region and $\varepsilon_i = k_{i-1} / k_i$. Thanks to $b_0 = 0$ and Eq. (47), we have $b_j = 0 \ (j = 1, 2, \cdots, n, n+1)$. In order to include the limiting case in an infinite isotropic homogeneous material, E_{0i} and P_{0i} must be unity. Then, all the unknown constants can be obtained by using Eq. (48). T-complete functions for each region $\Omega_j \ (j = 0, 1, 2, \cdots, n+1)$, $\mathbf{N}_{ej}$ can be written as

$$\mathbf{N}_{ej} = \left[\left(E_{ji} r^i + F_{ji} r^{-i} \right) \cos i\theta \quad \left(P_{ji} r^i + Q_{ji} r^{-i} \right) \sin i\theta \quad \cdots \right], \quad i = 1, 2, \cdots, \frac{\nu}{2} \tag{49}$$

NUMERICAL EXAMPLES

Example 1: Effects of Gauss Point and T-Complete Function

The intra-element and frame temperature fields are linked by the single hybrid functional given by Eq. (12) for the element stiffness matrix. All calculations are conducted by 1D Gaussian quadrature rule along the element outer-boundary. As is known, the number of Gauss points n_g depends on how many T-complete functions are truncated [43]. In the literature, unfortunately, only few studies focused on the effects of the number of Gauss points n_g and that of T-complete functions ν on the analysis. In order to understand the influence of Gauss point and T-complete function, a unit-cell model including a circular inclusion is investigeted (Fig. **5**). The unit cell is an equilateral triangle of side length *L*. The diameter of the inclusion is assumed to be $D = 2r_0$. The 12-node element with *D*=1mm and *L*=2mm is studied. In the analysis, the thermal conductivities of the inclusion and surrounding matrix are respectively taken as $k_1 = 0.53\,\text{W/mK}$ and $k_2 = 2\,\text{W/mK}$.

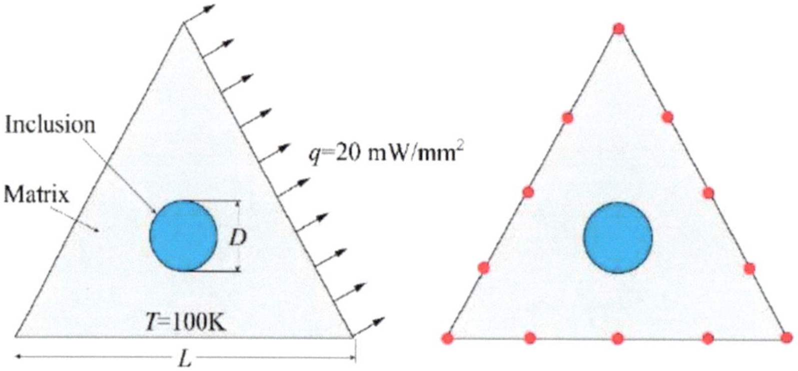

Fig. (5). Unit-cell triangular element with a circular inclusion: (a) problem discription, and (b) 12-node element.

It is noted that an increase in the number of elemental nodes leads to an increase in the number of truncated terms (the order) of T-complete functions. Owing to the nodal DOFs being twelve, ν should be an even number which is greater than or equal to 11. To determine the reasonable value of ν, 540 groups of temperautre data at 2809 points are calculated by changing ν from 12 to 70 and n_g from 3 to 20. Compared with ABAQUS (926 DC2D8 elements), the averaged relative error (*ARE*) [36] of 206 groups of data is lower than 1%. These data are used for further investigation. Fig. **(6a)** shows the variation of optimal maximum number of T-complete functions with the number of Gauss points. As can be seen, more truncated terms of T-complete functions requires more Gauss points seeding on each element side to keep accuracy. A conclusion can be drawn that the maximum number of T-complete functions (ν_{max}) varies linearly with the number of Gauss points under $ARE < 1\%$. Theoretically, this linear relationship reflects the condition of Gaussian quadrature rule, which is exact for univariate polynomials of up to degree $2n_g$-1 by suitably choosing integration points as well as weights. Fig. **(6b)** depicts the variation in *ARE* against the number of Gauss points for different values of ν. It is seen that an increase in the number of Gauss points n_g results in a decrease in *ARE* of the temperature. As expected, when n_g exceeds a critical value, *ARE* almost keep unchanged. This implies that for a fixed value of ν, further increasing the number of Gauss points is not beneficial for the accuracy. An optimal combination of ν and n_g can be determined by balancing the computational effort and accuracy. For the 12-node triangular element as shown in Fig. (**5**), an optimal choice is $\nu = 12$ and $n_g=11$.

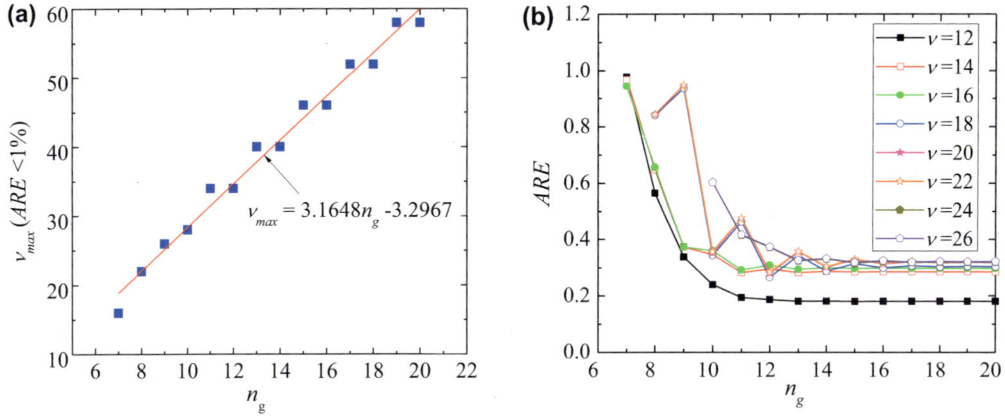

Fig. (6). Effects of ν_{max} and n_g on the accuracy (a) relationship and (b) optimal combination

Example 2: A Square Domain Containing Multiple Circular Inclusions

As the second example, a square domain including multiple circular inclusions is considered shown in Fig. (**7**). Fig. (**8**) depicts the mesh configuration for two cases of square edge packing. Five nodes are seeded on each element side with 11 Gauss points in the numerical implementation. 16, 20, 24 and 28 terms of T-complete functions are truncated for 8-, 10-, 12- and 14-sided elements, respectively. The thermal conductivities of the inclusion and surrounding matrix are set to be $k_1 = 0.115\,\text{W/mK}$ and $k_2 = 0.53\,\text{W/mK}$.

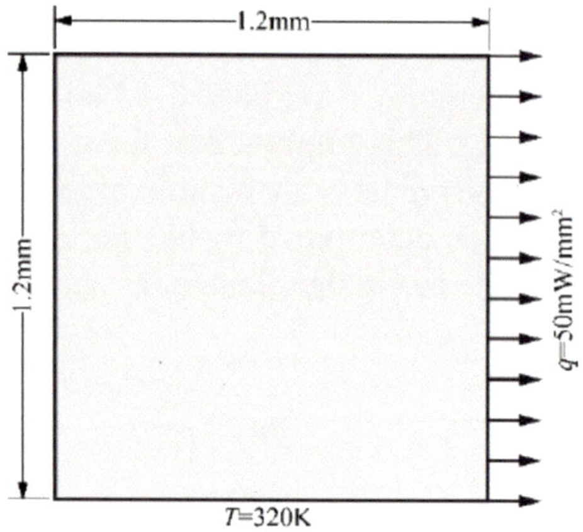

Fig. (7). A square domain embedded by multiple circular inclusions and boundary conditions.

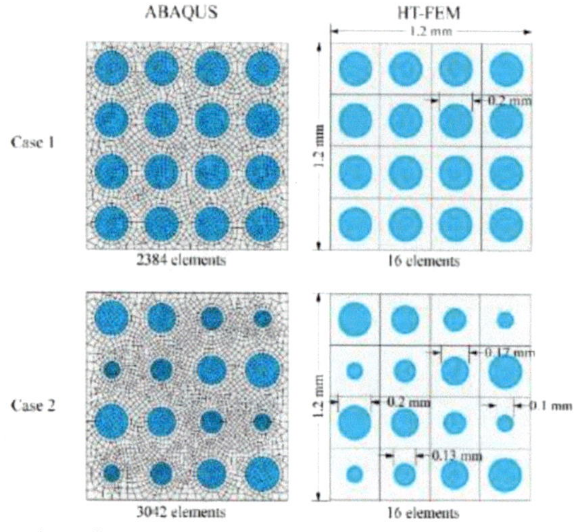

Fig. (8). Mesh configurations for two square packing cases.

In Case 1, 16 identically sized inclusions are regularly embedded in the matrix as shown in Fig. (**8**). A coarse mesh is discretized with 16 special square elements in the HT-FEM model while a refined mesh is discretized with 2384 CD2D8 elements in the ABAQUS model. As shown in Fig. (**9**) (left), identical results are obtained by HT-FEM without sacrificing any accuracy. The number of DOFs in Case 1 is reduced from 7133 to 145. This illustrates the high efficiency of the proposed method. As for Case 2, the same packing is used, but the inclusions are of different size. Fig. (**9**) (right) demonstrates a good agreement between HT-FEM and ABAQUS (with 3042 elements and 9089 nodes) again.

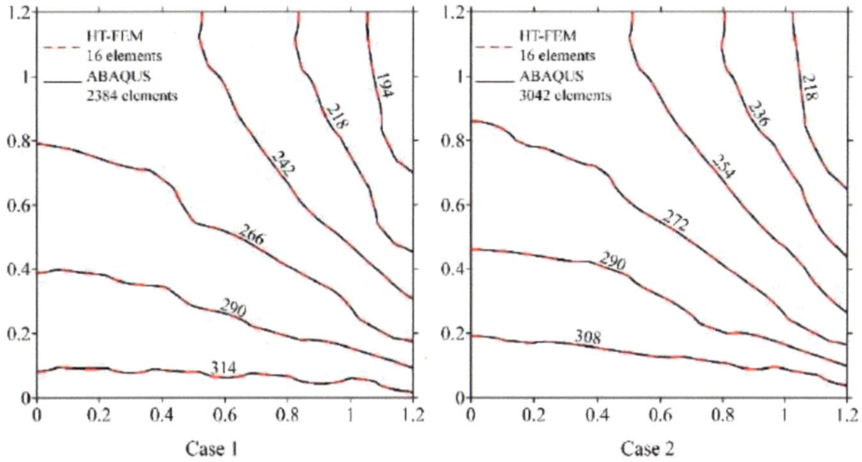

Fig. (9) Contour plots of the temperature.

Example 3: A Square Domain Containing 25 Elliptical Fibers

Generally, the effective thermal conductivity of fiber-reinforced composites is insensitive to microstructure in the longitudinal direction. In the transverse direction, however, the effective thermal conductivity is very different. In the third example, 25 identical elliptical fibers are embedded in a square domain as shown in Fig. (**10**). The fiber volume fraction is assumed to be 10%. The temperatures of 100K and 0K are exerted on the top and bottom surfaces while adiabatic boundary conditions are assumed on the remaining surfaces. Fig. **10(a)** displays two packing patterns: regular and random. The random packing includes two cases: (1) randomly-distributed fibers with the same inclination and (2) with random inclination. It is noted that the Voronoi tessellation technique is used to model the random packing with polygonal elements as shown in Fig. **10(b)**.

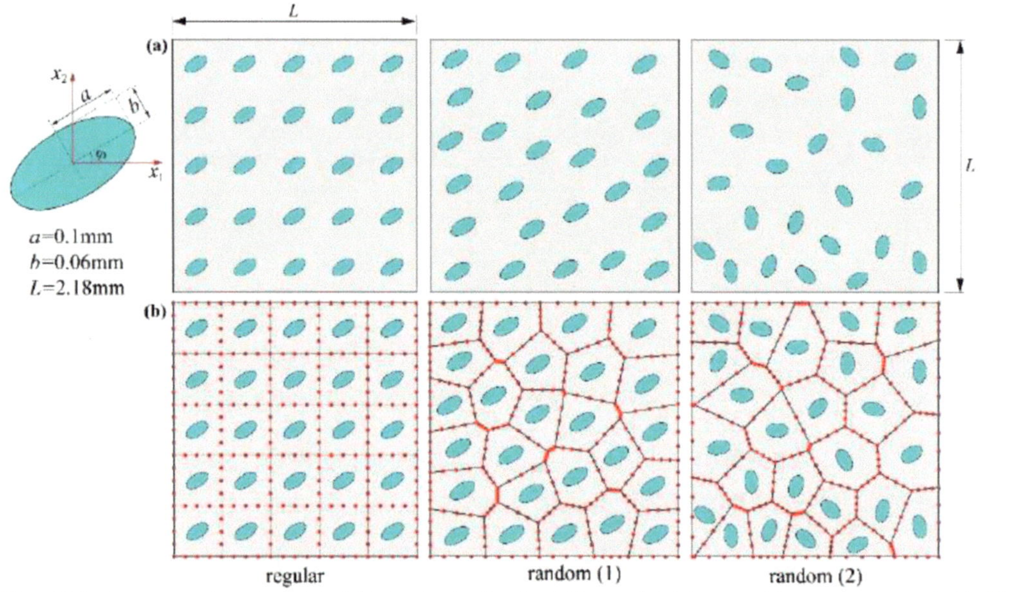

Fig. (10). A square domain with 25 identical elliptical fibers: (a) fiber packing pattern, and (b) HT-FEM mesh.

First, we consider the horizontal case ($\varphi = 0^{\circ}$). The generated square and polygonal meshes for the regular and random packing patterns are composed of 216 and 280 DOFs, respectively. Fig. (**11**) plots the heat flux distribution q_2 on the top surface. As is seen, the heat flow changes periodically in the regular packing pattern while this periodicity does not exist in the random case. This reveals that the thermal response depends greatly on the distribution of fibers in the matrix. It is also found that the proposed method exhibits higher efficiency and accuracy compared with ABAQUS. In the regular, random (1) and random (2) cases, ABAQUS uses 7886 DC2D6 elements (15861 DOFs), 8068 DC2D6 elements (16245 DOFs) and 8722 DC2D6 elements (17589 DOFs), respectively. Although the computational effort does not solely depend on the number of elements, at least HT-FEM achieves very similar accuracy with a coarse mesh.

Example 4: A Square Domain with 50 Randomly Dispersed Elliptical Fibers

In real composites, the fibers are randomly distributed in different sizes. Therefore, it is necessary to go into the more general model and try to understand the random effects. Fig. (**12**) depicts a 1mm square domain reinforced by 50 elliptical fibers. These fibers are randomly dispersed in different sizes and orientations. A temperature of 100K is applied on the left surface while a heat flux of 50mW/mm^2

is exerted on the right surface. The HT-FEM model is divided into 50 polygonal elements with 556 nodes. And the corresponding ABAQUS model includes 32670 DC2D6 elements with 65741 nodes. Fig. (**13**) displays the contour plots of temperature. It is concluded that HT-FEM can achieve similar results under a very coarse mesh. To illustrate the efficiency of the proposed method, Table **1** gives a comparison of computational time. It is seen that an increase in the number of Gauss points leads to an increase in the computational time. But *ARE* remains more or less constant at the value of 0.735% when n_g becomes greater than 6. Most importantly, HT-FEM consumes much less time than ABAQUS.

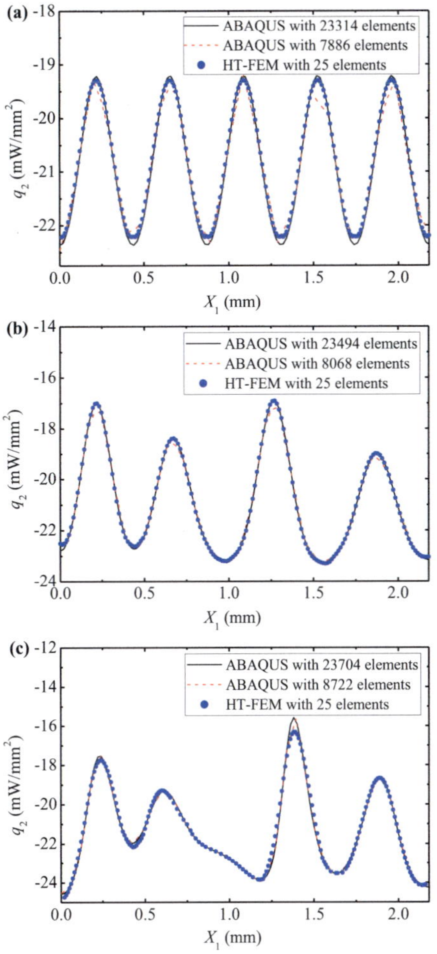

Fig. (11). Heat flux distribution on the top surface for **(a)** regular packing, **(b)** random packing (1) and **(c)** random packing (2).

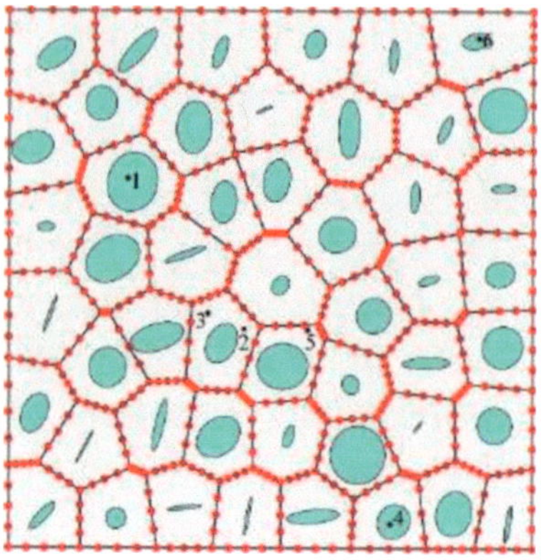

Fig. (12). Mesh configuration for the square domain with 50 randomly dispersed elliptical fibers.

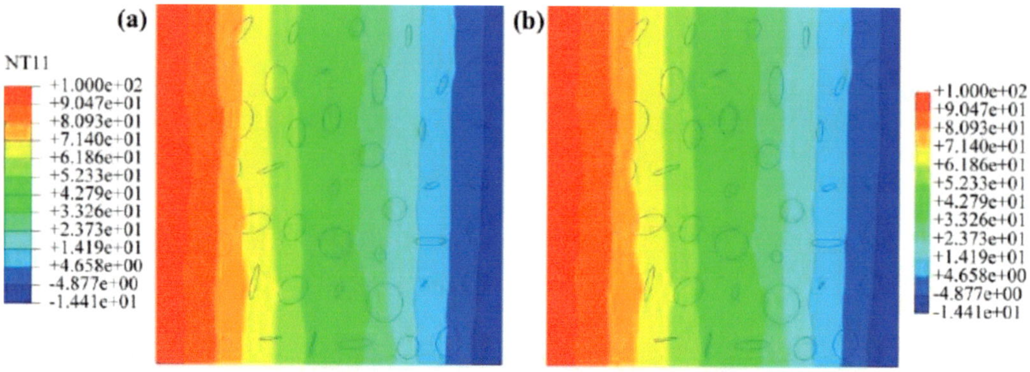

ABAQUS with 32670 elements (65741 DOFs)

HT-FEM with 50 elements (556 DOFs)

Fig. (13). Contour plots of temperature in the square domain with 50 randomly dispersed fibers: **(a)** ABAQUS, and **(b)** HT-FEM.

Table 1. Computational time and *ARE* for different numbers of Gauss points (n_g).

Type of method	HT-FEM						ABAQUS
	n_g=4	6	7	11	15	20	
Computational time (s)	0.094	0.110	0.119	0.140	0.157	0.172	3.8
***ARE* (%)**	0.773	0.736	0.735	0.735	0.735	0.735	---

Example 5: A Rectangular Domain with Four Coated Circular Fibers

We consider a rectangular domain reinforced by four coated circular fibers as shown in Fig. (**14**). The radii of four fibers are assumed to be 0.05mm. A temperature of 200K is applied on the upper surface while a temperature of 320K is specified on the lower surface. From left to right, the fibers are bonded by 1, 2, 3 and 4 layers of coating, respectively. The thickness of each coating for the first three fibers is 0.01mm. The fourth fiber has different coatings in thickness. From inside to outside, the thicknesses are assumed to be 0.006mm, 0.01mm, 0.013mm and 0.011mm. The thermal conductivities of the fibers and surrounding matrix are 0.115 W/mK and 1.06 W/mK. The coatings are assumed to be 0.23 W/mK, 0.3225 W/mK, 0.46 W/mK and 0.53 W/mK respectively. Fig. (**15**) plots a coarse HT-FEM mesh including four 24-node polygonal elements as well as a refined ABAQUS mesh including 10390 DC2D6 elements. To ensure numerical accuracy, each element side employs three nodal points so that the quadratic interpolation is used to approximate the temperature frame field [37,38]. Fig. (**16**) shows the temperature and heat flux distributions along the selected paths. A very good agreement is observed between ABAQUS and the proposed method.

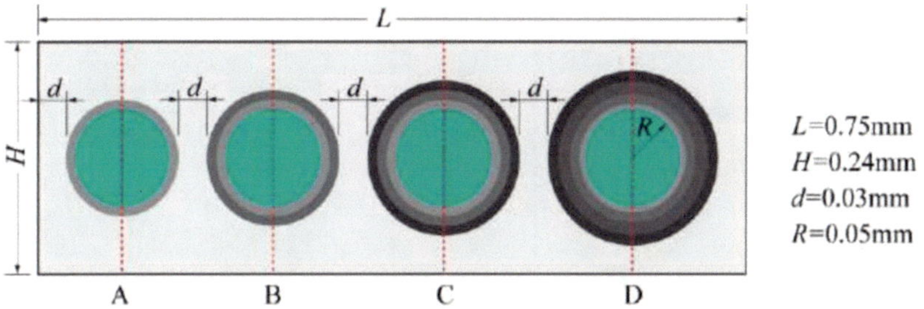

Fig. (14). A rectangular domain with four coated circular fibers.

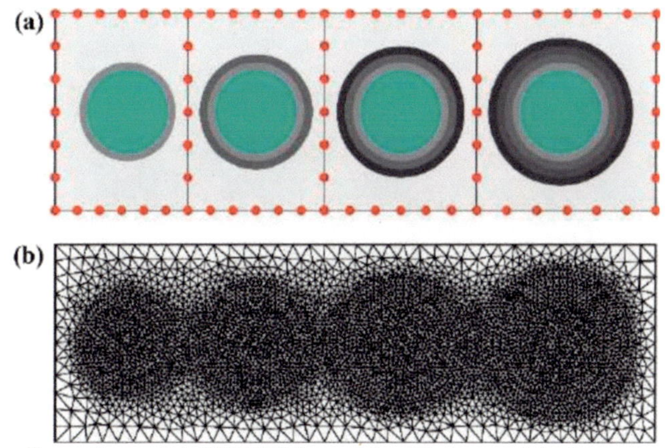

Fig. (15). Mesh configurations: **(a)** HT-FEM model with 4 special elements, and **(b)** ABAQUS model with 10390 DC2D6 elements.

Example 6: An Annular Domain with 22 Two-Layer Coated Circular Fibers

As the last example, an annular domain reinforced by 22 identical two-layer coated circular fibers of radius 0.05mm is considered. The geometry as well as thermal boundary conditions is illustrated in Fig. (**17**). The inner and outer coatings with thermal conductivities of 0.23W/mK and 0.53W/mK are assumed in the analysis. The fiber and surrounding matrix use the same thermal conductivities as in Example 5. Fig. (**18**) displays the finite element meshes for the annular domain. The ABAQUS model consists of 8102 DC2D6 elements and 16294 nodes. The HT-FEM model employs a polygonal mesh which includes only 22 special elements and 234 nodes. It can be seen that HT-FEM gives a unique advantage of grid partitioning. Fig. **(19)** depicts the contour plots of temperature, from which we can see that the HT-FEM results coincide well with ABAQUS ones. A conclusion can be drawn that HT-FEM is also suitable for more complex problems even on very coarse meshes.

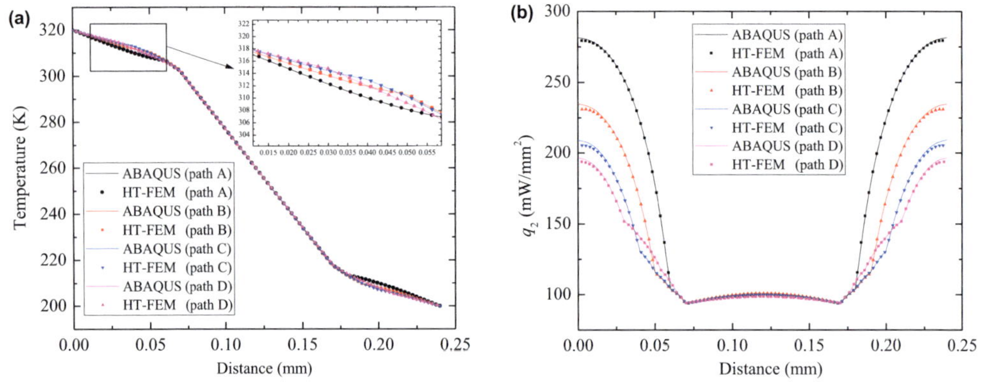

Fig. (16). Temperature and heat flux distributions along the selected paths.

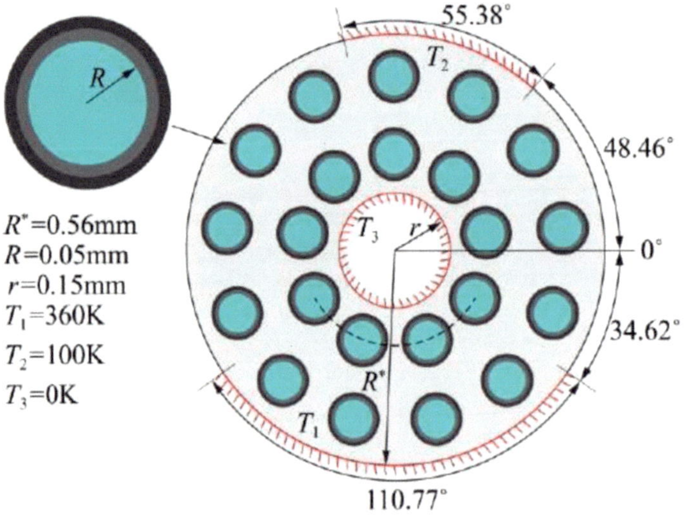

Fig. (17). An annular domain with 22 two-layer coated circular fibers.

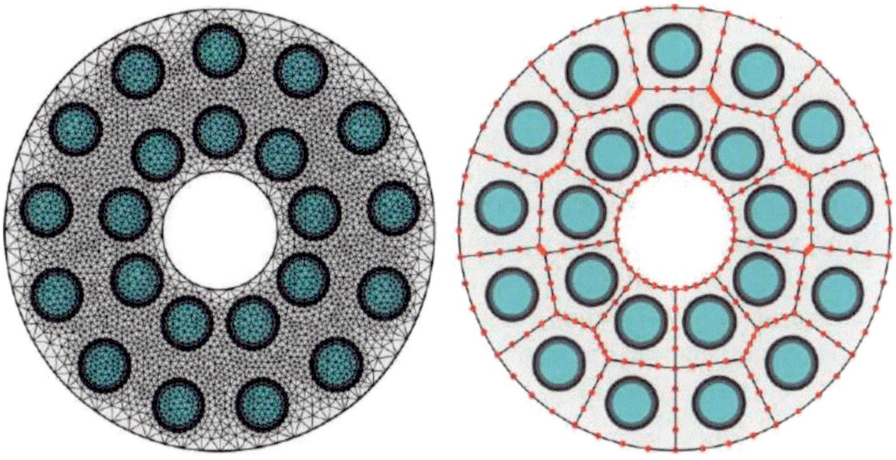

ABAQUS with 8102 elements (16294 DOFs) HT-FEM with 22 elements (234 DOFs)

Fig. (18). Mesh configurations of the annular domain.

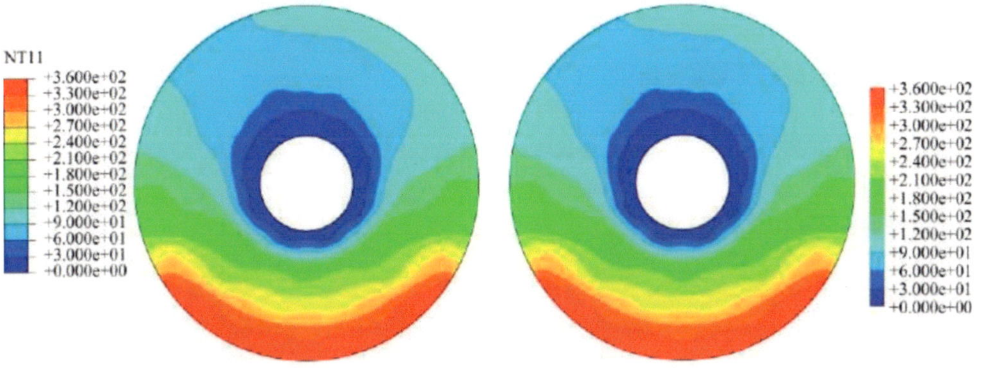

ABAQUS with 8102 elements HT-FEM with 22 elements

Fig. (19). Contour plots of the temperature in the annular domain.

CONCLUDING REMARKS

Owning to numerous fibers embedded in the matrix, seeking a numerical solution for heat conduction in a composite is intractable by using the conventional FEM. This stems from the poor ability of treating local effects induced by heterogeneous inclusions (fibers). By dint of the Voronoi tessellation technique, a polygonal mesh is produced to model heterogeneities. Two temperature fields are independently assumed inside the element domain (including the inclusion, interphase if any, and matrix) and along the element outer-boundary. To overcome the shortcoming of traditional treatment where each phase requires a functional, the current study focuses on the piecewise T-complete functions to link the inclusion, interphase (if any) and matrix. In doing so, only one hybrid functional is needed for establishing the finite element stiffness equation. The circular inclusion problem is first analyzed. And then, it is extended to study the elliptical inclusion problem by using the conformal mapping technique. Finally, the idea of single functional is applied to the coated circular inclusion problem. Numerical examples demonstrate that the proposed method reaches ease of simulating composites with two or more phases. Interestingly, there exists a linear relationship between the maximum number of T-complete functions and the number of Gauss points seeded on each element side. The computational accuracy may be further improved by increasing the number of sides for each element. Compared with ABAQUS, the proposed method exhibits a similar accuracy with high efficiency. An extension of the current work to other problems like elasticity, plate bending and piezoelectricity is possible.

CONSENT FOR PUBLICATION

Not applicable.

CONFLICT OF INTEREST

The author declares no conflict of interest, financial or otherwise.

ACKNOWLEDGEMENTS

Declared none

REFERENCES

[1] A.K. Mohanty, M. Misra, L.T. Drzal, S.E. Selke, B.R. Harte, and G. Hinrichsen. *Natural Fibers, Biopolymers, and Biocomposites: An Introduction*. In: A.K. Mohanty, M. Misra, L.T. Drzal, Eds. *Natural Fibers Biopolymers & Biocomposites*, Boca Raton: Taylor and Francis, 2005.

[2] S.T. Mileiko, "High temperature oxide-fibre/metal-matrix composites", *Mater. Chem. Phys.*, vol. 210, pp. 353-361, 2018.

[3] B. Xu, and F. Guo, "A micromechanics method for transverse creep behavior induced by interface diffusion in unidirectional fiber-reinforced metal matrix composites", *Int. J. Solids Struct.*, vol. 159, pp. 126-134.

[4] X. Jin, L. Yang, X. Du, and Y. Yang, "Modeling filtration performance of elliptical fibers with random distributions", *Advanced Powder Technology*, vol. 28, pp. 1193-1201, 2017.

[5] A. Tiwary, C. Hu, and S. Ghosh, "Numerical conformal mapping method based Voronoi cell finite element model for analyzing microstructures with irregular heterogeneities", *Finite Elem. Anal. Design*, vol. 43, pp. 504-520, 2007.

[6]L. Chen, Y.Z. Du, Y.D. Huang, F. Wu, H.M. Cheng, B. Fei, and J.H. Xin, "Hierarchical poly(p-phenylene benzobisoxazole)/graphene oxide reinforcement with multifunctional and biomimic middle layer", *Compos. Part A*, vol. 88, pp. 123-130, 2016.

[7] S.S. Chen, Y.W. Cao, and J.C. Feng, "Polydopamine as an efficient and robust platform to functionalize carbon fiber for high-performance polymer composites", *ACS Appl. Mater. Inter.*, vol. 6, pp. 349-56, 2014.

[8] S.L. Gao, E. Mäder, and R. Plonka, "Coatings for glass fibers in a cementitious matrix", *Acta Mater.*, vol. 52, pp. 4745-4755, 2004.

[9] J.H. Xiao, Y.L. Xu, and F.C. Zhang, "An analytical method for predicting the effective transverse thermal conductivity of nano coated fiber composites", *Compos. Struct.*, vol. 189, pp. 553-559, 2018.

[10] S. Kari, H. Berger, U. Gabbert, R. Guinovartdıaz, J. Bravocastillero, and R. Rodrıguezramos, "Evaluation of influence of interphase material parameters on effective material properties of three phase composites", *Compos. Scie. Tech.*, vol. 68, pp. 684-691, 2008.

[11] J.R. Yeh, "The effect of interface on the transverse properties of composites", Int. J. Solids Struct., vol. 29, pp. 2493-2502, 1992.

[12] G. Wacker, A.K. Bledzki, and A. Chate, "Effect of interphase on the transverse Young's modulus of glass/epoxy composites", *Compos. Part A*, vol. 29, pp. 619-626, 1998.

[13] S. Natarajan, M. Haboussi, and G. Manickam, "Application of higher-order structural theory to bending and free vibration analysis of sandwich plates with CNT reinforced composite facesheets", *Compos. Struct.*, vol. 113, pp. 197-207, 2014.

[14] C.M. Song, E.T. Ooi, A.L.N. Pramod, and S. Natarajan, "A novel error indicator and an adaptive refinement technique using the scaled boundary finite element method", *Eng. Anal. Bound. Elem.*, vol. 94, pp. 10-24, 2018.

[15] N. Sukumar, and A. Tabarraei, "Conforming polygonal finite elements", *Int. J. Numer. Method Eng.*, vol. 61, pp. 2045-2066, 2004.

[16] S. Ghosh, and S.N. Mukhopadhyay, "A material based finite element analysis of heterogeneous media involving Dirichlet tessellations", *Comput. Method Appl. Mech. Eng.*, vol. 104, pp. 211-247, 1993.

[17] P.L. Bishay, and S.N. Atluri, "High-performance 3D hybrid/mixed, and simple 3D Voronoi cell finite elements, for macro- & micro-mechanical modeling of solids, without using multi-field variational principles", *Comput. Model Eng. Sci.*, vol. 1, pp. 41-97, 2012.

[18] J. Jirousek, and N. Leon, "A powerful finite element for plate bending", *Comput. Method Appl. Mech.*, vol. 12, pp. 77-96, 1997.

[19] Q.H. Qin, "Hybrid-Trefftz finite element method for Reissner plates on an elastic foundation", *Comput. Method Appl. Mech.*, vol. 122, pp. 379-392, 1995.

[20] Q.H. Qin, "Hybrid Trefftz finite-element approach for plate bending on an elastic foundation", *Appl. Mathe. Model.*, vol. 18, pp.334-339, 1994.

[21] H. Wang and Q.H. Qin, "Some problems with the method of fundamental solution using radial basis functions", *Acta. Mech. Solida Sin.*, vol. 20, pp. 21-29, 2007.

[22] H. Wang and Q.H. Qin, "Meshless approach for thermo-mechanical analysis of functionally graded materials", *Eng. Anal. Bound. Elem.*, vol. 32, pp. 704-712, 2008.

[23] J. Zhang, and N. Katsube, "A hybrid finite element method for heterogeneous materials with randomly dispersed elastic inclusions", *Finite Elem. Anal. Des.*, vol. 19, pp. 45-55, 1995.

[24] J. Zhang, and N. Katsube, "A hybrid finite element method for heterogeneous materials with randomly dispersed rigid inclusions", *Int. J. Numer. Method Eng.*, vol. 38, pp. 1635-1653, 1995.

[25] J. Zhang, and N. Katsube, "A polygonal element approach to random heterogeneous media with rigid ellipses or elliptical voids", *Comput. Method Appl. Mech.* Eng., vol. 148, pp. 225-234, 1997.

[26] L. Dong, and S.N. Atluri, "T-Trefftz Voronoi cell finite elements with elastic/rigid inclusions or voids for micromechanical analysis of composite and porous materials", *CMES,* vol. 83, pp. 183-219, 2012.

[27] L. Dong, and S.N. Atluri, "Development of 3D T-Trefftz Voronoi cell finite elements with/without spherical voids &/or elastic/rigid inclusions for micromechanical modeling of heterogeneous materials", *CMC*, vol. 29, pp. 169-211, 2012.

[28] L. Dong, and S.N. Atluri, "Development of 3D Trefftz Voronoi cells with ellipsoidal voids &/or elastic/rigid inclusions for micromechanical modeling of heterogeneous materials", *CMC*, vol. 30, pp. 39-81, 2012.

[29] P.L. Bishay, and S.N. Atluri, "Trefftz-Lekhnitskii Grains (TLGs) for efficient Direct Numerical Simulation (DNS) of the micro/meso mechanics of porous piezoelectric materials", *Comput. Mater. Sci.*, vol. 83, pp. 235-249, 2014.

[30] P.L. Bishay, A. Alotaibi, and S.N. Atluri, "Multi-region Trefftz collocation grains (MTCGs) for modeling piezoelectric composite and porous materials in direct and inverse problems", *J. Mech. Mater. Struct.*, vol. 9, pp. 287-312, 2014.

[31] P.L. Bishay, L. Dong, and S.N. Atluri, "Multi-physics computational grains (MPCGs) for direct numerical simulation (DNS) of piezoelectric composite/porous materials and structures", *Comput. Mech.*, vol. 54: , pp. 1129-1139, 2014.

[32] M.C. Chen, and X.C. Ping, "A novel hybrid finite element analysis of inplane singular elastic field around inclusion corners in elastic media", *Int. J. Solids Struct.*, vol. 46, pp. 2527-2538, 2009.

[33] H. Wang H, and Q.H. Qin, "A new special coating/fiber element for analyzing effect of interface on thermal conductivity of composites", *Appl. Math. Comput.*, vol. 268, pp. 311-321, 2015.

[34] H. Wang H, Q.H. Qin, and Y. Xiao, "Special n-sided Voronoi fiber/matrix elements for clustering thermal effect in natural-hemp-fiber-filled cement composites", *Int. J. Heat Mass Tran.*, vol. 92, pp. 228-235, 2016.

[35] H. Wang, and Q.H. Qin, "Special fiber elements for thermal analysis of fiber-reinforced composites", *Eng. Comput.*, vol. 28, pp. 1079-1097, 2011.

[36] Z. She, K.Y. Wang, and P.C. Li, "Hybrid Trefftz polygonal elements for heat conduction problems with inclusions/voids", *Comput. Math. Appl.*, vol. 78, pp. 1978-1992, 2019.

[37] Q.H. Qin, and H. Wang, MATLAB and C Programming for Trefftz Finite Element Methods, Boca Raton: CRC, 2008.

[38] A. Portela, and A. Charafi, "Programming Trefftz boundary elements", *Adv. Eng. Softw.*, vol. 28, pp. 509-523, 1997.

[39] K.Y. Wang, L.Q. Zhang, and P.C. Li, "A four-node hybrid-Trefftz annular element for analysis of axisymmetric potential problems", *Finite Elem. Anal. Des.*, vol. 60, pp. 49-56, 2012.

[40] Z. She, K.Y. Wang, and H. Liu, "Thermal analysis of elliptical fiber-reinforced composites by the hybrid Trefftz finite element method", *Int. J. Heat. Mass Tran.*, vol. 144, pp. 118596, 2019.

[41] C.Y. Cao, A.B. Yu, and Q.H. Qin, "A new hybrid finite element approach for plane piezoelectricity with defects", *Acta Mech.*, vol. 224, pp. 41-61, 2013.

[42] Z. She, K.Y. Wang, and P.C. Li, "Thermal analysis of multilayer coated fiber-reinforced composites by the hybrid Trefftz finite element method", *Compos. Struct.*, vol. 224, pp. 110992, 2019.

[43] S. Natarajan, A.K.R. Kumar, S. Bordas, and E. Atroshchenko, "Trefftz polygonal finite element for linear elasticity: convergence, accuracy, and properties", *Asia. Pac. J. Comput. Eng.*, vol. 4, pp. 1-17, 2017.

CHAPTER 4

Hybrid-Trefftz Finite Element Method for Heat Conduction in Functionally Graded Materials

Zhuo-Jia Fu[1*], Wen-Zhi Xu[1], and Qiang Xi[1]

[1] *Center for Numerical Simulation Software in Engineering and Sciences, College of Mechanics and Materials, Hohai University Nanjing 211100, P. R. China*

Abstract: Accurate and efficient analysis of heat conduction behaviors in functionally graded materials (FGMs) is very important for the reliable design of equipment and structures under high temperature environments. The hybrid-Trefftz finite element method (HTFEM) for steady-state and transient heat conduction analysis in FGMs is introduced in this chapter. For transient heat conduction problems, Laplace transformation technique is adopted to deal with the time-dependent terms and then one of the popular numerical inverse Laplace transformations, Stehfest algorithm, is introduced to regain the time-dependent numerical solutions. In the HTFEM model, the Trefftz functions including the functionally graded features of the FGMs can be derived *via* various variable transformations, which are adopted to approximate the temperature/heat flux inside the element. The related element stiffness matrix can be obtained *via* a variational functional based on the developed Trefftz functions. Numerical accuracy and efficiency of the proposed HTFEM model is assessed through several benchmark examples. Compared with the standard FEM, the HTFEM is a semi-analytical and efficient computational method without the sensitivity issues of mesh distortion for heat conduction in FGMs.

Keywords: Functionally graded materials, Heat conduction, Hybrid FEM, Laplace transformation, Radial Trefftz function, T-complete function, Variable transformations.

INTRODUCTION

Functionally graded materials (FGMs) [1,2] belong to the two-component advanced composite materials, whose microstructure varies smoothly and gradiently from one component to another. These variation characteristics of material parameters are very suitable for engineering applications that require

[*]**Corresponding author Zhuo-Jia Fu:** Center for Numerical Simulation Software in Engineering and Sciences, College of Mechanics and Materials, Hohai University, Nanjing-211100, P. R. China; E-mail: 20130010@hhu.edu.cn

conflicting properties in one component [3]. In particular, the ceramic-metal FGMs combine the desirable properties of both metals and ceramics, such as thermal resistance from ceramic and fracture toughness from metal, which can be used in the equipment and structures under high temperature environments [4].

To design an appropriate FGM under high temperature environments, it is necessary to analyse and understand their heat conduction behaviours. Analytical methods are only available for simple regular domains with simple boundary conditions. For the cases with arbitrary geometric domains subjected to complex boundary conditions, various numerical methods, such as the finite element method (FEM) [5-8], the boundary element method (BEM) [9,10], the domain-type meshless methods [11,12] and the boundary-type meshless methods [13-16] and so on, have been carried out for analysing heat conduction behaviors of FGMs. However, the standard FEM has to introduce additional techniques, such as the stepwise constant model [5,6] and the graded element model [7,8], for dealing with the inhomogeneous materials with continuously varied physical quantities, which makes it inefficient for simulating heat conduction in FGMs. The BEM can avoid adopting this inefficient additional technique due to the use of semi-analytical fundamental solutions as basis functions. However, one has to pay the price of numerical calculation of singular or hyper-singular integrals, which is a mathematically complex and time-consuming issue.

In contrast to the aforementioned numerical methods, HTFEM [17] proposed in 1977 [18] is one kind of advanced FEMs associated with Trefftz methods [19-23], which has inherent advantages in the numerical simulation of heat conduction behaviours in FGMs, because the Trefftz functions including the specific graded material features are introduced to represent the elemental fields. The mathematical concepts of Trefftz functions (also called as T-complete functions) are proposed by Herrera and his collaborators [24,25]. So far, the HTFEM has been successfully applied to varied science and engineering fields, for examples, 2D elasticity [26], Kirchhoff plates [27], thick plates [28,29], general 3D solid mechanics [30], potential problems [31,32], Helmholtz problems [33], transient heat conduction analysis [34], and piezoelectric materials [35] and contact problems [14]. In addition, radial Trefftz functions (nonsingular general solutions) [36] have been used in Trefftz methods for solving various boundary value problems, which are also called as boundary knot method [37,38] and boundary collocation method [39] in literature. Naturally, the radial Trefftz functions (nonsingular general solutions) are considered as the trial functions of intra-element fields in the HTFEM, which is also called as HGS-FEM [40]. Unlike the standard FEM, HTFEM introduces the interface continuity conditions between two adjacent elements to combine the fields inside the elements with the fields along with the element frame [41]. These temperature/heat flux fields in the element are represented by a linear combination of the derived Trefftz functions satisfying the considered homogeneous governing equation in advance. As mentioned in the literature [41], the main benefits of the HTFEM are: 1) it only requires the numerical integrals along with the boundaries of the element and releases the restriction of the element geometry to avoid the boresome mesh adjustment; 2) it provides the accurate performance with the rapid convergence rate [42].

On the other hand, for transient heat conduction problems, the following two schemes can be introduced into the HTFEM: 1) time stepping method [43,44], it transforms the time-dependent heat conduction equations into several time-independent nonhomogeneous equations, and then employs the HTFEM coupled with several additional particular solution techniques to solve these nonhomogeneous problems. The numerical accuracy and stability may be sensitive to the time-stepping size. 2) Laplace transformation technique [10,13,38,45], the time derivative terms of the considered governing equation vanish and results in a series of time-independent PDEs in Laplace-space domain, and then employs numerical inverse Laplace transformation to regain the original time-dependent solutions from the corresponding HTFEM solutions in Laplace-space domain. Since the numerical inverse Laplace transformation (NILT) is an ill-posed process, the numerical accuracy and stability rely on the selection of efficient and accurate NILT.

This chapter presents a HTFEM in conjunction with variable transformations and Laplace transformation for heat conduction analysis in FGMs. Section 2 presents the heat conduction model and the derivation of the related Trefftz functions, including the T-complete functions and the radial Trefftz functions, by means of variable transformations, and constructs the corresponding variational functional to generate the hybrid Trefftz finite element formulation, and introduces Laplace transformation and Stehfest numerical inverse Laplace transformation to cope with transient heat conduction problems in FGMs. Several benchmark examples have been tested to show the numerical accuracy and efficiency of the proposed HTFEM in Section 3. Finally, Section 4 summarizes some conclusions and potential extensions.

STATEMENT OF HEAT CONDUCTION ANALYSIS IN FGMS

Heat Conduction Model

Consider a heat conduction problem [15] under a 2D arbitrary geometrical shape FGM $\Omega \subset \Re^2$, whose boundary is $\Gamma=\partial\Omega$. The governing differential equation is:

$$\rho(\mathbf{x})c(\mathbf{x})\frac{\partial u(\mathbf{x},t)}{\partial t}=\nabla\cdot\left[\mathbf{K}(\mathbf{x})\nabla u(\mathbf{x},t)\right]+Q(\mathbf{x},t), \quad \mathbf{x}\in\Omega \tag{1}$$

subjected to the boundary conditions:

Dirichlet boundary condition

$$u(\mathbf{x},t)=g_D(\mathbf{x},t), \quad \mathbf{x}\in\Gamma_D \tag{2a}$$

Neumann boundary condition

$$q(\mathbf{x},t)=g_N(\mathbf{x},t), \quad \mathbf{x}\in\Gamma_N \tag{2b}$$

and the initial condition

$$u(\mathbf{x},0)=u_0(\mathbf{x}),\mathbf{x}\in\Omega \tag{3}$$

in which $u(\mathbf{x},t)$ stands for temperature on node $\mathbf{x}$ at time instant t, $\Gamma=\Gamma_D+\Gamma_N$, $\mathbf{n}=[n_1,n_2]$ is the outward unit normal vector at boundary $\mathbf{x}\in\Gamma$, $g_D(\mathbf{x},t)$, $g_N(\mathbf{x},t)$ and $u_0(\mathbf{x})$ stand for the given functions. $Q(\mathbf{x},t)$ denotes the known heat source, $\mathbf{K}=\begin{bmatrix}K_{11}(\mathbf{x}) & K_{12}(\mathbf{x})\\ K_{21}(\mathbf{x}) & K_{22}(\mathbf{x})\end{bmatrix}$, $\rho(\mathbf{x})$ and $c(\mathbf{x})$ denote the matrix of heat conductivity, the mass density and the specific heat, respectively. Herein the heat conductivity and the product of mass density and specific heat are assumed to have the following forms

$$K_{ij}(\mathbf{x})=k_{ij}f(\mathbf{x}), \quad \mathbf{x}\in\Omega, \ i=1,2; j=1,2 \tag{4}$$

$$\rho(\mathbf{x})c(\mathbf{x}) = \rho_0 c_0 f(\mathbf{x}), \quad \mathbf{x} \in \Omega \tag{5}$$

Let the following variable transformation

$$v(\mathbf{x},t) = \sqrt{f(\mathbf{x})}u(\mathbf{x},t) \tag{6}$$

Eqs. (1)-(3) are represented as

$$\nabla \cdot \left[\mathbf{k}\nabla v(\mathbf{x},t)\right] + \eta v(\mathbf{x},t) + \frac{Q(\mathbf{x},t)}{\sqrt{f(\mathbf{x})}} = \rho_0 c_0 \frac{\partial v(\mathbf{x},t)}{\partial t} \tag{7}$$

$$v(\mathbf{x},t) = \sqrt{f(\mathbf{x})}g_D(\mathbf{x},t), \quad \mathbf{x} \in \Gamma_D \tag{8a}$$

$$q_v(\mathbf{x},t) = h_v\left(v(\mathbf{x},t) + \bar{g}_N(\mathbf{x},t)\right), \mathbf{x} \in \Gamma_N \tag{8b}$$

$$v(\mathbf{x},0) = \sqrt{f(\mathbf{x})}u_0(\mathbf{x}), \mathbf{x} \in \Omega \tag{9}$$

where $\quad q_v(\mathbf{x},t) = -\mathbf{k}\left[\nabla v(\mathbf{x},t)\right]\cdot\mathbf{n}, \; h_v = -\mathbf{k}\left[\frac{\nabla\left(\sqrt{f(\mathbf{x})}\right)}{\sqrt{f(\mathbf{x})}}\right]\cdot\mathbf{n},$

and

$$\bar{g}_N(\mathbf{x},t) = \frac{g_N(\mathbf{x},t)}{h_v\sqrt{f(\mathbf{x})}}, \quad \eta = \sum_{i,j=1}^{2} k_{ij}\left(\frac{\dfrac{\partial f(\mathbf{x})}{\partial x_i}\dfrac{\partial f(\mathbf{x})}{\partial x_j}}{\left(2f(\mathbf{x})\right)^2} - \frac{\dfrac{\partial^2 f(\mathbf{x})}{\partial x_i \partial x_j}}{2f(\mathbf{x})}\right) \tag{10}$$

It should be mentioned that, with the following specific forms of $f(\mathbf{x})$, the parameter η of Eq. (10) can be simplified:

(i) Quadratic function $f(\mathbf{x}) = (b_0 + b_1x_1 + b_2x_2)^2$ with arbitrary constants b_0, b_1, b_2, then $\eta = 0$ can be derived according to Eq. (10).

(ii) Exponential function $f(\mathbf{x}) = c_1 e^{\varsigma_1 x_1 + \varsigma_2 x_2}$ with arbitrary constants ς_1, ς_2, c_1, then $\eta = -\sum_{i=1}^{2}\sum_{j=1}^{2} \varsigma_i k_{ij} \varsigma_j$ can be derived according to Eq. (10).

(iii)Trigonometric function $f(\mathbf{x}) = \prod_{i=1}^{2}\left(d_i \cos(\beta_i x_i) + e_i \sin(\beta_i x_i)\right)^2$ with arbitrary constants $d_1, e_1, \beta_1, d_2, e_2, \beta_2$, then $\eta = k_{11}\beta_1^2 + k_{22}\beta_2^2$ can be derived according to Eq. (10) when $k_{12} = k_{21} = 0$.

To solve the transformed PDEs (7)-(9), the following two forms are discussed.

a) Steady State PDE

Without the time-dependent term, the following steady state PDE is considered,

$$\nabla \cdot \left[\mathbf{k} \nabla v(\mathbf{x})\right] + \eta v(\mathbf{x}) = -\frac{Q(\mathbf{x})}{\sqrt{f(\mathbf{x})}} \tag{11}$$

$$v(\mathbf{x}) = \sqrt{f(\mathbf{x})} g_D(\mathbf{x}), \quad \mathbf{x} \in \Gamma_D \tag{12a}$$

$$q_v(\mathbf{x},t) = h_v\left(v(\mathbf{x},t) + \bar{g}_N(\mathbf{x},t)\right), \mathbf{x} \in \Gamma_N \tag{12b}$$

By implementing the variable transformation

$$\begin{pmatrix} y_1 \\ y_2 \end{pmatrix} = \begin{pmatrix} 1/\sqrt{k_{11}} & 0 \\ -k_{12}/\sqrt{k_{11}\Delta_k} & \sqrt{k_{11}}/\sqrt{\Delta_k} \end{pmatrix} \begin{pmatrix} x_1 \\ x_2 \end{pmatrix}, \ \mathbf{y} = \mathbf{k}^{-}\mathbf{x}, \tag{13}$$

the related Trefftz functions of governing equation (11) without the heat source ($Q(\mathbf{x}) = 0$) can be derived according to the parameter η,

when $\eta > 0$, the T-complete functions

$$\begin{aligned}\phi_{T1}=J_0\left(\sqrt{\eta}r\right),\cdots,\phi_{T(2m)}=J_m\left(\sqrt{\eta}r\right)\cos\left(m\theta\right),\\ \phi_{T(2m+1)}=J_m\left(\sqrt{\eta}r\right)\sin\left(m\theta\right),\ \ldots\end{aligned}\quad ,\ m=1,2,\ldots,\ \left(r,\theta\right)\in\Omega \tag{14}$$

and the radial Trefftz functions

$$\phi_{RTj}=J_0\left(\sqrt{\eta}R_j\right), \tag{15}$$

when $\eta<0$, the T-complete functions

$$\begin{aligned}\phi_{T1}=I_0\left(\sqrt{-\eta}r\right),\cdots,\phi_{T(2m)}=I_m\left(\sqrt{-\eta}r\right)\cos\left(m\theta\right),\\ \phi_{T(2m+1)}=I_m\left(\sqrt{-\eta}r\right)\sin\left(m\theta\right),\ \ldots\end{aligned}\quad ,\ m=1,2,\ldots,\ \left(r,\theta\right)\in\Omega \tag{16}$$

and the radial Trefftz functions

$$\phi_{RTj}=I_0\left(\sqrt{-\eta}R_j\right), \tag{17}$$

when $\eta=0$, the T-complete functions

$$\phi_{T1}=1,\cdots,\phi_{T(2m)}=r^m\cos\left(m\theta\right),\phi_{T(2m+1)}=r^m\sin\left(m\theta\right),\ \ldots,\ m=1,2,\ldots,\ \left(r,\theta\right)\in\Omega \tag{18}$$

and the radial Trefftz function is an constant, which cannot be used. Then the following nonsingular kernel functions can be employed as

1) Asymptotic radial Trefftz functions [46]

$$\phi_{ARTj}=J_0\left(\sqrt{\eta_\varepsilon}R_j\right)\text{ or }I_0\left(\sqrt{-\eta_\varepsilon}R_j\right)\text{ with a very small value }\eta_\varepsilon \tag{19}$$

2) Harmonic functions [47]

$$\phi_{Hj}=e^{-c_H\left(R_{jx}^2-R_{jy}^2\right)}\cos\left(2c_HR_{jx}R_{jy}\right) \tag{20}$$

where J_m, I_m denote, respectively, the m-order Bessel and modified Bessel function of first kind, c_H is a problem-dependent parameter, and $r=\sqrt{y_1^2+y_2^2}$, $\theta=\arctan\left(y_2/y_1\right)$, $R_j=\sqrt{R_{jx}^2+R_{jy}^2}$, $R_{jx}=y_1-\bar{y}_{j1}$, $R_{jy}=y_2-\bar{y}_{j2}$, in which $\bar{\mathbf{y}}_j=\left(\bar{y}_{j1},\bar{y}_{j2}\right)$ stands for the source node on the boundary.

b) Transient PDE

Define Laplace transformation of $v\left(\mathbf{x},t\right)$ as

$$L\left(v(\mathbf{x},t)\right)=\tilde{v}(\mathbf{x},\varepsilon)=\int_0^{\infty}v(\mathbf{x},t)e^{-\varepsilon t}dt \tag{21}$$

Then the transient PDEs (7)-(9) can be converted to a series of time-independent PDEs in Laplace-space domain *via* the above Laplace transformation, namely,

$$\nabla\cdot\left[\mathbf{k}\nabla\tilde{v}\left(\mathbf{x},\varepsilon\right)\right]+\omega\tilde{v}\left(\mathbf{x},\varepsilon\right)=Q_L\left(\mathbf{x},\varepsilon\right) \tag{22}$$

$$\tilde{v}\left(\mathbf{x},\varepsilon\right)=\sqrt{f\left(\mathbf{x}\right)}\tilde{g}_1\left(\mathbf{x},\varepsilon\right),\quad \mathbf{x}=\left(x_1,x_2\right)\in\Gamma_D \tag{23a}$$

$$q_v(\mathbf{x},\varepsilon)=h_v\left(\tilde{v}(\mathbf{x},\varepsilon)+\bar{g}_N\left(\mathbf{x},\varepsilon\right)\right),\mathbf{x}\in\Gamma_N \tag{23b}$$

where generalized heat source $Q_L\left(\mathbf{x},\varepsilon\right)=-\rho_0c_0v\left(\mathbf{x},0\right)-\dfrac{Q\left(\mathbf{x},\varepsilon\right)}{\sqrt{f\left(\mathbf{x}\right)}}$. By using the same variable transformation (13), the related Trefftz functions of governing equation (22) without the heat source ($Q_L\left(\mathbf{x},\varepsilon\right)=0$) can be derived according to the parameter $\omega=\eta-\rho_0c_0\varepsilon$,

when $\omega>0$, the T-complete functions

$$\begin{matrix}\phi_{T1}=J_0\left(\sqrt{\omega}r\right),\cdots,\phi_{T(2m)}=J_m\left(\sqrt{\omega}r\right)\cos\left(m\theta\right),\\ \phi_{T(2m+1)}=J_m\left(\sqrt{\omega}r\right)\sin\left(m\theta\right),\ \dots\end{matrix},\ m=1,2,\dots,\ \left(r,\theta\right)\in\Omega \tag{24}$$

and the radial Trefftz functions

$$\phi_{RTj}=J_0\left(\sqrt{\omega}R_j\right), \tag{25}$$

when $\omega<0$, the T-complete functions

$$\begin{matrix}\phi_{T1}=I_0\left(\sqrt{-\omega}r\right),\cdots,\phi_{T(2m)}=I_m\left(\sqrt{-\omega}r\right)\cos\left(m\theta\right),\\ \phi_{T(2m+1)}=I_m\left(\sqrt{-\omega}r\right)\sin\left(m\theta\right),\ \dots\end{matrix},\ m=1,2,\dots,\ \left(r,\theta\right)\in\Omega \tag{26}$$

and the radial Trefftz functions

$$\phi_{RTj}=I_0\left(\sqrt{-\omega}R_j\right), \tag{27}$$

when $\omega=0$, the related T-complete functions are the same to Eq. (18), and the corresponding asymptotic radial Trefftz functions and harmonic functions are the same to Eqs. (19) and (20), respectively.

HTFEM Formulation

To implement the HTFEM, a group of non-overlapping elements are first generated to discretize the whole domain Ω. Consider an intra-element domain Ω_e bounded by the element boundary Γ_e in a particular element (element e), two different independent fields are given [40,48] as follows:

1) An intra-element field is defined by

$$v_e\left(\mathbf{x}\right)=\sum_{j=1}^{m}N_{ej}\left(\mathbf{x}\right)c_{ej}=\mathbf{N}_e\left(\mathbf{x}\right)\mathbf{c}_e \qquad \forall\mathbf{x}\in\Omega_e \tag{28}$$

where $\mathbf{c}_e$ represents unknown coefficients and m stands for the number of Trefftz function terms. N_{ej} denotes the Trefftz functions (T-complete function ϕ_{Tj} or radial

Trefftz function ϕ_{RTj} or asymptotic radial Trefftz function ϕ_{ARTj} or harmonic function ϕ_{Hj}) satisfying homogeneous governing Eq. (11) for steady state case and homogeneous governing Eq. (22) for transient case. For the considered element e, it is vital to select the appropriate number of Trefftz function terms to obtain a good solution. The basic rule is chosen the minimum number of Trefftz functions as $m = N_{dof} - 1$, where N_{dof} denotes the number of nodal degrees of freedom (DOFs) for the considered element.

As shown in Fig. (**1**), the temperature field [48] in the considered element is represented *via* the Trefftz functions defined in a local coordinate system $\mathbf{x} = (x_1, x_2)$ paralleling to the global coordinate system $\mathbf{X} = (X_1, X_2)$. The related heat flux on Γ_e is represented as follow

$$q_{ve} = -\sum_{i,j=1}^{2} k_{ij} \left(\sqrt{f(\mathbf{x})} \frac{\partial v_e}{\partial x_j} - v_e \frac{\partial \sqrt{f(\mathbf{x})}}{\partial x_j} \right) n_i = \mathrm{Q}_{ve} \mathrm{c}_e \qquad \textbf{(29)}$$

where

$$\mathbf{Q}_{ve} = -\sum_{i,j=1}^{2} k_{ij} \left(\sqrt{f(\mathbf{x})} \frac{\partial \mathbf{N}_e}{\partial x_j} - \mathbf{N}_e \frac{\partial \sqrt{f(\mathbf{x})}}{\partial x_j} \right) n_i(\mathbf{x}) = -\mathbf{AkT}_e \qquad \textbf{(30)}$$

with

$$\mathbf{A} = [n_1 \quad n_2], \quad \mathbf{T}_e = \left[\sqrt{f(\mathbf{x})} \frac{\partial \mathbf{N}_e}{\partial x_1} - \mathbf{N}_e \frac{\partial \sqrt{f(\mathbf{x})}}{\partial x_1} \quad \sqrt{f(\mathbf{x})} \frac{\partial \mathbf{N}_e}{\partial x_2} - \mathbf{N}_e \frac{\partial \sqrt{f(\mathbf{x})}}{\partial x_2} \right]^{\mathrm{T}}$$

and the undetermined coefficients $\mathbf{c}_e$ could be calculated *via* varied approaches, such as variational approach and least square etc., to approximately fulfill both the given boundary conditions and the interface continuity conditions between two adjacent elements. In this study, the following inter-element interface continuity conditions are implemented,

$$v_e = v_f \quad on\ \Gamma_e \cap \Gamma_f \text{ conformity} \tag{31a}$$

$$q_{ve} + q_{vf} = 0 \quad on\ \Gamma_e \cap \Gamma_f \text{ reciprocity} \tag{31b}$$

Then in order to obtain the relation expression between the unknown coefficients $\mathbf{c}_e$ and the nodal parameters, the second independent temperature field can be defined as follow

2) An auxiliary conforming frame field [48]

$$\hat{v}_e(\mathbf{x}) = \hat{\mathbf{N}}_e(\mathbf{x})\mathbf{d}_e, \qquad \mathbf{x} \in \Gamma_e \tag{32}$$

is defined in terms of nodal degrees of freedom (DOF) $\mathbf{d}_e$ along with the boundary Γ_e of the considered element, where $\hat{\mathbf{N}}_e$ stands for the polynomial basis functions adopted in the standard FEM. On each side of the considered element, the frame field discretized by *Nf* nodes can be represented as follow

$$\hat{v} = \sum_{i=1}^{Nf} \hat{N}_i(\xi) v_i \tag{33}$$

where $\tilde{N}_i(\xi)$ stands for the shape functions based on the coordinate ξ. Then the DOF number of the considered element can be calculated as $N_{dof} = Nf \times N_{side}$, where N_{side} denotes the side number of the considered element. By using the properties of the shape functions in 1D line element [49]:

$$\hat{N}_i(\xi_i) = 1,\ \hat{N}_j(\xi_i) = 0 \quad for\ i \neq j,\ \sum_{i=1}^{Nf} \hat{N}_i(\xi) = 1,$$

the following shape functions with varied nodes can be given:

(1) shape function of an element side with 2 nodes

$$\xi = \{-1, 1\},\ \hat{N}_1(\xi) = \frac{1-\xi}{2},\ \hat{N}_2(\xi) = \frac{1+\xi}{2}$$

(2) shape function of an element side with 3 nodes

$$\xi = \{-1,0,1\},\ \hat{N}_1(\xi) = -\frac{\xi(1-\xi)}{2},\ \hat{N}_2(\xi) = 1-\xi^2,\ \hat{N}_3(\xi) = \frac{\xi(1+\xi)}{2}$$

(3) shape function of an element side with 4 nodes

$$\xi = \left\{-1,-\frac{1}{3},\frac{1}{3},1\right\}$$

$$\hat{N}_1(\xi) = \frac{1}{16}(1-\xi)\left(-10+9\left(\xi^2+1\right)\right),\ \hat{N}_2(\xi) = \frac{9}{16}\left(1-\xi^2\right)(1-3\xi),$$

$$\hat{N}_3(\xi) = \frac{9}{16}\left(1-\xi^2\right)(1+3\xi),\ \hat{N}_4(\xi) = \frac{1}{16}(1+\xi)\left(-10+9\left(\xi^2+1\right)\right).$$

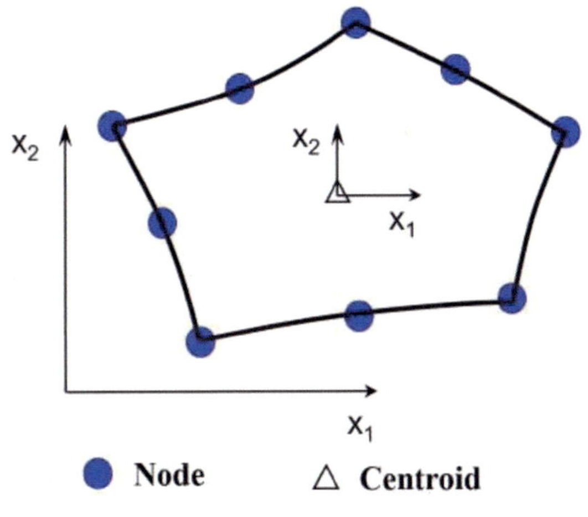

Fig. (1). The field inside a particular element.

By using the variational method [17,48], the HTFEM formulation for analyzing the heat conduction behavior in FGMs can be constructed. In the present variational method, a modified variational principle is established by combining the traditional potential functional with several additional terms, where the additional terms are constructed to guarantee the satisfaction of intereface continuity conditions between two adjacent elements (Eq. (31)) and Neumann boundary conditions on the physical boundary. By introducing the procedure from Ref. [50], the functional for the homogeneous problem (Eqs. (11) and (12) with $Q(\mathbf{x})=0$) can be established as follow

$$\Pi_m = \sum_e \Pi_{me} \quad \textbf{(34)}$$

with

$$\Pi_{me} = -\frac{1}{2}\int_{\Omega_e}\left(k_{ij}v_{,i}v_{,j} - \eta v^2\right)d\Omega - \frac{1}{2}\int_{\Gamma_{qe}} h_v\left(\hat{v}+\bar{g}_N\right)^2 d\Gamma + \int_{\Gamma_e} q_v\left(\hat{v}-v\right)d\Gamma \quad \textbf{(35)}$$

The boundary Γ_e of the considered element (element *e*) is made up of the following parts:

$$\Gamma_e = \Gamma_{ue} \cup \Gamma_{qe} \cup \Gamma_{Ie} \text{ and } \Gamma_{ue} \cap \Gamma_{qe} = \Gamma_{qe} \cap \Gamma_{Ie} = \Gamma_{ue} \cap \Gamma_{Ie} = \varnothing \quad \textbf{(36)}$$

where Γ_{Ie} denotes the intereface boundaries between the considered element '*e*' and its adjacent elements.

Then we derive the governing equation (11) with $Q(\mathbf{x})$=0 subjected to boundary conditions (12) and interface continuity conditions (31) between the adjacent elements from the stationary condition of the functional (34). The following first-order variational of the functional (35) can be presented

$$\begin{aligned}\delta\Pi_{me} = &-\int_{\Omega_e}\left(k_{ij}v_{,j}\delta v_{,i} - \eta v\delta v\right)\mathrm{d}\Omega - \int_{\Gamma_{qe}} h_v\left(\hat{v}+\bar{g}_N\right)\delta\hat{v}d\Gamma \\ &+\int_{\Gamma_e} q_v\left(\delta\hat{v}-\delta v\right)d\Gamma + \int_{\Gamma_e}\delta q_v\left(\hat{v}-v\right)\mathrm{d}\Gamma\end{aligned} \quad \textbf{(37)}$$

By using the divergence theorem

$$\int_\Omega h_{,i}\mathrm{d}\Omega = \int_\Gamma h n_i \mathrm{d}\Gamma \quad \textbf{(38)}$$

For arbitrary smooth-enough function h, the functional (37) can be rewritten

$$\begin{aligned}\delta\Pi_{me} &= \int_{\Omega_e}\left[\left(k_{ij}v_{,j}\right)_{,i}+\eta v\right]\delta v\mathrm{d}\Omega-\int_{\Gamma_e}k_{ij}v_{,j}n_{,i}\delta v d\Gamma-\int_{\Gamma_{qe}}h_v\left(\hat{v}+\bar{g}_N\right)\delta\hat{v}d\Gamma \\ &\quad+\int_{\Gamma_e}q_v\left(\delta\hat{v}-\delta v\right)d\Gamma+\int_{\Gamma_e}\delta q_v\left(\hat{v}-v\right)\mathrm{d}\Gamma \\ &=\int_{\Omega_e}\left(\left(k_{ij}v_{,j}\right)_{,i}+\eta v\right)\delta v\mathrm{d}\Omega+\int_{\Gamma_e}\delta q_v\left(\hat{v}-v\right)\mathrm{d}\Gamma+\int_{\Gamma_{ue}}q_v\delta\hat{v}d\Gamma+\int_{\Gamma_{Ie}}q_v\delta\hat{v}d\Gamma \\ &\quad+\int_{\Gamma_{qe}}\left[q_v-h_v\left(\hat{v}+\bar{g}_N\right)\right]\delta\hat{v}d\Gamma\end{aligned} \tag{39}$$

Since the potential conformity is forced to satisfy, one yields

$$\begin{aligned}\delta\hat{v}=\delta v=0 \qquad & on\ \Gamma_{ue}\ \left(v=\hat{v}\right)\\ \delta\hat{v}^e=\delta\hat{v}^f \qquad & on\ \Gamma_{Ief}\ \left(\hat{v}^e=\hat{v}^f\right)\end{aligned} \tag{40}$$

Next Eq. (39) is rewritten as

$$\begin{aligned}\delta\Pi_{me}=&\int_{\Omega_e}\left(\left(k_{ij}v_{,j}\right)_{,i}+\eta v\right)\delta v\mathrm{d}\Omega+\int_{\Gamma_e}\delta q_v\left(\hat{v}-v\right)\mathrm{d}\Gamma\\ &+\int_{\Gamma_{qe}}\left[q_v-h_v\left(\hat{v}+\bar{g}_N\right)\right]\delta\hat{v}d\Gamma+\int_{\Gamma_{Ie}}q_v\delta\hat{v}d\Gamma\end{aligned} \tag{41}$$

By employing the stationary condition $\delta\Pi_{me}=0$, we can obtain

$$\left(k_{ij}v_{,j}\right)_{,i}+\eta v=0, \quad \text{in } \Omega_e \tag{42}$$

$$\hat{v}=v, \quad \text{on } \Gamma_e \tag{43}$$

$$q_v-h_v\hat{v}=g_N, \quad \text{on } \Gamma_{qe} \tag{44}$$

Then the field continuity Eq. (31) can be imposed to assemble two adjacent elements. Consider two adjacent particular elements ‘e’ and ‘f’, one may obtain

$$\begin{aligned}\delta\Pi_{m(e+f)}=&\int_{\Omega_{e+f}}\left(\left(k_{ij}v_{,j}\right)_{,i}+\eta v\right)\delta v\mathrm{d}\Omega+\int_{\Gamma_e}\delta q_v\left(\hat{v}-v\right)\mathrm{d}\Gamma++\int_{\Gamma_f}\delta q_v\left(\hat{v}-v\right)\mathrm{d}\Gamma\\ &+\int_{\Gamma_{qe+qf}}\left[q_v-h_v\left(\hat{v}+\bar{g}_N\right)\right]\delta\hat{v}d\Gamma+\int_{\Gamma_{Ief}}q_v\delta\hat{v}^{ef}d\Gamma+\cdots\end{aligned} \tag{45}$$

To vanish the functional $\delta\Pi_{m(e+f)} = 0$, the reciprocity condition (31b) $q_{ve} + q_{vf} = 0$ should be satisfied on the interface boundary Γ_{Ief} between two adjacent particular elements 'e' and 'f'. Therefore, the functional (35) is used to produce the element stiffness matrix *via* the variational method presented in [41]. By using the divergence theorem to the functional (35) again, the following functional for the HTFEM model can be obtained

$$\begin{aligned}\Pi_{me} &= \frac{1}{2}\left(\int_{\Omega_e}\left[\left(k_{ij}v_{,j}\right)_{,i} + \eta v\right]v\mathrm{d}\Omega + \int_{\Gamma_e} q_v v d\Gamma\right) \\ &- \frac{1}{2}\int_{\Gamma_{qe}} h_v\left(\hat{v} + \overline{g}_N\right)^2 d\Gamma + \int_{\Gamma_e} q_v\left(\hat{v} - v\right) d\Gamma\end{aligned} \quad \textbf{(46)}$$

In functional (46), the governing equation (11) with $Q(\mathbf{x}) = 0$ is satisfied in advance since the semi-analytical Trefftz functions are used as the basis functions in the HTFEM model. Then the final functional for the HTFEM model can be only defined on the boundary of the considered element *e*

$$\Pi_{me} = -\frac{1}{2}\int_{\Gamma_e} q_v v d\Gamma + \int_{\Gamma_e} q_v \hat{v} d\Gamma - \frac{1}{2}\int_{\Gamma_{qe}} h_v\left(\hat{v} + \overline{g}_N\right)^2 d\Gamma \quad \textbf{(47)}$$

Substituting Eqs. (28), (29) and (32) into the functional (47), we can obtain

$$\Pi_e = -\frac{1}{2}\mathbf{c}_e^{\mathrm{T}}\mathbf{H}_e\mathbf{c}_e + \mathbf{c}_e^{\mathrm{T}}\mathbf{G}_e\mathbf{d}_e - \frac{1}{2}\mathbf{d}_e^{\mathrm{T}}\mathbf{F}_e\mathbf{d}_e - \mathbf{d}_e^{\mathrm{T}}\mathbf{f}_e - \mathbf{a} \quad \textbf{(48)}$$

in which

$$\mathbf{H}_e = \int_{\Gamma_e} \mathbf{Q}_{ve}^{\mathrm{T}}\mathbf{N}_e \mathrm{d}\Gamma \quad \mathbf{G}_e = \int_{\Gamma_e} \mathbf{Q}_{ve}^{\mathrm{T}}\hat{\mathbf{N}}_e \mathrm{d}\Gamma \quad \mathbf{F}_e = \int_{\Gamma_{eq}} h_v \hat{\mathbf{N}}_e^{\mathrm{T}}\hat{\mathbf{N}}_e \mathrm{d}\Gamma$$

$$\mathbf{f}_e = \int_{\Gamma_e} h_v \overline{g}_N \hat{\mathbf{N}}_e^{\mathrm{T}} \mathrm{d}\Gamma \qquad \mathbf{a}_e = \frac{1}{2}\int_{\Gamma_{eq}} h_v \overline{g}_N^2 \mathrm{d}\Gamma$$

The relation expression between unknown coefficient vector $\mathbf{c}_e$ and the nodal DOF $\mathbf{d}_e$ can be derived by minimizing the functional Π_e with respect to $\mathbf{c}_e$ and $\mathbf{d}_e$

$$\frac{\partial \Pi_e}{\partial \mathbf{c}_e^{\mathrm{T}}} = -\mathbf{H}_e \mathbf{c}_e + \mathbf{G}_e \mathbf{d}_e = \mathbf{0} \qquad \textbf{(49a)}$$

$$\frac{\partial \Pi_e}{\partial \mathbf{d}_e^{\mathrm{T}}} = \mathbf{G}_e^T \mathbf{c}_e - \mathbf{F}_e \mathbf{d}_e - \mathbf{f}_e = \mathbf{0} \qquad \textbf{(49b)}$$

which leads to

$$\mathbf{c}_e = \mathbf{H}_e^{-1} \mathbf{G}_e \mathbf{d}_e \qquad \textbf{(50)}$$

and

$$\mathbf{K}_e \mathbf{d}_e = \mathbf{f}_e \qquad \textbf{(51)}$$

where $\mathbf{K}_e = \mathbf{G}_e^{\mathrm{T}} \mathbf{H}_e^{-1} \mathbf{G}_e - \mathbf{F}_e$ denotes the generalized element stiffness matrix. Due to the property of the Trefftz functions, the missing constant temperature should be recovered by using the reference method [41], in which the intra-element thermal field of the considered element e can be represented as

$$v_e = \mathbf{N}_e \mathbf{c}_e + c_0 \qquad \textbf{(52)}$$

where the constant temperature c_0 is computed by introducing the following least square fitting of v_e and $\hat{v}_e$ at all the nodes on the element boundary

$$\sum_{i=1}^{n} \left(\mathbf{N}_e \mathbf{c}_e + c_0 - \hat{v}_e \right)^2 \Big|_{\text{node } i} = \min \qquad \textbf{(53)}$$

which finally gives

$$c_0 = \frac{1}{n} \sum_{i=1}^{n} \Delta v_{ei} \qquad \textbf{(54)}$$

in which $\Delta v_{ei} = \left(\hat{v}_e - \mathbf{N}_e \mathbf{c}_e\right)\Big|_{\text{node } i}$ and n is the element node number. Once the nodal DOF $\mathbf{d}_e$ is obtained by solving Eq. (51), the unknown coefficient vector $\mathbf{c}_e$ and constant temperature c_0 can be calculated from Eqs. (50) and (54). Then one can employ Eq. (52) to compute the thermal field v at any nodes in an element.

Next consider the nonhomogeneous problem (11) and (12), the following appoximation solution can be written by

$$v = v_h + v_p \tag{55}$$

where v_h and v_p stand for, respectively, the homogeneous solution and the particular solution. Assuming that the particular solution v_p satisfies

$$\nabla \cdot \left[\mathbf{k} \nabla v_p(\mathbf{x})\right] + \eta v_p(\mathbf{x}) = -\frac{Q(\mathbf{x})}{\sqrt{f(\mathbf{x})}} = Q_p(\mathbf{x}) \tag{56}$$

then the new homogeneous equation

$$\nabla \cdot \left[\mathbf{k} \nabla v_h(\mathbf{x})\right] + \eta v_h(\mathbf{x}) = 0 \tag{57}$$

with the updated boundary conditions

$$v_h(\mathbf{x}) = \sqrt{f(\mathbf{x})} g_D(\mathbf{x}) - v_p(\mathbf{x}) = g_{Dp}(\mathbf{x}), \quad \mathbf{x} \in \Gamma_D \tag{58a}$$

$$q_{vh}(\mathbf{x}) - h_v \left[v_h(\mathbf{x}) + g_{Np}(\mathbf{x})\right] = 0, \mathbf{x} \in \Gamma_N \tag{58b}$$

where

$$g_{Np}(\mathbf{x}) = \overline{g}_N(\mathbf{x}) - q_{vp}(\mathbf{x})/h_v + v_p(\mathbf{x}) \tag{59}$$

can be solved by using the HTFEM formulation presented above.

Now we go back to determine the particular solution v_p in Eq. (56). In the past decade, the dual reciprocity method (DRM) [44,51] and multiple reciprocity method (MRM) [52-54] have been considered as as two of the most popular techniques to compute the particular solution. Since the MRM can be considered as an extension of the DRM, this study only gives a brief introduction on the DRM.

In the DRM implementation, the nonhomogeneous term $Q_p(\mathbf{x})$ in Eq. (56) can be evaluated by a linear combination of linearly-independent interpolation basis functions, such as trigonometric polynomials, Fourier series functions, Chebyshev polynomials and radial basis functions (RBFs) etc. Taking the RBFs as an example in the DRM, the expression can be represented as

$$Q_p(\mathbf{x}) \cong \sum_{i=1}^{n_p} \alpha_i^p \boldsymbol{\varphi}^p(r_i) \text{ with its matrix form } \mathbf{Q}_p = \boldsymbol{\varphi}^p \boldsymbol{\alpha}^p \tag{60}$$

where $\boldsymbol{\alpha}^p = \{\alpha_i^p\}$ are the underdetermined coefficient vector, n_p represents the number of additional inner collocation nodes $\{\mathbf{x}^p\}$ in the DRM. By solving the linear system of equations (60), it yields

$$\boldsymbol{\alpha}^p = (\boldsymbol{\varphi}^p)^{-1} \mathbf{Q}_p, \tag{61}$$

Then the particular solution $\mathbf{v}_p(\mathbf{x})$ at $\mathbf{x} \in \Omega$ can be obtained as follow

$$\mathbf{v}_p(\mathbf{x}) = \sum_{i=1}^{n_p} \alpha_i^p \boldsymbol{\psi}^p(r_i), \tag{62}$$

where $\boldsymbol{\psi}^p(r_i)$ satisfies the expression

$$\boldsymbol{\varphi}^p(r_i) = (\nabla \cdot \mathbf{k} \nabla + \eta)\{\boldsymbol{\psi}^p(r_i)\}. \tag{63}$$

For the selection of the radial basis function pair ($\boldsymbol{\psi}^p$ and $\boldsymbol{\varphi}^p$), one may find more details in the literature [36]. Once the particular solutions $\mathbf{v}_p(\mathbf{x})$ are determined, the corresponding homogeneous potential can be obtained directly by using Eq.

(58a), and then the right-hand side vector $\mathbf{f}_e$ in the element stiffness Eq. (51) can be replaced by the following updated form

$$\mathbf{f}_e^h = \int_{\Gamma_e} h_v g_{Np} \hat{\mathbf{N}}_e^{\mathrm{T}} \mathrm{d}\Gamma = \int_{\Gamma_e} h_v \left(\bar{g}_N - q_{vp}/h_v + v_p\right) \hat{\mathbf{N}}_e^{\mathrm{T}} \mathrm{d}\Gamma \tag{64}$$

Eq. (64) can be further presented by using the Gaussian numerical integral

$$\begin{aligned} \mathbf{f}_e^h &= \int_{-1}^{+1} h_v \left(\bar{g}_N - q_{vp}/h_v + v_p\right) \hat{\mathbf{N}}_e^{\mathrm{T}} J d\xi \\ &\approx \sum_{k=1}^{n_G} w_k \hat{\mathbf{N}}_{ei}^{\mathrm{T}}(\xi_k)\left(\bar{g}_N(\xi_k) - q_{vp}(\xi_k)/h_v(\xi_k) + v_p(\xi_k)\right) J(\xi_k) \end{aligned} \tag{65}$$

It should be mentioned that the right-hand side verctor $\mathbf{f}_e^h$ is updated by computing the particular solutions at Gaussian integral points instead of nodal points. Finally, the full solution of nonhomogeneous problem (11) and (12) is the sum of the particular solution and the homogeneous solution.

Of course, the HTFEM formulation for solving Laplace-transformed nonhomogeneous problem (22) and (23) can be similarly derived. Once the Laplace-space solutions $\tilde{v}(\mathbf{x},\varepsilon)$ are determined by using the HTFEM, the time-dependent solutions $v(\mathbf{x},t)$ can be regained by the numerical inverse Laplace transformation. In the present numerical implementation the well-known Stehfest algorithm [55] is employed. Assuming that $\tilde{f}(\varepsilon)$ is the Laplace transformation of $f(t)$, the following approximation solution f_s for a specific time instant *t* ican be expressed by

$$f_s(t) = \frac{\ln 2}{t} \sum_{i=1}^{M} V_i \tilde{f}\left(\frac{\ln 2}{t} i\right) \tag{66}$$

where $V_i = (-1)^{\frac{M}{2}+i} \sum_{k=\left[\frac{i+1}{2}\right]}^{\min\left(i,\frac{M}{2}\right)} \frac{k^{\frac{M}{2}}(2k)!}{\left(\frac{M}{2}-k\right)!k!(k-1)!(i-k)!(2k-i)!}$, in which $[C]$ stands for the nearest integers less than or equal to *C*. According to the suggestion of the literature [10], *M=10* is used in the present numerical implementation. Therefore, for a specific time instant t, M boundary value problems related to the parameters $\varepsilon = \frac{\ln 2}{t} i,\ i = 1, 2, \cdots, M$ need to be solved.

Finally the inverse variable transformation $u(\mathbf{x},t) = \frac{v(\mathbf{x},t)}{\sqrt{f(\mathbf{x})}}$ can be employed to provide the numerical solutions of the original heat conduction problems (1)-(3) in FGMs.

NUMERICAL RESULTS AND DISCUSSIONS

This section verifies the accuracy and efficiency of the HTFEM for the heat conduction analysis in a square FGM plate. Then the comparison between the HTFEM with T-complete functions (TCF) and radial Trefftz functions (RTF) for heat conduction problems in a L-shaped FGM plate is made. The numerical accuracy is measured by the average relative error $Rerr(U)$ and normalized error $Nerr(U)$, whose definitions are respectively presented as:

$$Rerr(U) = \frac{\sqrt{\frac{1}{NT}\sum_{i=1}^{NT}\left|U_{num}(\mathbf{x}_i) - U_{exact}(\mathbf{x}_i)\right|^2}}{\sqrt{\frac{1}{NT}\sum_{i=1}^{NT}\left|U_{exact}(\mathbf{x}_i)\right|^2}}, \tag{67}$$

$$Nerr(U) = \frac{\left|U_{num}(\mathbf{x}_i) - U_{exact}(\mathbf{x}_i)\right|}{\max\limits_{1 \le i \le NT}\left|U_{exact}(\mathbf{x}_i)\right|}, \tag{68}$$

where NT represents the total number of evenly-distributed test nodes in the considered domain, $U_{num}(\mathbf{x}_i)$ and $U_{exact}(\mathbf{x}_i)$ stand for the numerical and analytical solutions (temperature u or heat flux in x_1 direction q_{x1} or heat flux in x_2 direction q_{x2}) at node $\mathbf{x}_i$, respectively. Unless otherwise specified, NT=100 and Gauss-Legendre quadrature formula with five points is adopted for the numerical evaluation of the integrals encountered in the present HTFEM numerical implementation.

Example 1: Consider the heat conduction problem in an exponential FGM under a square domain $\Omega = (-1,1) \times (-1,1)$, whose heat conductivity coefficients are defined by Eq. (4) with $\mathbf{k}=\begin{pmatrix} 1 & 0.2 \\ 0.2 & 4 \end{pmatrix}$ and $f(\mathbf{x})=e^{\sum_{i=1}^{2}\varsigma_i x_i}$ ($\varsigma_1=0.1, \varsigma_1=0.4$). The analytical solution can be derived as follows:

$$u(\mathbf{x})=e^{\frac{(Tx+Ty)}{\tau}\sum_{i=1}^{2}\sum_{j=1}^{2}\varsigma_i k_{ij}\varsigma_j-\sum_{i=1}^{2}\varsigma_i x_i} \quad \textbf{(69)}$$

in which

$$\tau=\sqrt{k_{11}\left(\frac{\sqrt{\Delta_{\mathbf{k}}}-k_{12}}{k_{11}}\right)^2+2k_{12}\left(\frac{\sqrt{\Delta_{\mathbf{k}}}-k_{12}}{k_{11}}\right)+k_{22}}$$

$$Tx=\frac{x_1\sqrt{\Delta_{\mathbf{k}}}}{k_{11}}, Ty=-\frac{x_1 k_{12}}{k_{11}}+x_2 .$$

Table **1** list the convergence rate of both temperature and heat flux as well as the condition numbers of the final stiffness matrix K and the matrix H_e for element 1 against the number of elements in the present HTFEM implementation with T-complete function term m=9. It can be found from Table **1** that the proposed HTFEM with the derived T-complete functions (16) by using only 4 elements (2×2 meshes, see Fig. **2a**) can provide acceptable numerical results. The HTFEM converges rapidly to the analytical solution at the beginning of the mesh refinement. While with the further mesh refinement, its convergence rate becomes very slow caused by the ill-conditioning interpolation matrix H_e. And the errors of the temperature are averagely one order of magnitude smaller than those of the heat flux.

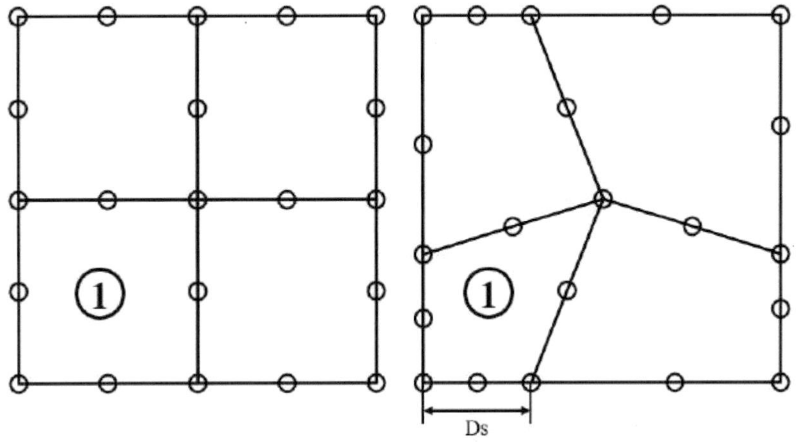

Fig. (2). Configuration of 2×2 regular meshes and distorted meshes adopted in the proposed HT-FEM method for Example 1.

Table 1. Numerical results of Example 1 obtained by the HTFEM with different meshes.

Elem (Node)	2×2 (21)	4×4 (65)	6×6 (133)	8×8 (225)
Cond (K)	3.293	10.534	22.853	40.848
Cond (H_e)	5.286E+10	1.310E+13	3.328E+14	2.620E+15
Rerr of u	2.183E-4	2.831E-5	1.590E-5	1.322E-5
Rerr of q_{x1}	3.695E-3	9.073E-4	5.615E-4	2.524E-4
Rerr of q_{x2}	7.256E-3	2.517E-3	1.484E-3	1.114E-3

Table 2. Numerical results of Example 1 obtained by the HTFEM with different terms *m* of T-complete functions in 2×2 meshes.

m	8	9	10	11
Cond (K)	3.219	3.293	3.356	3.367
Cond (H_e)	1.262E+8	5.286E+10	3.208E+11	5.953E+13
Rerr of u	2.013E-4	2.183E-4	2.575E-4	2.010E-4
Rerr of q_{x1}	3.317E-3	3.695E-3	4.799E-3	3.295E-3
Rerr of q_{x2}	8.419E-3	7.256E-3	7.271E-3	8.406E-3

The effect of the T-complete function term *m* on the numerical implementation of HTFEM are investigated and presented in Tables **2** and **3**. We can observe that the numerical accuracy of the HTFEM is insensitive to the T-complete function term, while the condition number of the matrix H_e increases remarkably with an increase of the T-complete function term number. To balance the accuracy and stability of the proposed HTFEM, m=9 is adopted in the following numerical implementation.

Table 3. Numerical results of Example 1 obtained by the HTFEM with different terms *m* of T-complete functions in 4×4 meshes.

m	**8**	**9**	**10**	**11**
Cond(K)	10.062	10.534	11.047	11.047
Cond(H$_e$)	7.843E+9	1.310E+13	7.932E+13	1.108E+16
***Rerr* of u**	2.832E-5	2.831E-5	2.831E-5	2.831E-5
***Rerr* of q$_{x1}$**	9.096E-4	9.073E-4	9.114E-4	9.115E-4
***Rerr* of q$_{x2}$**	2.520E-3	2.517E-3	2.523E-3	2.523E-3

To investigate the effect of the mesh distortion on the sensitivity of the HTFEM model, the distorted meshes as presented in Fig. (**2b**) are considered, where *Ds* stands for mesh distortion parameter. Table **4** lists the numerical results obtained by the HTFEM with different mesh distortion parameters. From Table **4**, it can be found that, unlike the traditional FEM, the proposed HTFEM is not so sensitive to mesh distortion, which releases the restriction of the element geometry to avoid the boresome mesh adjustment.

Table 4. Numerical results of Example 1 obtained by the HTFEM with/without distorted mesh (*Ds*=0.5).

Distorted Parameter	**Undistorted (*Ds*=0.5)**	***Ds*=0.3**	***Ds*=0.4**	***Ds*=0.6**	***Ds*=0.7**
Cond (K)	3.293	3.687	3.383	3.362	3.671
Cond (H$_e$)	5.286E+10	2.241E+11	1.159E+11	2.200E+10	9.012E+9
Rerr of u	2.183E-4	5.505E-4	3.590E-4	2.474E-4	4.315E-4
Rerr of q$_{x1}$	3.695E-3	7.372E-3	5.434E-3	4.278E-3	5.555E-3
Rerr of q$_{x2}$	7.256E-3	1.236E-2	9.757E-3	6.970E-3	9.686E-3

Example 2: This example is the heat conduction problem in a L-shape FGM with the heat conductivity defined by Eq. (4) with $\mathbf{k}=\begin{pmatrix}2 & 0\\ 0 & 1\end{pmatrix}$ and $f(\mathbf{x})=e^{\sum_{i=1}^{2}\varsigma_i x_i}$ ($\varsigma_1=0,\varsigma_1=1$). The following analytical solution can be derived

$$u(\mathbf{x})=\sqrt{\frac{1-Tx/Tr}{2Tr}}\sinh(Tr)e^{-Ty} \quad \textbf{(70)}$$

where $Tx=\frac{x_1}{\sqrt{2}}-1, Ty=x_2, Tr=\sqrt{Tx^2+Ty^2}$. The corresponding full Dirichlet boundary conditions are easily obtained from the analytical solution.

We compare the proposed HTFEM with both T-complete function and radial Trefftz function in this case. The related numerical results of example 2 are shown in Tables **5** and **6**. It can be found from Tables **5** and **6** that both the proposed HTFEM based on T-complete function and radial Trefftz function perform the acceptable numerical results with only three elements. In addition, both these two types of the HTFEM converge fast to the analytical solutions with an increase of the element number up to 75 in this case. Then the HTFEM results converges slowly and even diverges with the further increase of the element number (larger than 75) due to the ill-conditioned matrix H_e.

Fig. (**3**) represents the the normalized error distribution of temperature u and heat flux components q_{x1} and q_{x2} obtained by the HTFEM (75 elements) with T-complete function (TCF) and radial Trefftz function (RTF). We can observe from Fig. (**3**) that the proposed HTFEM with both T-complete function and radial Trefftz function provides accurate temperature and heat flux fields.

Table 5. Numerical results of Example 2 by HTFEM based on T-complete function with different numbers of elements (nodes).

Element (Node)	3 (18)	12 (53)	27 (106)	48 (177)	75 (266)	108 (373)
Cond (K)	1.867	6.568	14.237	24.999	38.645	65.912
Cond (H_e)	4.012E+9	9.642E+11	2.443E+13	2.479E+14	1.466E+15	5.736E+15
Rerr of u	8.807E-3	1.493E-3	3.893E-4	1.200E-4	5.152E-5	4.832E-5
Rerr of q_{x1}	6.199E-2	1.659E-2	8.847E-3	4.906E-3	2.499E-3	2.710E-3
Rerr of q_{x2}	8.478E-2	1.741E-2	1.035E-2	3.934E-3	2.681E-3	2.232E-3

Table 6. Numerical results of Example 2 by HTFEM based on radial Trefftz function with different numbers of elements (nodes).

Element (Node)	3 (18)	12 (53)	27 (106)	48 (177)	75 (266)	108 (373)
Cond (K)	1.869	6.409	14.959	26.260	112.090	4413.629
Cond (H_e)	6.616E+15	3.155E+16	4.056E+16	4.118E+16	1.168E+17	1.131E+17
Rerr of u	7.557E-3	1.245E-3	4.207E-4	1.355E-4	8.098E-5	1.787E-4
Rerr of q_{x1}	6.248E-2	1.794E-2	9.175E-3	4.839E-3	2.793E-3	4.550E-3
Rerr of q_{x2}	5.054E-2	1.270E-2	8.886E-3	3.816E-3	3.091E-3	3.947E-3

CONCLUDING REMARKS

This paper derives the semi-analytical Trefftz functions (T-complete functions and radial Trefftz functions) for heat conduction analysis in FGMs by means of variable transformations. The Trefftz functions are then employed to establish the HTFEM model for heat conduction analysis under two-dimensional FGMs. Laplace transformation technique and Stehfest numerical inverse Laplace transformation are adopted to cope with the time-dependent terms of transient heat conduction problems in FGMs. Besides the dual reciprocity method has been introduced into the HTFEM for simulating the heat conduction beaviors containing the inner heat sources.

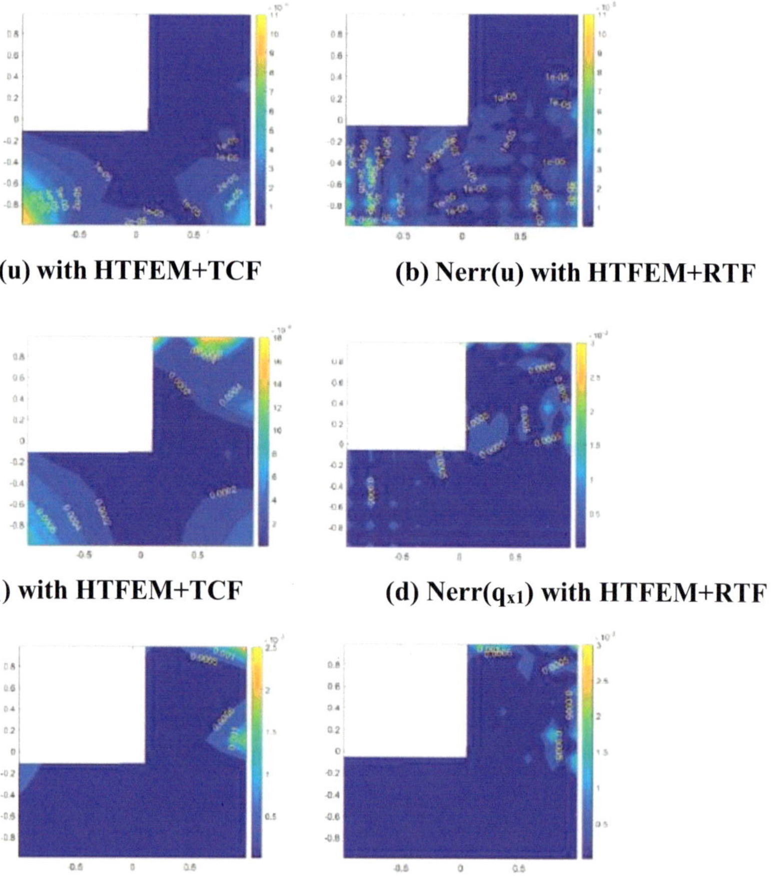

(a) Nerr(u) with HTFEM+TCF **(b) Nerr(u) with HTFEM+RTF**

(c) Nerr(q_{x1}) with HTFEM+TCF **(d) Nerr(q_{x1}) with HTFEM+RTF**

(e) Nerr(q_{x2}) with HTFEM+TCF **(f) Nerr(q_{x2}) with HTFEM+RTF**

Fig. (3). Normalized error distributions of temperature u and heat flux components q_{x1} and q_{x2} computed by HTFEM (75 elements) based on T-complete function (TCF) and radial Trefftz function (RTF).

Numerical investigation demonstrates that the proposed HTFEM is one of the most competitive numerical schemes for heat conduction analysis in 2D FGMs. It provides the accurate numerical solutions and converges to the analytical solution with an increase of the element number. Numerical results verify that the HTFEM is not sensitive to mesh distortion. In addition, the proposed HTFEM with radial Trefftz function (RTF) can work equally well with the one with traditional T-complete functions (TCF). The benefit of the HTFEM based on RTF is to remove the troublesome issue on constructing appropriate T-complete functions in the HTFEM based on TCF.

Future work on this research topic could be the development of the proposed HTFEM to 3D composite materials [56, 57] and be coupled with boundary discretization schemes for structural-acoustic coupling analysis [58, 59].

CONSENT FOR PUBLICATION

Not applicable.

CONFLICT OF INTEREST

The author declares no conflict of interest, financial or otherwise.

ACKNOWLEDGEMENTS

The work described in this paper was supported by the National Science Fund of China (Grant No. 11772119), the Fundamental Research Funds for the Central Universities (Grant No. B200202124), the Foundation for Open Project of State Key Laboratory of Mechanics and Control of Mechanical Structures (Nanjing University Of Aeronautics And Astronautics) (Grant No. MCMS-E-0519G01), Alexander von Humboldt Research Fellowship (ID: 1195938) and the Six Talent Peaks Project in Jiangsu Province of China (Grant No. 2019-KTHY-009).

REFERENCES

[1] Y. Koike, "High-bandwidth graded-index polymer optical fibre," *Polymer,* vol. 32, no. 10, pp. 1737-1745, 1991.

[2] Y. Miyamoto, W. A. Kaysser, B. H. Rabin, A. Kawasaki, and R. G. Ford, *Functionally graded materials: Design, processing and applications*. Springer, 1999.

[3] L. J. Gray, T. Kaplan, J. D. Richardson, and G. H. Paulino, "Green's functions and boundary integral analysis for exponentially graded materials: Heat conduction," *J. Appl. Mech.-Trans. ASME,* vol. 70, no. 4, pp. 543-549, 2003.

[4] F. Erdogan, "Fracture mechanics of functionally graded materials," *Composites Engineering,* vol. 5, no. 7, pp. 753-770, 1995.

[5] G. Anlas, M. H. Santare, and J. Lambros, "Numerical Calculation of Stress Intensity Factors in Functionally Graded Materials," *International Journal of Fracture,* vol. 104, no. 2, pp. 131-143, 2000.

[6] H. Li, J. Lambros, B. A. Cheeseman, and M. H. Santare, "Experimental investigation of the quasi-static fracture of functionally graded materials," *Int. J. Solids Struct.,* vol. 37, no. 27, pp. 3715-3732, 2000.

[7] M. H. Santare and J. Lambros, "Use of Graded Finite Elements to Model the Behavior of Nonhomogeneous Materials," *Journal of Applied Mechanics,* vol. 67, no. 4, pp. 819-822, 2000.

[8] J. H. Kim and G. H. Paulino, "Isoparametric Graded Finite Elements for Nonhomogeneous Isotropic and Orthotropic Materials," *Journal of Applied Mechanics,* vol. 69, pp. 502-514, 2002.

[9] A. Sutradhar and G. H. Paulino, "The simple boundary element method for transient heat conduction in functionally graded materials," *Computer Methods in Applied Mechanics and Engineering,* vol. 193, no. 42-44, pp. 4511-4539, 2004.

[10] A. Sutradhar, G. H. Paulino, and L. J. Gray, "Transient heat conduction in homogeneous and non-homogeneous materials by the Laplace transform Galerkin boundary element method," *Engineering Analysis with Boundary Elements,* vol. 26, no. 2, pp. 119-132, 2002.

[11] J. Sladek, V. Sladek, C. L. Tan, and S. N. Atluri, "Analysis of transient heat conduction in 3D anisotropic functionally graded solids by the MLPG," *CMES-Computer Modeling in Engineering & Sciences,* vol. 32, pp. 161-174, 2008.

[12] V. Sladek, J. Sladek, M. Tanaka, and C. Zhang, "Transient heat conduction in anisotropic and functionally graded media by local integral equations," *Engineering Analysis with Boundary Elements,* vol. 29, no. 11, pp. 1047-1065, 2005.

[13] Z. J. Fu, L. W. Yang, Q. Xi, and C. S. Liu, "A boundary collocation method for anomalous heat conduction analysis in functionally graded materials," *Computers & Mathematics with Applications,* 2020. doi: https://doi.org/10.1016/j.camwa.2020.02.023.

[14] H. Wang, Q. H. Qin, and Y. L. Kang, "A meshless model for transient heat conduction in functionally graded materials," *Computational Mechanics,* vol. 38, no. 1, pp. 51-60, 2006.

[15] Q. Xi, Z. J. Fu, and T. Rabczuk, "An efficient boundary collocation scheme for transient thermal analysis in large-size-ratio functionally graded materials under heat source load," *Computational Mechanics,* vol. 64, no. 5, pp. 1221-1235, 2019.

[16] Z. J. Fu, W. Chen, and Q. H. Qin, "Three Boundary Meshless Methods for Heat Conduction Analysis in Nonlinear FGMs with Kirchhoff and Laplace Transformation," *Advances in Applied Mathematics and Mechanics,* vol. 4, no. 5, pp. 519-542, 2012.

[17] Q. H. Qin, "Trefftz Finite Element Method and Its Applications," *Applied Mechanics Reviews,* vol. 58, no. 5, pp. 316-337, 2005.

[18] J. Jirousek and N. Leon, "A powerful finite element for plate bending," *Computer Methods in Applied Mechanics and Engineering,* vol. 12, no. 1, pp. 77-96, 1977.

[19] Z. C. Li, T. T. Lu, H. T. Huang, and A. H. D. Cheng, "Trefftz, collocation, and other boundary methods - A comparison," *Numerical Methods for Partial Differential Equations,* vol. 23, no. 1, pp. 93-144, 2007.

[20] Z. J. Fu, L. W. Yang, H. Q. Zhu, and W. Z. Xu, "A semi-analytical collocation Trefftz scheme for solving multi-term time fractional diffusion-wave equations," *Engineering Analysis with Boundary Elements,* vol. 98, pp. 137-146, 2019.

[21] B. Pluymers, B. van Hal, D. Vandepitte, and W. Desmet, "Trefftz-Based Methods for Time-Harmonic Acoustics," *Archives of Computational Methods in Engineering,* vol. 14, no. 4, pp. 343-381, 2007.

[22] W. Chen, Z. J. Fu, and Q. H. Qin, "Boundary Particle Method with High-Order Trefftz Functions," *CMC-Computers Materials & Continua,* vol. 13, no. 3, pp. 201-217, 2010.

[23] Q. Xi, Z. Fu, W. Wu, H. Wang, and Y. Wang, "A novel localized collocation solver based on Trefftz basis for potential-based inverse electromyography," *Applied Mathematics and Computation,* vol. 390, p. 125604, 2021.

[24] I. Herrera and F. J. Sabina, "Connectivity as an alternative to boundary integral equations: Construction of bases," *Proceedings of the National Academy of Sciences,* vol. 75, no. 5, p. 2059, 1978.

[25] I. Herrera, "Boundary methods: A criterion for completeness," *Proceedings of the National Academy of Sciences,* vol. 77, no. 8, p. 4395, 1980.

[26] M. Dhanasekar, J. J. Han, and Q. H. Qin, "A hybrid-Trefftz element containing an elliptic hole," *Finite Elements in Analysis and Design,* vol. 42, no. 14-15, pp. 1314-1323, 2006.

[27] Q. H. Qin, "Hybrid Trefftz Finite-Element Approach for Plate-Bending on an Elastic-Foundation," *Applied Mathematical Modelling,* vol. 18, no. 6, pp. 334-339, 1994.

[28] J. Petrolito, "Hybrid-trefftz quadrilateral elements for thick plate analysis," *Computer Methods in Applied Mechanics and Engineering,* vol. 78, no. 3, pp. 331-351, 1990.

[29] Q. H. Qin, "Hybrid-Trefftz finite element method for Reissner plates on an elastic foundation," *Computer Methods in Applied Mechanics and Engineering,* vol. 122, no. 3-4, pp. 379-392, 1995.

[30] K. Peters, E. Stein, and W. Wagner, "A new boundary-type finite element for 2-D- and 3-D-elastic structures," *International Journal for Numerical Methods in Engineering,* vol. 37, no. 6, pp. 1009-1025, 1994.

[31] A. P. Zielinski and O. C. Zienkiewicz, "Generalized finite element analysis with T-complete boundary solution functions," *International Journal for Numerical Methods in Engineering,* vol. 21, no. 3, pp. 509-528, 1985.

[32] H. Wang, Q. H. Qin, and D. Arounsavat, "Application of hybrid Trefftz finite element method to non-linear problems of minimal surface," *International Journal for Numerical Methods in Engineering,* vol. 69, no. 6, pp. 1262-1277, 2007.

[33] K. Sze and G. Liu, "Hybrid-Trefftz six-node triangular finite element models for Helmholtz problem," *Computational Mechanics,* vol. 46, no. 3, pp. 455-470, 2010.

[34] J. Jirousek and Q. H. Qin, "Application of hybrid-Trefftz element approach to transient heat conduction analysis," *Computers & Structures,* vol. 58, no. 1, pp. 195-201, 1996.

[35] Q. H. Qin, "Variational formulations for TFEM of piezoelectricity," *Int. J. Solids Struct.,* vol. 40, no. 23, pp. 6335-6346, 2003.

[36] W. Chen, Z. J. Fu, and C. S. Chen, *Recent advances in radial basis function collocation methods*. Springer, 2014.

[37] Z. J. Fu, W. Chen, and Q. H. Qin, "Boundary knot method for heat conduction in nonlinear functionally graded material," *Engineering Analysis with Boundary Elements,* vol. 35, no. 5, pp. 729-734, 2011.

[38] Z. J. Fu, Q. Xi, W. Chen, and A. H. D. Cheng, "A boundary-type meshless solver for transient heat conduction analysis of slender functionally graded materials with exponential variations," *Computers & Mathematics with Applications,* vol. 76, no. 4, pp. 760-773, 2018.

[39] J. T. Chen, M. H. Chang, K. H. Chen, and I. L. Chen, "Boundary collocation method for acoustic eigenanalysis of three-dimensional cavities using radial basis function," *Computational Mechanics,* vol. 29, no. 4, pp. 392-408, 2002.

[40] Z. J. Fu, W. Chen, and Q. H. Qin, "Hybrid Finite Element Method Based on Novel General Solutions for Helmholtz-Type Problems," *CMC-Computers Materials & Continua,* vol. 21, no. 3, pp. 187--208, 2011.

[41] Q. H. Qin, *The Trefftz Finite and Boundary Element Method.* Southampton: WIT Press, 2000.

[42] J. Jirousek, A. Venkatesh, A. P. Zielinski, and H. Rabemanantsoa, "Comparative study of p-extensions based on conventional assumed displacement and Hybrid-Trefftz fe models," *Computers & Structures,* vol. 46, no. 2, pp. 261-278, 1993.

[43] L. L. Cao, Q. H. Qin, and N. Zhao, "An RBF-MFS model for analysing thermal behaviour of skin tissues," *International Journal of Heat and Mass Transfer,* vol. 53, no. 7-8, pp. 1298-1307, 2010.

[44] Z. J. Fu, J. H. Shi, W. Chen, and L. W. Yang, "Three-dimensional transient heat conduction analysis by boundary knot method," *Mathematics and Computers in Simulation,* vol. 165, pp. 306-317, 2019.

[45] Z. J. Fu, W. Chen, and H. T. Yang, "Boundary particle method for Laplace transformed time fractional diffusion equations," *Journal of Computational Physics,* vol. 235, pp. 52-66, 2013.

[46] W. Chen, L. J. Shen, Z. J. Shen, and G. W. Yuan, "Boundary knot method for Poisson equations," *Engineering Analysis with Boundary Elements,* vol. 29, no. 8, pp. 756-760, 2005.

[47] Z. J. Fu, and W. Chen, "A Truly Boundary-Only Meshfree Method Applied to Kirchhoff Plate Bending Problems," *Advances in Applied Mathematics and Mechanics,* vol. 1, no. 3, pp. 341-352, 2009.

[48] Z. J. Fu, Q. H. Qin, and W. Chen, "Hybrid-Trefftz finite element method for heat conduction in nonlinear functionally graded materials," *Engineering Computations,* vol. 28, no. 5, pp. 578-599, 2011.

[49] Q. H. Qin and H. Wang, *Matlab and C Programming for Trefftz Finite Element Methods*. Boca Raton: CRC Press, 2009.

[50] H. Wang and Q. H. Qin, "Hybrid FEM with fundamental solutions as trial functions for heat conduction simulation," *Acta Mechanica Solida Sinica,* vol. 22, no. 5, pp. 487-498, 2009.

[51] P.W. Partridge, C.A. Brebbia, and L.C. Wrobel, *The dual reciprocity boundary element method.* Computational Mechanics Publications, 1992.

[52] A.J. Nowak and A.C. Neves, *The Multiple Reciprocity Boundary Element Method.* Computational Mechanics Publication, 1994.

[53] Q. Xi, Z. J. Fu, C. Alves, and H. L. Ji, "A semi-analytical boundary collocation solver for the inverse Cauchy problems in heat conduction under 3D FGMs with heat source," *Numerical Heat Transfer, Part B: Fundamentals,* vol. 76, no. 5, pp. 311-327, 2019.

[54] Z. J. Fu, W. Chen, and C. Z. Zhang, "Boundary particle method for Cauchy inhomogeneous potential problems," *Inverse Problems in Science and Engineering,* vol. 20, no. 2, pp. 189-207, 2012.

[55] D. P. Gaver, "Observing Stochastic Processes, and Approximate Transform Inversion," *Operations Research,* vol. 14, no. 3, pp. 444-459, 1966.

[56] J. R. Berger, P. A. Martin, V. Mantic, and L. J. Gray, "Fundamental solutions for steady-state heat transfer in an exponentially graded anisotropic material," *Zeitschrift Fur Angewandte Mathematik Und Physik,* vol. 56, no. 2, pp. 293-303, 2005.

[57] W. Z. Qu, C. M. Fan, and Y. M. Zhang, "Analysis of three-dimensional heat conduction in functionally graded materials by using a hybrid numerical method," *International Journal of Heat and Mass Transfer*, vol. 145, p. 118771, 2019.

[58] Z. J. Fu, Q. Xi, Y. D. Li, H. Huang, and T. Rabczuk, "Hybrid FEM-SBM solver for structural vibration induced underwater acoustic radiation in shallow marine environment," *Computer Methods in Applied Mechanics and Engineering,* vol. 369, p. 113236, 2020.

[59] Q. Xi, Z. J. Fu, Y. D. Li, and H. Huang, "A Hybrid GFDM–SBM Solver for Acoustic Radiation and Propagation of Thin Plate Structure Under Shallow Sea Environment," *Journal of Theoretical and Computational Acoustics,* vol. 28, no. 02, p. 2050008, 2020.

CHAPTER 5

Introduction to Fundamental Solution Based Finite Element Methods

Qing-Hua Qin[1*]

[1] *Department of Engineering, Shenzhen MSU-BIT University, 518172 Shenzhen, China*

Abstract: Further to the discussion of hybrid Trefftz FEM in Chapter 1, this chapter presents an overview of applications of the fundamental solution based FEM for heat conduction and seepage problems. The fundamental solution based FEM discussed here is significantly different from the previous three FEM types mentioned in Chapter 1. In this method, a linear combination of the fundamental solution at different points approximates the field variable within the element. The independent frame field defined along the element boundary and the modified variational functional are employed to guarantee inter-element continuity, generate the final stiffness equation and establish linkage between the boundary frame field and the internal field in the element.

Keywords: Finite element method, Functionally graded material, Fundamental solution, Heat conduction, Seepage.

INTRODUCTION

The fundamental solution based FEM was initiated in 2008 [1, 2] and has now become a very popular and powerful computational method in mechanical engineering. Unlike analytical solutions, which are available only for a few problems with simple geometries and boundary conditions [3-13], the fundamental solution based FEM is a highly efficient computational tool for the solution of various complex boundary value problems [1, 14-16]. So far, this method has been applied to potential problems [17-21], plane elasticity [22-24], wave propagation [25], composites [26-32], piezoelectric materials [33-35], three-dimensional problems [36, 37], functionally graded materials [38-41], bioheat transfer problems [42-48], thermal elastic problems [49], hole problems [50, 51], homogenize heterogeneous elastic medium [52], heat conduction problems [2, 53], crack problems [54], topology optimization [41], axisymmetric potential problems [55], micromechanics problems [56, 57], and anisotropic elastic problems [58-60].

[*]**Corresponding author Qing-Hua Qin:** Department of Engineering, Shenzhen MSU-BIT University, 518172 Shenzhen, China; E-mail: qinghua.qin@smbu.edu.cn

Following the introduction presented above, basic formulations of fundamental solution based FEM and their applications to heat conduction problems are presented in the section "heat conduction problems". Extension to time-dependent heat conduction problems in functionally graded materials is presented in the section "Transient heat conduction in functionally graded materials". Further extension to seepage problems is presented in the section "orthotropic seepage problems". Finally, a brief summary of the approach is provided.

HEAT CONDUCTION PROBLEMS

This section is concerned with the application of fundamental solution (FS) based FEM to the solution of heat conduction problems. The section is written is based on the work presented in [2]. Beginning with a description of its basic equation, the construction of FS based FEM is discussed. It includes the derivation of the assumed intra-element field, auxiliary conforming frame field, modified variational principle, and recovery of rigid-body motion.

Basic Equation of Heat Conduction

Consider a heat conduction problem in a general plane domain Ω. Its field equation is written as [2]:

$$\frac{\partial}{\partial X_1}\left(k\frac{\partial u(\mathbf{x})}{\partial X_1}\right)+\frac{\partial}{\partial X_2}\left(k\frac{\partial u(\mathbf{x})}{\partial X_2}\right)=0 \qquad \forall\mathbf{x}\in\Omega \tag{1}$$

and with the boundary conditions:

—Dirichlet boundary condition related to unknown temperature field

$$u=\bar{u} \qquad \text{on } \Gamma_u \tag{2}$$

—Neumann boundary condition for the boundary heat flux

$$q=-ku_{,i}n_i=\bar{q} \qquad \text{on } \Gamma_q \tag{3}$$

where k stands for the thermal conductivity, u is the sought field variable and q represents the boundary heat flux. n_i is the ith component of the outward normal vector to the boundary $\Gamma=\Gamma_u\cup\Gamma_q$, and $\bar{u}$ and $\bar{q}$ are specified functions on the related boundaries, respectively. Space derivatives are indicated by a comma, *i.e.* $u_{,i} = \partial u / \partial X_i$, and the subscript index i takes values 1 and 2 in our analysis. Additionally, the repeated subscript indices stand for summation convention. For convenience, Eq. (3) is rewritten in matrix form as:

$$q = -k\mathbf{A}\begin{bmatrix} u_{,1} \\ u_{,2} \end{bmatrix} = \bar{q} \qquad \textbf{(4)}$$

where $\mathbf{A} = \begin{bmatrix} n_1 & n_2 \end{bmatrix}$.

Basic Formulation of FS-FEM Finite Element Approach

In this section, the procedure for developing a hybrid finite element model with the FS as an interior trial function is described based on the boundary value problem defined by Eqs (1)-(3).

As in the hybrid Trefftz-FEM, the main idea of the proposed approach is to establish a hybrid finite element formulation whereby intra-element continuity is enforced on a nonconforming internal displacement field formed by a linear combination of fundamental solutions at points outside the element domain under consideration, while an auxiliary frame field is independently defined on the element boundary to enforce field continuity across inter-element boundaries. But unlike in the HT FEM, the intra-element field is constructed based on the fundamental solution, rather than T-functions. Consequently, a variational functional corresponding to the new trial function is required to derive the related stiffness matrix equation. With the problem domain divided into some subdomains or elements denoted by Ω_e with the element boundary Γ_e, additional continuity is usually required on the common boundary Γ_{Ief} between any two adjacent elements '*e*' and '*f*' (Fig. **1**):

$$\left.\begin{array}{ll} u_e = u_f & \text{(conformity)} \\ q_e + q_f = 0 & \text{(reciprocity)} \end{array}\right\} \text{ on } \Gamma_{Ief} = \Gamma_e \cap \Gamma_f \qquad \textbf{(5)}$$

in the proposed hybrid FE approach.

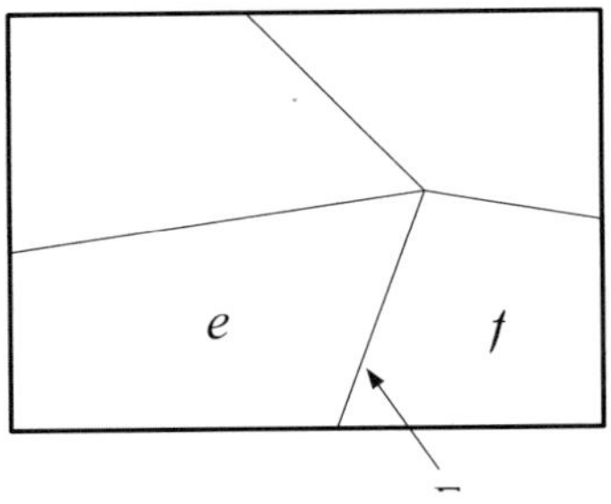

Fig. (1). Illustration of continuity between two adjacent elements '*e*' and '*f*'.

Ω_e

Non-Conforming Intra-Element Field

Activated by the idea of fundamental solution's method (MFS) [61], to remove the singularity of the fundamental solution, for a particular element, say element e, which occupies the subdomain, we assume that the field variable defined in the element domain is extracted from a linear combination of fundamental solutions centred at different source points (Fig. **2**), that is:

c_{ej}

$$u_e(\mathbf{x}) = \sum_{j=1}^{n_s} N_e(\mathbf{x}, \mathbf{y}_j) c_{ej} = \mathbf{N}_e(\mathbf{x}) \mathbf{c}_e^{n_s} \qquad \forall \mathbf{x} \in \Omega_e, \mathbf{y}_j \notin \Omega_e \tag{6}$$

where are undetermined coefficients and is the number of virtual sources outside the element e. $N_e(\mathbf{x}, \mathbf{y}_j)$ is the fundamental solution to the 2D heat conduction and generally satisfies

$$k\nabla^2 N_e(\mathbf{x}, \mathbf{y}) + \delta(\mathbf{x}, \mathbf{y}) = 0 \qquad \forall \mathbf{x}, \mathbf{y} \in \mathbb{R}^2 \tag{7}$$

which gives

$$N_e(\mathbf{x}, \mathbf{y}) = -\frac{1}{2\pi k} \ln r(\mathbf{x}, \mathbf{y}) \tag{8}$$

clearly, Eq. (6) analytically satisfies Eq. (1) due to the solution property of $N_e(\mathbf{x}, \mathbf{y}_j)$.

In practice, the generation of virtual sources is usually done by means of the formulation employed in the MFS [61, 62]:

$$\mathbf{y} = \mathbf{x}_b + \gamma(\mathbf{x}_b - \mathbf{x}_c) \tag{9}$$

where γ is a dimensionless coefficient, $\mathbf{x}_b$ is the elementary boundary point and $\mathbf{x}_c$ the geometrical centroid of the element. For a particular element shown in Fig. (**2**), we can use the nodes of the element to generate related source points for simplicity.

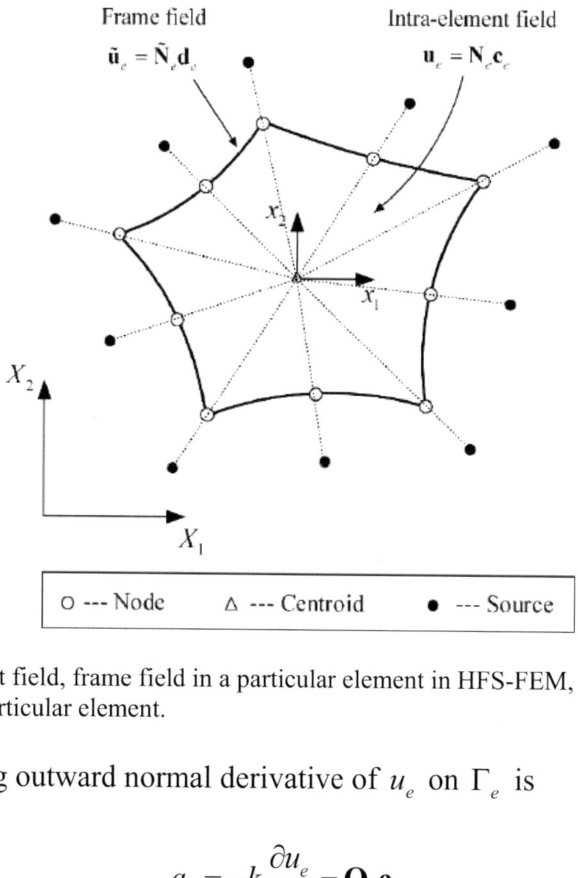

Fig. (2). Intra-element field, frame field in a particular element in HFS-FEM, and the generation of source points for a particular element.

The corresponding outward normal derivative of u_e on Γ_e is

$$q_e = -k\frac{\partial u_e}{\partial n} = \mathbf{Q}_e\mathbf{c}_e \tag{10}$$

where

$$\mathbf{Q}_e = -k\frac{\partial \mathbf{N}_e}{\partial n} = -k\mathbf{A}\mathbf{T}_e \tag{11}$$

with

$$\mathbf{T}_e = \left\{ \frac{\partial \mathbf{N}_e}{\partial x_1} \quad \frac{\partial \mathbf{N}_e}{\partial x_2} \right\}^T \tag{12}$$

Auxiliary Conforming Frame Field

To enforce conformity on the field variable u, for instance, $u_e = u_f$ on $\Gamma_e \cap \Gamma_f$ of any two neighbouring elements *e* and *f*, an auxiliary inter-element frame field $\tilde{u}$ is used and expressed in terms of the same degrees of freedom (DOF) $\mathbf{d}$, as used in the conventional finite element method. In this case, $\tilde{u}$ is confined to the entire element boundary, that is,

$$\tilde{u}_e(\mathbf{x}) = \tilde{\mathbf{N}}_e(\mathbf{x})\mathbf{d}_e \tag{13}$$

which is independently assumed along the element boundary in terms of nodal DOF $\mathbf{d}_e$, where $\tilde{\mathbf{N}}_e$ represents the conventional finite element interpolating functions. For example, a simple interpolation of the frame field on the side with three nodes of a particular element (Fig. (**2**) can be given in the form:

$$\tilde{u} = \tilde{N}_1 u_1 + \tilde{N}_2 u_2 + \tilde{N}_3 u_3 \tag{14}$$

where $\tilde{N}_i$ ($i = 1,2,3$) stands for shape functions in terms of the natural coordinate ξ defined in Fig. (**3**).

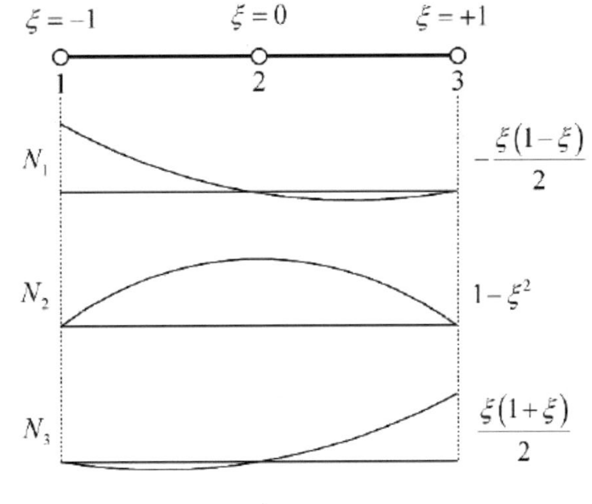

Fig. (3). Typical quadratic interpolation for the frame field.

Modified Variational Principle and Stiffness Equation

Modified Functional

For the boundary value problem defined in Eqs (1)-(3) and (5), because the stationary conditions of the traditional potential or complementary variational functional cannot guarantee satisfaction for the inter-element continuity condition required in the proposed FS finite element model, a modified potential functional is developed as:

$$\Pi_m = \sum_e \Pi_{me} = \sum_e \left\{ \Pi_e + \int_{\Gamma_e} (\tilde{u} - u) q \mathrm{d}\Gamma \right\} \tag{15}$$

where

$$\Pi_e = -\frac{1}{2} \int_{\Omega_e} k u_{,i} u_{,i} \mathrm{d}\Omega - \int_{\Gamma_{qe}} \bar{q} \tilde{u} \mathrm{d}\Gamma \tag{16}$$

in which the governing Eq. (1) is assumed to be satisfied, a priori, for deriving the FS finite element model. The boundary Γ_e of a particular element consists of the following parts

$$\Gamma_e = \Gamma_{ue} \cup \Gamma_{qe} \cup \Gamma_{Ie} \tag{17}$$

where Γ_{Ie} represents the inter-element boundary of element '*e*' shown in Fig. (**1**).

Stiffness Equation

Having independently defined the intra-element field and frame field in a particular element (Fig. **2**), the next step is to generate the element stiffness equation through a variational approach.

Following the approach described in [1], the variational functional Π_e corresponding to a particular element e of the present problem can be written as

$$\Pi_{me} = -\frac{1}{2} \int_{\Omega_e} k u_{,i} u_{,i} \mathrm{d}\Omega - \int_{\Gamma_{qe}} \bar{q} \tilde{u} \mathrm{d}\Gamma + \int_{\Gamma_e} q (\tilde{u} - u) \mathrm{d}\Gamma \tag{18}$$

Applying the divergence theorem described in [1] to the above functional, we have the final functional for the FS finite element model:

$$
\begin{aligned}
\Pi_{me} &= \frac{1}{2}\left[\int_{\Gamma_e} qu\mathrm{d}\Gamma + \int_{\Omega_e} uk\nabla^2 u\mathrm{d}\Omega\right] - \int_{\Gamma_{qe}} \bar{q}\tilde{u}\mathrm{d}\Gamma + \int_{\Gamma_e} q(\tilde{u}-u)\mathrm{d}\Gamma \\
&= -\frac{1}{2}\int_{\Gamma_e} qu\mathrm{d}\Gamma - \int_{\Gamma_{qe}} \bar{q}\tilde{u}\mathrm{d}\Gamma + \int_{\Gamma_e} q\tilde{u}\mathrm{d}\Gamma
\end{aligned}
\tag{19}
$$

Then, substituting Eqs (6), (10) and (13) into the functional (19) finally produces

$$
\Pi_e = -\frac{1}{2}\mathbf{c}_e^{\mathrm{T}}\mathbf{H}_e\mathbf{c}_e - \mathbf{d}_e^{\mathrm{T}}\mathbf{g}_e + \mathbf{c}_e^{\mathrm{T}}\mathbf{G}_e\mathbf{d}_e \tag{20}
$$

in which

$$
\mathbf{H}_e = \int_{\Gamma_e} \mathbf{Q}_e^{\mathrm{T}}\mathbf{N}_e\mathrm{d}\Gamma = \int_{\Gamma_e} \mathbf{N}_e^{\mathrm{T}}\mathbf{Q}_e\mathrm{d}\Gamma
$$

$$
\mathbf{G}_e = \int_{\Gamma_e} \mathbf{Q}_e^{\mathrm{T}}\tilde{\mathbf{N}}_e\mathrm{d}\Gamma, \quad \mathbf{g}_e = \int_{\Gamma_{qe}} \tilde{\mathbf{N}}_e^{\mathrm{T}}\bar{q}\mathrm{d}\Gamma
$$

The symmetry of $\mathbf{H}_e$ is obvious from the scale definition (20) of the variational functional Π_e.

To enforce inter-element continuity on the common element boundary, the unknown vector $\mathbf{c}_e$ must be expressed in terms of nodal DOF $\mathbf{d}_e$. The minimization of the functional Π_e with respect to $\mathbf{c}_e$ and $\mathbf{d}_e$ respectively yields

$$
\begin{aligned}
\frac{\partial \Pi_e}{\partial \mathbf{c}_e^{\mathrm{T}}} &= -\mathbf{H}_e\mathbf{c}_e + \mathbf{G}_e\mathbf{d}_e = \mathbf{0} \\
\frac{\partial \Pi_e}{\partial \mathbf{d}_e^{\mathrm{T}}} &= \mathbf{G}_e^{\mathrm{T}}\mathbf{c}_e - \mathbf{g}_e = \mathbf{0}
\end{aligned}
\tag{21}
$$

from which the optional relationship between $\mathbf{c}_e$ and $\mathbf{d}_e$ and the stiffness equation can be produced:

$$
\mathbf{c}_e = \mathbf{H}_e^{-1}\mathbf{G}_e\mathbf{d}_e \quad \text{and} \quad \mathbf{K}_e\mathbf{d}_e = \mathbf{g}_e \tag{22}
$$

where $\mathbf{K}_e = \mathbf{G}_e^{\mathrm{T}} \mathbf{H}_e^{-1} \mathbf{G}_e$ stands for the element stiffness matrix.

It is worth pointing out that the evaluation of the right-handed vector $\mathbf{g}_e$ in Eq. (22) is the same as that in the conventional FEM, which is obviously convenient for implementation of the FS-FEM into the existing FEM program.

Recovery of Rigid-Body Motion

Considering the physical definition of the fundamental solution, it is necessary to recover the missing rigid-body motion modes from the above results.

Following the method presented in [5], the missing rigid-body motion can be recovered by writing the internal potential field of a particular element *e* as:

$$u_e = \mathbf{N}_e \mathbf{c}_e + c_0 \tag{23}$$

where the undetermined rigid-body motion parameter c_0 can be calculated using the least square matching of u_e and $\tilde{u}_e$ at element nodes [63]:

$$\sum_{i=1}^{n} \left(\mathbf{N}_e \mathbf{c}_e + c_0 - \tilde{u}_e \right)^2 \Big|_{\text{node } i} = \min \tag{24}$$

which finally gives:

$$c_0 = \frac{1}{n} \sum_{i=1}^{n} \Delta u_{ei} \tag{25}$$

in which $\Delta u_{ei} = \left(\tilde{u}_e - \mathbf{N}_e \mathbf{c}_e \right)\Big|_{\text{node } i}$ and n is the number of element nodes.

Once the nodal field is determined by solving the final stiffness equation, the coefficient vector $\mathbf{c}_e$ can be evaluated from Eq. (22), and then c_0 is evaluated from Eq. (25). Finally, the potential field u at any internal point in an element can be obtained by means of Eq. (6).

TRANSIENT HEAT CONDUCTION IN FUNCTIONALLY GRADED MATERIALS

In the previous section, applications of FS-FEM to heat conduction problems were presented. Extension of the procedure to transient heat conduction in functionally graded materials (FGMs) is now described. In the process, a Laplace transform approach is used to handle the time variable. Then, a fundamental solution in Laplace space for FGMs is constructed. Next, a hybrid graded element is formulated based on the obtained fundamental solution and a frame field. The majority of this section was presented in [39].

Basic Formulations

Statement of Heat Conduction Problems on FGMs

Consider a two-dimensional (2D) transient heat conduction problem:

$$\nabla\cdot(k(X)\nabla u(X,t)) = \rho(X)c(X)\frac{\partial u(X,t)}{\partial t} \tag{26}$$

with the boundary conditions:

—Dirichlet boundary condition

$$u(X,t)=\bar{u}\,(X,t) \qquad \text{on } \Gamma_u \tag{27}$$

—Neumann boundary condition

$$q(X,t)=\bar{q}\,(X,t) \qquad \text{on } \Gamma_q \tag{28}$$

where t denotes the time variable ($t>0$). k is the thermal conductivity dependent on the special variables $X\in\Omega\subset R^d$. d is the number of dimensions of the solution domain Ω ($d=2$ in this study). ρ is the mass density. c is the specific heat, and u is the unknown temperature field. q represents the boundary heat flux defined by $q=-k\partial u/\partial n$, where n is the unit outward normal to the boundary $\Gamma=\Gamma_u\cup\Gamma_q$. $\bar{u}$ and $\bar{q}$ are specified temperature and heat flow, respectively, on the related boundaries. In addition, an initial condition must be given for the time-dependent problem. In this section, a zero initial temperature distribution is considered, *i.e.*

$$u(X,0) = u_0(X) = 0 \tag{29}$$

The composition and the volume fraction of FGM constituents vary gradually with the coordinate X, giving a non-uniform microstructure with continuously graded macroproperties (conductivity, specific heat, density). In the present work, to make the derivation tractable, the mass density is assumed to be constant within each element and takes the value of ρ at the centroid of the element. The thermal conductivity and specific heat have been chosen to have the same functional variation so that the thermal diffusivity η is constant, that is:

$$k(X) = k_0 f(X) \tag{30}$$

$$c(X) = c_0 f(X) \tag{31}$$

and

$$\eta = \frac{k_0}{c_0 \rho} \tag{32}$$

It should be mentioned that the foregoing assumption in FGMs leads to a class of solvable problems and can provide benchmark solutions for other numerical methods such as the FEM, meshless method [64], and BEM [65-67]. Moreover, it can provide valuable insight into the thermal behaviour of FGMs [7]. Therefore this assumption has been adopted by many researchers for solving transient thermal problems in FGMs, see [68, 69].

Laplace Transformation and Fundamental Solution in Laplace Space

The Laplace transformation (LT) of a function $u(X,t)$ is defined by

$$L(u(X,t)) = U(X,s) = \int_0^\infty u(X,t) e^{-st} dt \tag{33}$$

where s is the Laplace parameter. Using integration by parts, it is possible to show that:

$$L[\frac{\partial u(X,t)}{\partial t}] = sU(X,s) - u_0(X) \tag{34}$$

The boundary conditions (27) and (28) become:

$$U(X,s)=\frac{\bar{u}(X,t)}{s} \qquad \text{on } \Gamma_u \tag{35}$$

$$P(X,s)=\frac{\bar{q}}{s}(X,t) \qquad \text{on } \Gamma_q \tag{36}$$

Exponentially Graded Material

First, we consider a FGM with thermal conductivity and specific heat varying exponentially in just one Cartesian coordinate, direction X_2:

$$k(X_2)=k_0 e^{2\beta X_2} \tag{37}$$

$$c(X_2)=c_0 e^{2\beta X_2} \tag{38}$$

where β is the non-homogeneity graded parameter.

Substituting Eq. (37) and Eq. (38) into Eq. (26) yields:

$$\nabla^2 u+2\beta u_{X_2}=\frac{1}{\alpha}\frac{\partial u}{\partial t} \tag{39}$$

where u_{X_2} denotes the derivative of u with respect to X_2 ($u_{X_2}=\partial u/\partial X_2$)

After performing the LT, Eq. (39) becomes

$$\nabla^2 U+2\beta U_{X_2}-\frac{s}{\alpha}U=0 \tag{40}$$

in LT space ,where $u_0(X)=0$ (at $t=0$) is considered (see Eq. (29)).

To obtain the fundamental solution of Eq. (40), the following substitution is used here:

$$U=e^{-\beta X_2}G \tag{41}$$

In this case, the differential Eq. (40) in Laplace space becomes:

$$\nabla^2 G - (\beta^2 + \frac{s}{\alpha})G = 0 \tag{42}$$

Obviously, Eq. (42) is the modified Helmholz equation, whose fundamental solution is:

$$G = \frac{1}{2\pi} K_0(\sqrt{\beta^2 + \frac{s}{\alpha}} r) \tag{43}$$

Making use of Eq. (41), we obtain the fundamental solution of Eq. (40) in Laplace space:

$$N(X, X_S) = \frac{1}{2\pi} e^{-\beta(X_2 - X_{S_2})} K_0(\sqrt{\beta^2 + \frac{s}{\alpha}} r) \tag{44}$$

where $r = \|X - X_S\|$, X and X_S denote an arbitrary field point and source point in the infinite domain, respectively. K_0 is the modified Bessel function of the second kind of zero order.

General Method for FGMs with Different Variation of Properties

The method can be extended to a broader range of FGMs, not only exponential but also quadratic and trigonometric material variation, by variable transformations [70]. By defining a variable [70]:

$$v(X, t) = \sqrt{k(X)} u(X, t) \tag{45}$$

Eq. (26) can be rewritten as:

$$\nabla^2 v + (\frac{\nabla k(X) \cdot \nabla k(X)}{4k^2(X)} - \frac{\nabla^2 k(X)}{2k(X)})v = \frac{\rho c(X)}{k(X)} \frac{\partial v}{\partial t} \tag{46}$$

For simplicity, define:

$$k^{'}(X) = \frac{\nabla k(X) \cdot \nabla k(X)}{4k^2(X)} - \frac{\nabla^2 k(X)}{2k(X)} \tag{47}$$

Then, Eq. (46) can be rewritten as:

$$\nabla^2 v + k^{'}(X) v = \frac{1}{\alpha} \frac{\partial v}{\partial t} \tag{48}$$

After performing the LT, the differential Eq. (48) becomes:

$$\nabla^2 V + k^{'} V - \frac{s}{\alpha} V = 0 \tag{49}$$

When $k^{'}$ is a constant, Eq. (49) is a modified Helmholz equation whose fundamental solution is known. Then, the fundamental solution of Eq. (26) in Laplace space can be obtained by inverse transformation:

$$N(X, X_S) = \frac{1}{2\pi} \frac{K_0(\sqrt{-k^{'} + \frac{s}{\alpha}} r)}{k(X)^{1/2} k(X_s)^{1/2}} \tag{50}$$

For quadratic material:

$$k(X) = k_0 (a_1 + \beta X_2)^2 \tag{51}$$

In this case, $k^{'} = 0$ in Eq. (49).

For trigonometric material,

$$k(X) = k_0 (a_1 \cos \beta X_2 + a_2 \sin \beta X_2)^2 \tag{52}$$

In this case, $k^{'} = \beta^2$ in Eq. (49).

For exponential material,

$$k(X) = k_0 (a_1 e^{\beta X_2} + a_2 e^{-\beta X_2})^2 \tag{53}$$

In this case, $k' = -\beta^2$ in Eq. (49). Substituting $k' = -\beta^2$ into Eq. (50) and using the exponential law, the fundamental solution given by Eq. (50) reduces to Eq. (44).

Note that for quadratic, trigonometric, and exponential variations of both heat conductivity and specific heat, the FGM transient problem can be transformed into the same differential equation, which has a simple and standard form (Eq. (48)) [70].

Generation of Graded Element

In this section, an element formulation is presented to deal with materials with continuous variation of physical properties. Such an element model is usually known as a hybrid graded element, and can be used for solving the boundary value problem in Laplace space. The governing equation is the transformed governing equation of Eq. (26) in Laplace space. The corresponding boundary conditions are defined in Eqs (35)-(36).

The proposed approach is based on a hybrid finite element formulation in which fundamental solutions are taken as internal interpolation functions (FS-FEM) [2]. As in the HT-FEM, the main idea of FS-FEM is to establish an appropriate hybrid FE formulation whereby intra-element continuity is enforced on a nonconforming internal field formed by a linear combination of fundamental solutions at points outside the element domain under consideration, while an auxiliary frame field is independently defined on the element boundary to enforce field continuity across the inter-element boundaries. But unlike the proce$\Omega_e$in HT FEM, the intra-element fields are constructed based on the fundamental solution, rather than T-functions. Consequently, a variational functional corresponding to the new trial function is required to derive the related stiffness matrix equation. As in the conventional FEM, the solution domain is divided into subdomains or elements. For a particular element, say element *e*, its domain is denoted by and bounded by Γ_e. Since a nonconforming function is used for modelling the internal fields, additional continuities are usually required over the common boundary Γ_{Ief} between any two adjacent elements '*e*' and '*f*' (Fig.**1**) [71].

$$\left.\begin{array}{ll} U_e = U_f & \text{(conformity)} \\ P_e + P_f = 0 & \text{(reciprocity)} \end{array}\right\} \quad \text{on } \Gamma_{Ief} = \Gamma_e \cap \Gamma_f \tag{54}$$

in the proposed hybrid FE approach.

Non Conforming Intra-Element Field

For a particular element, say element *e*, which occupies the subdomain Ω_e, the field variable within the element is extracted from a linear combination of fundamental solutions centred at different source points (see Fig. **2**), that is,

$$U_e(x)=\sum_{j=1}^{n_s} N_e(x,x_S)c_{ej}=\mathbf{N}_e(x)\mathbf{c}_e \qquad \forall x\in\Omega_e\subset R^d, x_S\notin\Omega_e \tag{55}$$

where c_{ej} is undetermined coefficients and n_s is the number of virtual sources outside the element *e*. $N_e(x,x_S)$ is the required fundamental solution expressed in local element coordinates (x_1,x_2) rather than global coordinates (X_1,X_2) (see Fig. 2). Clearly, Eq. (55) analytically satisfies the transformed governing equation of Eq. (26) in Laplace space due to the inherent property of $N_e(x,x_S)$.

The fundamental solution for FGMs (N_e in Eq. (55)) is used to approximate the intra-element field for a FGM. The smooth variation of material properties throughout an element can be achieved by using the fundamental solution which can reflect the impact of a concentrated unit source acting at a point on any other points. The model inherits the essence of a FGM, and so can simulate FGMs more naturally than the stepwise constant approximation which has been frequently used in conventional FEM. Fig. (**4**) illustrates the difference between the two models when the thermal conductivity varies along direction X_2.

Note that the thermal conductivity in Eq. (26) is defined in the global coordinate system. When contriving the intra-element field for each element, this formulation must be transferred into the local element coordinate system defined at the centre of the element, and the graded heat conductivity $k(X)$ in Eq. (30) can then be expressed by:

$$k_e(X)=k_C(X)f(X) \tag{56}$$

for a particular element *e*, where $k_C(X)$ denotes the value of conductivity at the centroid of each element and can be calculated as follows:

$$k_C(X) = k_0 f(X_C) \tag{57}$$

where X_C is the global coordinates of the element centroid.

Accordingly, k_C is used to replace k (see Eq. (50)) in the formulation of the fundamental solution for the FGM and to construct the intra-element field in the local element coordinate system for each element.

In practice, the generation of virtual sources is usually achieved by means of the following formulation employed in the MFS [69]

$$y = x_b + \gamma\left(x_b - x_c\right) \tag{58}$$

where γ is a dimensionless coefficient, x_b and x_c are, respectively, boundary point and geometrical centroid of the element. For a particular element as shown in Fig. (**2**), for simplicity we can use the nodes of the element to generate related source points.

The corresponding normal heat flux on Γ_e is given by:

$$P_e = -k_e \frac{\partial U_e}{\partial X_j} n_i = \mathbf{Q}_e \mathbf{c}_e \tag{59}$$

where

$$\mathbf{Q}_e = -k_e \frac{\partial \mathbf{N}_e}{\partial X_j} n_i = -k_e \mathbf{A} \mathbf{T}_e \tag{60}$$

with

$$\mathbf{T}_e = \left[\mathbf{N}_{e,1} \quad \mathbf{N}_{e,2}\right]^{\mathrm{T}} \qquad \mathbf{A} = \left[n_1 \quad n_2\right] \tag{61}$$

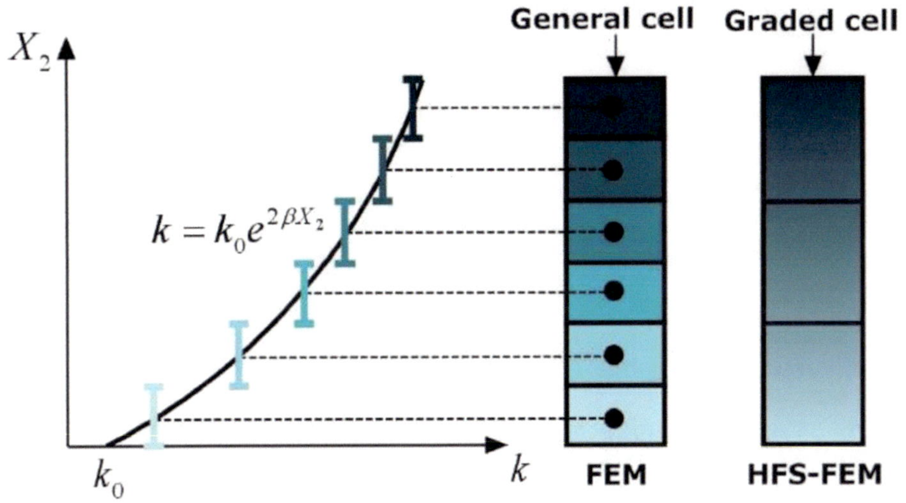

Fig. (4). Comparison of computational cells in conventional FEM and the proposed hybrid fundamental solution based FEM (HFS-FEM).

Auxiliary Conforming Frame Field

In order to enforce conformity on the field variable U, for instance, $U_e = U_f$ on $\Gamma_e \cap \Gamma_f$ of any two neighbouring elements *e* and *f*, an auxiliary inter-element frame field $\tilde{U}$ is used and expressed in terms of nodal degrees of freedom (DOF), $\mathbf{d}$, as used in conventional FEM as:

$$\tilde{U}_e(x) = \tilde{\mathbf{N}}_e(x)\mathbf{d}_e \tag{62}$$

which is independently assumed along the element boundary, where $\tilde{\mathbf{N}}_e$ represents the conventional FE interpolating functions. For example, a simple interpolation of the frame field on the side with three nodes of a particular element can be given in the form:

$$\tilde{U} = \tilde{N}_1\Psi_1 + \tilde{N}_2\Psi_2 + \tilde{N}_3\Psi_3 \tag{63}$$

where $\tilde{N}_i$ ($i = 1,2,3$) stands for shape functions, which are the same as those in conventional FEM.

Modified Variational and Stiffness Equation

When the intra-element field and frame field in a particular element have been independently defined (see Fig. **2**), the element stiffness equation can be generated through a variational approach. Here we present the results directly; details of the derivation can be found in [1, 63].

The final functional defined only on the element boundary is:

$$\Pi_{me} = -\frac{1}{2}\int_{\Gamma_e} PU\mathrm{d}\Gamma - \int_{\Gamma_{qe}} \frac{\bar{q}}{S}\tilde{U}\mathrm{d}\Gamma + \int_{\Gamma_e} P\tilde{U}\mathrm{d}\Gamma \qquad \textbf{(64)}$$

Substituting Eqs (6), (10) and (55) into the functional (64) yields:

$$\Pi_e = -\frac{1}{2}\mathbf{c}_e^{\mathrm{T}}\mathbf{H}_e\mathbf{c}_e - \mathbf{d}_e^{\mathrm{T}}\mathbf{g}_e + \mathbf{c}_e^{\mathrm{T}}\mathbf{G}_e\mathbf{d}_e \qquad \textbf{(65)}$$

where

$$\mathbf{H}_e = \int_{\Gamma_e} \mathbf{Q}_e^{\mathrm{T}}\mathbf{N}_e\mathrm{d}\Gamma \quad \mathbf{G}_e = \int_{\Gamma_e} \mathbf{Q}_e^{\mathrm{T}}\tilde{\mathbf{N}}_e\mathrm{d}\Gamma \quad \mathbf{g}_e = \int_{\Gamma_{qe}} \tilde{\mathbf{N}}_c^{\mathrm{T}}\frac{\bar{\mathrm{q}}}{S}\mathrm{d}\Gamma \qquad \textbf{(66)}$$

Next, to enforce inter-element continuity on the common element boundary, the unknown vector $\mathbf{c}_e$ must be expressed in terms of nodal DOF $\mathbf{d}_e$. The minimization of the functional Π_e with respect to $\mathbf{c}_e$ and $\mathbf{d}_e$, respectively, yields:

$$\begin{cases} \dfrac{\partial \Pi_e}{\partial \mathbf{c}_e^{\mathrm{T}}} = -\mathbf{H}_e\mathbf{c}_e + \mathbf{G}_e\mathbf{d}_e = \mathbf{0} \\ \dfrac{\partial \Pi_e}{\partial \mathbf{d}_e^{\mathrm{T}}} = \mathbf{G}_e^{\mathrm{T}}\mathbf{c}_e - \mathbf{g}_e = \mathbf{0} \end{cases} \qquad \textbf{(67)}$$

from which the optional relationship between $\mathbf{c}_e$ and $\mathbf{d}_e$, and the stiffness equation can be produced in the form:

$$\mathbf{c}_e = \mathbf{H}_e^{-1}\mathbf{G}_e\mathbf{d}_e \quad \text{and} \quad \mathbf{K}_e\mathbf{d}_e = \mathbf{g}_e \qquad \textbf{(68)}$$

where $\mathbf{K}_e = \mathbf{G}_e^{\mathrm{T}}\mathbf{H}_e^{-1}\mathbf{G}_e$ stands for the element stiffness matrix.

Numerical Inversion of Lt

In this section, we present a brief review of the inversion of the LT used in this work. In general, once the solution for $U(X,s)$ in the Laplace space is found numerically by the method proposed here, inversion of the LT is required to obtain the solution for $u(X,t)$ in the original physical domain. There are many inversion approaches for LT algorithms available in the literature [68]. A comprehensive review of those approaches can be found in [72]. In terms of numerical accuracy, computational efficiency and ease of implementation, Davies and Martin showed that Stehfest's algorithm provides good accuracy with a fairly wide range of functions [73]. Therefore, Stehfest's algorithm is chosen in our study.

If $F(s)$ is the LT of $f(t)$, an approximate value f_a of the function $f(t)$ for a specific time $t=T$ is given by:

$$f_a = \frac{\ln 2}{T}\sum_{i=1}^{N} V_i F\left(\frac{\ln 2}{T} i\right) \tag{69}$$

where

$$V_i = (-1)^{N/2+i} \sum_{k=\frac{i+1}{2}}^{\min(i,N/2)} \frac{k^{N/2}(2k)!}{(N/2-k)!k!(k-1)!(i-k)!(2k-i)!} \tag{70}$$

in which N must be an even number. In implementation, one should compare the results for different N to investigate whether the function is smooth enough, and determine an optimum N [68]. Stehfest suggested $N=10$; other researchers have found no significant change for $6 \le N \le 10$ [73]. Therefore, $N=10$ is adopted here. That means that for each specific time T it is necessary to solve different boundary value problems with different corresponding Laplace parameters $s = \frac{\ln 2}{T} i, i = 1, 2 \cdots 10$ in Laplace space 10 times, then weight and sum the solutions obtained in the Laplace space.

ORTHOTROPIC SEEPAGE PROBLEM

This section presents a further extension of the procedure in the section "HEAT CONDUCTION PROBLEM" to the orthotropic seepage problem. In the process, fundamental solutions for orthotropic seepage problems are employed to construct the element interior hydraulic head field and the conventional shape functions are used for the element frame hydraulic head field. Then these two independent fields are connected by a hybrid functional. The major contents presented in this section was based on the work in [18].

Mathematical Model

Basic Formulation of Orthotropic Seepage Problems

Considering a bounded domain Ω in a two-dimensional space $\Re^2$ which refers to the rectangular coordinates (X_1, X_2), the generalized steady-state water flow in homogeneous orthotropic solids is expressed as [74]:

$$k_1 \frac{\partial^2 H}{\partial X_1^2} + k_2 \frac{\partial^2 H}{\partial X_2^2} = 0 \qquad (X_1, X_2) \in \Omega \subset \Re^2 \tag{71}$$

with the boundary conditions:

$$\begin{cases} H = \bar{H} & \text{on } \Gamma_H \\ q = \bar{q} & \text{on } \Gamma_q \end{cases} \tag{72}$$

where H is the hydraulic head in the domain Ω, and $\bar{H}$ and $\bar{q}$ are the specified hydraulic head and seepage flow, respectively. $k_i > 0$ denotes orthotropic hydraulic conductivities of materials. As well, in Eq. (72), $\Gamma = \Gamma_H \cup \Gamma_q$.

The normal flow flux in Eq. (72) is defined by:

$$q \equiv q_1 n_1 + q_2 n_2 = -\left(k_1 \frac{\partial H}{\partial X_1}\right) n_1 - \left(k_2 \frac{\partial H}{\partial X_2}\right) n_2 \tag{73}$$

or in matrix form:

$$q \equiv \{n_1 \quad n_2\}\begin{Bmatrix} q_1 \\ q_2 \end{Bmatrix} = -\{n_1 \quad n_2\}\begin{bmatrix} k_1 & 0 \\ 0 & k_2 \end{bmatrix}\begin{Bmatrix} \dfrac{\partial H}{\partial X_1} \\ \dfrac{\partial H}{\partial X_2} \end{Bmatrix} \tag{74}$$

where n_i is the unit outward normal to the boundary Γ.

Fig. (**5**) shows a classic problem of two-dimensional seepage flow in a dam, in which the material is defined to be orthotropic along the two coordinate axes, and the bottom surface of the dam is assumed to be impermeable.

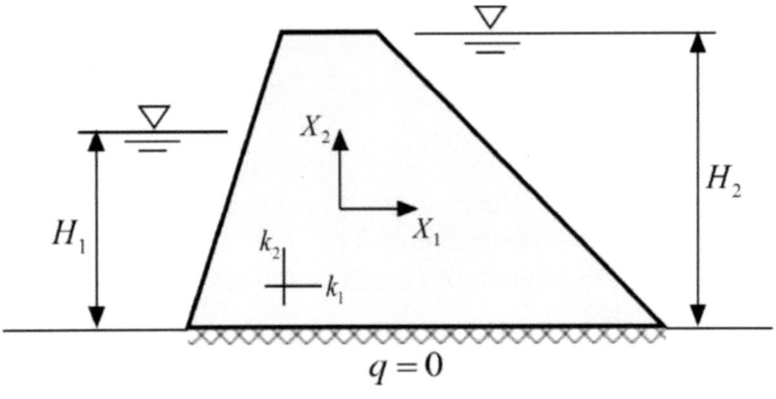

Fig. (5). Classical seepage problems of an orthotropic dam.

Fundamental Solution for Orthotropic Seepage Problems

The fundamental solution for a two-dimensional orthotropic seepage problem is obtained by considering:

$$k_1 \frac{\partial^2 H^*}{\partial X_1^2} + k_2 \frac{\partial^2 H^*}{\partial X_2^2} + \delta(P,Q) = 0 \tag{75}$$

where $P(X_1, X_2)$ and $Q(X_1^s, X_2^s)$ are the field point and source point, respectively.

By means of variables transformation, the induced seepage hydraulic head field H^* at the field point P can be given by [75-77]:

$$H^*(P,Q)=\frac{1}{2\pi\sqrt{k_1k_2}}\ln\left(\sqrt{k_2r_1^2+k_1r_2^2}\right) \tag{76}$$

in which $r_i = X_i - X_i^s \quad (i=1,2)$.

Making use of Eq. (70), the derivatives of the hydraulic head fundamental solution are written as:

$$\left.\begin{aligned}\frac{\partial H^*}{\partial X_1}&=\frac{1}{2\pi\sqrt{k_1k_2}}\frac{k_2r_1}{k_2r_1^2+k_1r_2^2}\\ \frac{\partial H^*}{\partial X_2}&=\frac{1}{2\pi\sqrt{k_1k_2}}\frac{k_1r_2}{k_2r_1^2+k_1r_2^2}\end{aligned}\right\} \tag{77}$$

Hybrid Finite Element Formulation

In the present hybrid finite element formulation, two independent element fields are independently defined and are linked by a hybrid functional, from which the stiffness equation in terms of unknown nodal potential and the relationship of these two fields are established.

Hydraulic Head within the Element

In the interior of an element, say element e, the field of the hydraulic head can be approximated by a linear combination of fundamental solutions of the governing equation (75), that is,

$$H(P)=\sum_{i=1}^{N_s}c_iH^*(P,Q_i)=\{\mathbf{N}\}\{\mathbf{c}_e\} \tag{78}$$

where N_s is the number of source points located outside the element domain, c_i represents an unknown coefficient, and

$$\begin{aligned}\{\mathbf{N}\}&=\left\{H^*(P,Q_1)\quad H^*(P,Q_2)\quad \cdots \quad H^*\left(P,Q_{N_s}\right)\right\}\\ \{\mathbf{c}_e\}&=\left\{c_1\quad c_2\quad \cdots \quad c_{N_s}\right\}^{\mathrm{T}}\end{aligned} \tag{79}$$

Differentiating Eq. (78) yields the normal seepage flow:

$$q(P)=\{\mathbf{Q}\}\{\mathbf{c}_e\} \tag{80}$$

where

$$\{\mathbf{Q}\}=-\{n_1 \quad n_2\}\begin{bmatrix} k_1 & 0 \\ 0 & k_2 \end{bmatrix}\begin{bmatrix} \dfrac{\partial\{\mathbf{N}\}}{\partial X_1} \\ \dfrac{\partial\{\mathbf{N}\}}{\partial X_2} \end{bmatrix} \tag{81}$$

with

$$\frac{\partial\{\mathbf{N}\}}{\partial X_i}=\left\{\frac{\partial H^*(P,Q_1)}{\partial X_i} \quad \frac{\partial H^*(P,Q_2)}{\partial X_i} \quad \cdots \quad \frac{\partial H^*\left(P,Q_{N_s}\right)}{\partial X_i}\right\} \qquad (i=1,2) \tag{82}$$

Hydraulic Head along the Element Boundary

The frame flow field $\tilde{\mathbf{H}}$ is independently defined in terms of the element nodal hydraulic head $\{\mathbf{d}_e\}$ by:

$$\tilde{H}(P)=\{\tilde{\mathbf{N}}\}\{\mathbf{d}_e\} \tag{83}$$

where $\{\tilde{\mathbf{N}}\}$ is an interpolation function vector consisting of the conventional shape functions used in the FEM or BEM. For example, we consider a hybrid element with 5 edges, each edge having 3 nodes. For the edge defined by nodes 3, 4, 5, the hydraulic head distribution on the quadratic edge can be expressed in terms of the interpolation vector $\{\tilde{\mathbf{N}}\}$ and the nodal hydraulic head vector $\{\mathbf{d}_e\}$:

$$\begin{aligned} &\{\tilde{\mathbf{N}}\}=\{0 \quad 0 \quad \tilde{N}_1 \quad \tilde{N}_2 \quad \tilde{N}_3 \quad 0 \quad 0 \quad 0 \quad 0 \quad 0\} \\ &\{\mathbf{d}_e\}=\{d_1 \quad d_2 \quad d_3 \quad d_4 \quad d_5 \quad d_6 \quad d_7 \quad d_8 \quad d_9 \quad d_{10}\}^{\mathrm{T}} \end{aligned} \tag{84}$$

where $\tilde{N}_i$ $(i=1,2,3)$ stands for conventional quadratic shape functions in terms of the local natural coordinate ξ:

$$\tilde{N}_1 = -\frac{\xi(1-\xi)}{2}, \quad \tilde{N}_2 = 1-\xi^2, \quad \tilde{N}_3 = \frac{\xi(1+\xi)}{2} \tag{85}$$

Hybrid Functional

For a given element e, which occupies a subdomain Ω_e with boundary Γ_e, the hybrid functional is defined as [1, 78]:

$$\Pi_e = -\frac{1}{2}\int_{\Omega_e}\left[k_1\left(\frac{\partial H}{\partial X_1}\right)^2 + k_2\left(\frac{\partial H}{\partial X_2}\right)^2\right]\mathrm{d}\Omega - \int_{\Gamma_{qe}} \bar{q}\tilde{H}\mathrm{d}\Gamma + \int_{\Gamma_e} q\left(\tilde{H} - H\right)\mathrm{d}\Gamma \tag{86}$$

Applying the Gauss theorem, we have:

$$\Pi_e = \frac{1}{2}\int_{\Omega_e}\left(k_1\frac{\partial^2 H}{\partial X_1^2} + k_2\frac{\partial^2 H}{\partial X_2^2}\right)H\mathrm{d}\Omega - \int_{\Gamma_{qe}} \bar{q}\tilde{H}\mathrm{d}\Gamma - \frac{1}{2}\int_{\Gamma_e} qH\mathrm{d}\Gamma + \int_{\Gamma_e} q\tilde{H}\mathrm{d}\Gamma \tag{87}$$

Due to the physical definition of orthotropic fundamental solutions, the internal hydraulic head analytically satisfies the governing equation (71), thus the domain integral in Eq. (87) is equal to zero. As a result, we have

$$\Pi_e = -\frac{1}{2}\int_{\Gamma_e} qH\mathrm{d}\Gamma - \int_{\Gamma_{qe}} \bar{q}\tilde{H}\mathrm{d}\Gamma + \int_{\Gamma_e} q\tilde{H}\mathrm{d}\Gamma \tag{88}$$

in which only element boundary integrals are involved.

Substitution of the interior field (78) and the frame field (83) into the functional (88) yields:

$$\Pi_e = -\frac{1}{2}\{\mathbf{c}_e\}^{\mathrm{T}}[\mathbf{H}_e]\{\mathbf{c}_e\} - \{\mathbf{d}_e\}^{\mathrm{T}}\{\mathbf{g}_e\} + \{\mathbf{c}_e\}^{\mathrm{T}}[\mathbf{G}_e]\{\mathbf{d}_e\} \tag{89}$$

Subsequently, minimization of the functional Π_e with respect to $\{\mathbf{c}_e\}$ and $\{\mathbf{d}_e\}$ respectively gives:

$$\left.\begin{aligned}\frac{\partial \Pi_e}{\partial \{\mathbf{c}_e\}^{\mathrm{T}}} &= -[\mathbf{H}_e]\{\mathbf{c}_e\}+[\mathbf{G}_e]\{\mathbf{d}_e\}=\{\mathbf{0}\} \\ \frac{\partial \Pi_e}{\partial \{\mathbf{d}_e\}^{\mathrm{T}}} &= [\mathbf{G}_e]^{\mathrm{T}}\{\mathbf{c}_e\}-\{\mathbf{g}_e\}=\{\mathbf{0}\}\end{aligned}\right\} \tag{90}$$

from which the optional relationship between $\{\mathbf{c}_e\}$ and $\{\mathbf{d}_e\}$:

$$\{\mathbf{c}_e\}=[\mathbf{H}_e]^{-1}[\mathbf{G}_e]\{\mathbf{d}_e\} \tag{91}$$

and the element stiffness equation in terms of the nodal hydraulic head:

$$[\mathbf{K}_e]\{\mathbf{d}_e\}=\{\mathbf{g}_e\} \tag{92}$$

are obtained. In Eq. (92), $[\mathbf{K}_e]=[\mathbf{G}_e]^{\mathrm{T}}[\mathbf{H}_e]^{-1}[\mathbf{G}_e]$.

Assembling the element stiffness equation (92) over all elements gives the final global stiffness equation, from which the nodal hydraulic head can be obtained, and then the unknown coefficients $\{\mathbf{c}_e\}$ are determined using Eq. (91).

Numerical Experiments

Verification of the Proposed Model

As the first example, an orthotropic medium with the domain $\Omega=(0,10)\times(0,11)$ is considered to verify the present algorithm. Under the assumption of orthotropic hydraulic conductivities $k_1 = 1\text{m/s}$ and $k_2 = 5\text{m/s}$, the exact solution of the problem is given by:

$$H(X_1,X_2)=-10X_1^2+2X_2^2+3X_1X_2 \tag{93}$$

which is also used to apply the mixed boundary conditions on the boundary of the rectangular plate. Fig. (**6**) shows the geometry of the domain and the specified boundary conditions for this example.

In the computation, 36 hybrid elements with 4 quadratic edges are used to model the computing domain, as shown in Fig. (**7**). Numerical solutions from the present method are compared with the exact results and listed in Table **1**, from which we can see that the present method has a very high accuracy.

As well, the convergence of the present algorithm is investigated and the results are plotted in Fig. (**8**), in which the average relative error ($Arerr$) is defined as:

$$Arerr(H)=\sqrt{\sum_{1}^{N}(H_{numerical}-H_{exact})_i^2 \Big/ \sum_{i}^{N}(H_{exact})_i^2} \tag{94}$$

where N is the number of sampling points. It is obvious that the error decreases with the increase in the number of elements.

Finally, the entire distribution of the hydraulic head in the rectangular domain is plotted in Fig. (**9**) for reference.

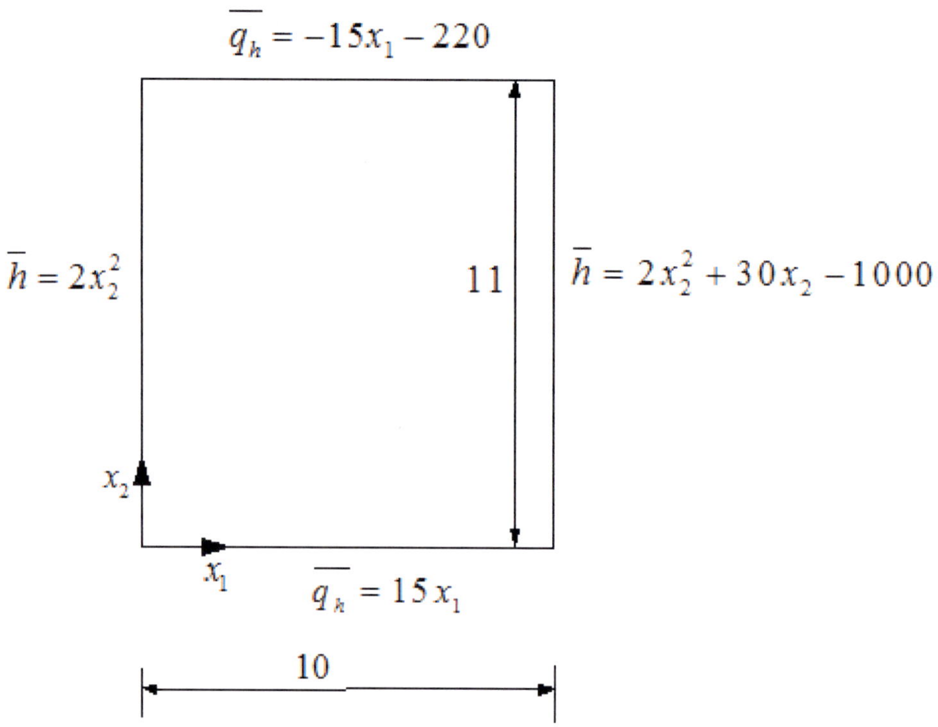

Fig. (6). Rectangular domain and boundary conditions.

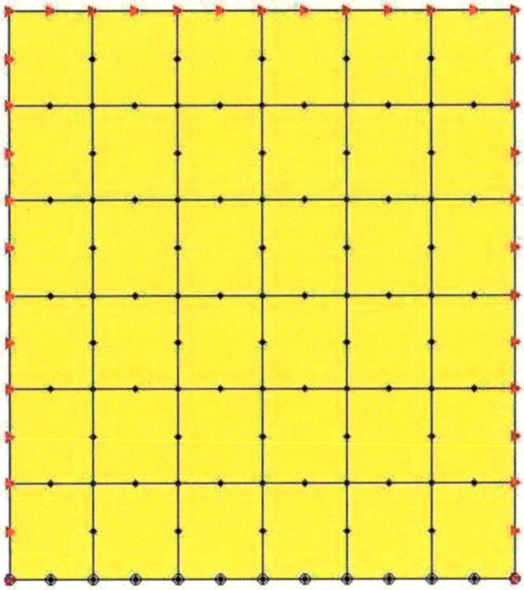

Fig. (7). Illustration of division of elements.

Table 1. Hydraulic head along $X_2 = 0$.

Coordinates		Hydraulic Head		Difference
X_1	X_2	Numerical solutions	Exact solutions	
0.0000	0.0000	0.0000	0.0000	0.00E+00
0.8333	0.0000	-6.9428	-6.9444	-1.64E-03
1.6667	0.0000	-27.7826	-27.7778	4.82E-03
2.5000	0.0000	-62.4978	-62.5000	-2.20E-03
3.3333	0.0000	-111.1167	-111.1111	5.59E-03
4.1667	0.0000	-173.6087	-173.6111	-2.40E-03
5.0000	0.0000	-250.0062	-250.0000	6.20E-03
5.8333	0.0000	-340.2750	-340.2778	-2.74E-03

(Table 1) cont.....

6.6667	0.0000	-444.4513	-444.4444	6.88E-03
7.5000	0.0000	-562.4968	-562.5000	-3.20E-03
8.3333	0.0000	-694.4517	-694.4444	7.31E-03
9.1667	0.0000	-840.2749	-840.2777	-2.76E-03
10.0000	0.0000	-1000.0000	-1000.0000	0.00E+00

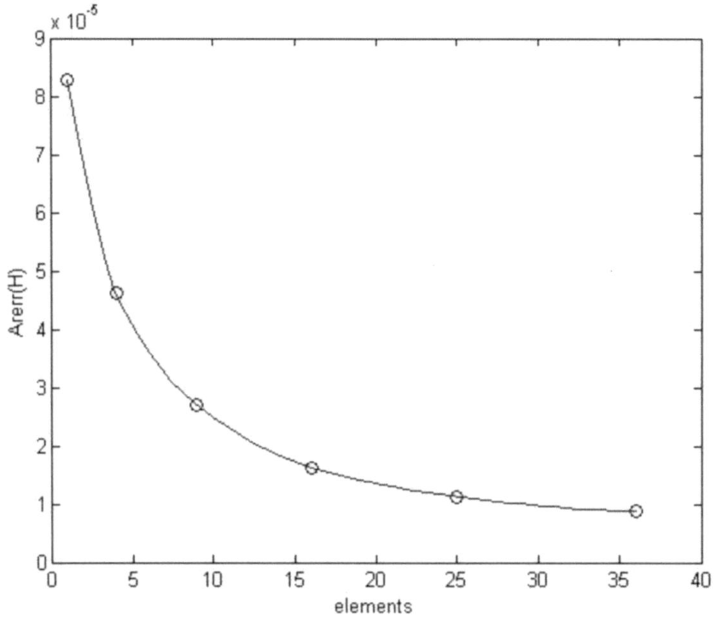

Fig. (8). Convergence demonstration of the proposed approach.

Seepage in a Trapezoidal Dam

In this example, a typical two-dimensional model of water flow through a dam shown in Fig. (**10**) is considered. The hydraulic conductivities of two coordinate directions are $k_1 = 4.0\times10^{-5}$ m/s , $k_2 = 1.0\times10^{-5}$ m/s . The hydraulic head is 15 m at the upstream and 6 m at the downstream. The other edges of the dam are assumed to be impermeable.

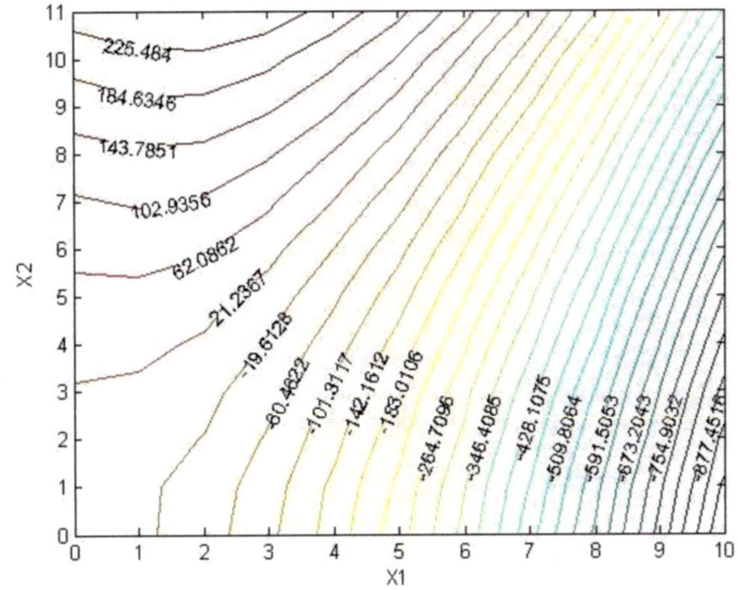

Fig. (9). The hydraulic head obtained from the proposed approach.

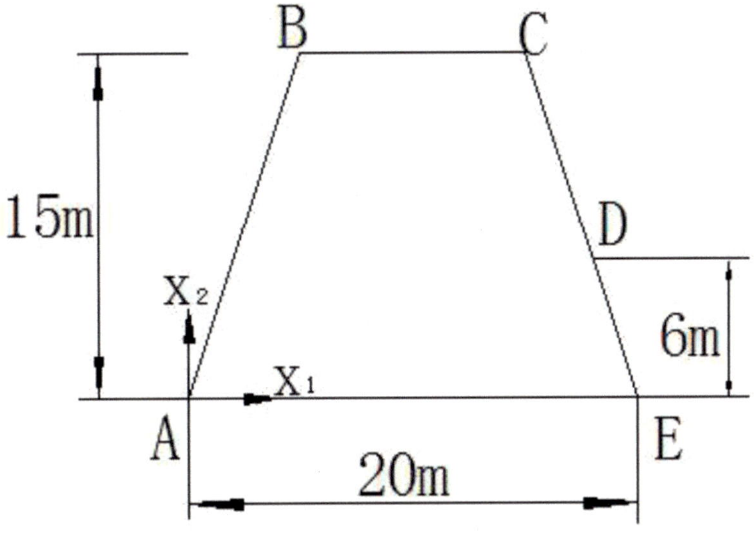

Fig. (10). Geometrical configuration of the dam.

In the computation, 25 quadratic elements are employed to discretize the dam domain, as shown in Figs. (**11**). Fig. (**12**) displays the contour map of the hydraulic head. It is found that there is a large hydraulic head gradient in the interaction region of the right slope and the water level, while in the region close to the upper surface of the dam, the hydraulic head gradient becomes small.

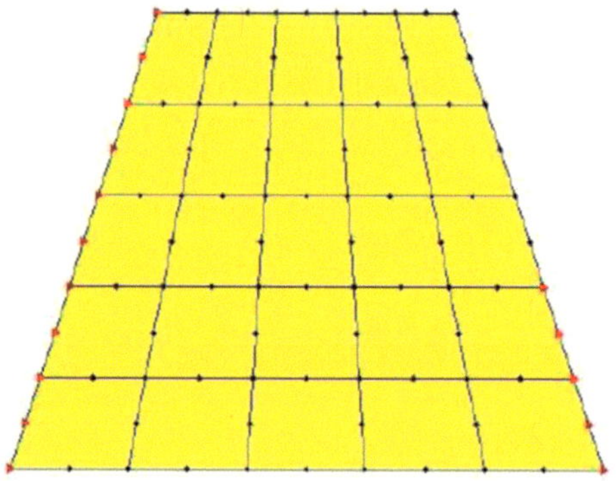

Fig. (11). Mesh division for the dam.

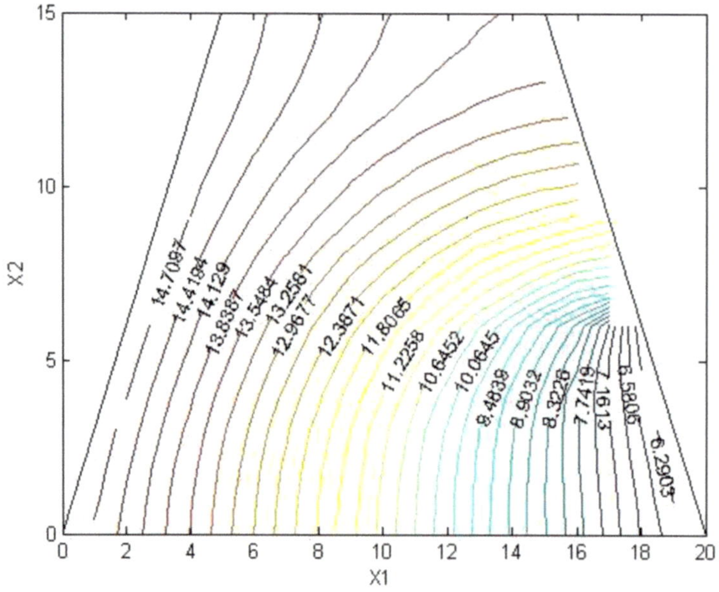

Fig. (12). Contour map of the hydraulic head in the dam.

CONCLUSIONS

A new type of fundamental solution based FEM formulation has been presented for analysing two-dimensional heat conduction problems, with applications to functionally graded materials and seepage problems. In the model, a linear combination of the fundamental solution at points outside the element domain is used to approximate the field variable within an element domain, and a frame field defined on the elementary boundary is introduced to guarantee inter-element continuity. To adapt the new trial function (fundamental solution) a modified variational functional is introduced to establish the corresponding stiffness matrix equation. Typically, the proposed

FS-FEM has the following features:

- Compared to conventional FEM, the formulation requires integration along the element boundaries only, which simplifies calculation of the stiffness matrix and easily generates arbitrary shaped elements.
- In contrast to the T-complete function used in HT-FEM, the fundamental solution in FS-FEM is easy to derive, and further, the determination of source points is easier to operate than selecting appropriate terms from T-complete series in HT-FEM.
- FS-FEM can define the fundamental solution at element level and thus can be flexibly used to analyse problems with different material properties. In contrast, BEM usually requires fundamental solutions to be defined in the full domain, which is not convenient for such problems with different materials. Moreover, the non-singular boundary integrals are used in the FS-FEM, rather than singular or hyper-singular boundary integrals as in the formulation of the conventional BEM.

CONSENT FOR PUBLICATION

Not applicable.

CONFLICT OF INTEREST

The author declares no conflict of interest, financial or otherwise.

ACKNOWLEDGEMENTS

Declared none.

REFERENCES

[1] Q. H. Qin and H. Wang, *Matlab and C programming for Trefftz finite element methods*. New York: CRC Press, 2008.

[2] H. Wang and Q. H. Qin, "Hybrid FEM with fundamental solutions as trial functions for heat conduction simulation," *Acta Mechanica Solida Sinica,* vol. 22, no. 5, pp. 487-498, 2009.

[3] Q. H. Qin and Y. Mai, "Crack growth prediction of an inclined crack in a half-plane thermopiezoelectric solid," *Theoretical and Applied Fracture Mechanics,* vol. 26, no. 3, pp. 185-191, 1997.

[4] Q. H. Qin, "Thermoelectroelastic Green's function for a piezoelectric plate containing an elliptic hole," *Mechanics of Materials,* vol. 30, no. 1, pp. 21-29, 1998.

[5] Q. H. Qin and Y.-W. Mai, "Thermoelectroelastic Green's function and its application for bimaterial of piezoelectric materials," *Archive of Applied Mechanics,* vol. 68, no. 6, pp. 433-444, 1998.

[6] Q. H. Qin, Y.-W. Mai, and S.-W. Yu, "Some problems in plane thermopiezoelectric materials with holes," *International Journal of Solids and Structures,* vol. 36, no. 3, pp. 427-439, 1999.

[7] Q. H. Qin, "Green's functions of magnetoelectroelastic solids with a half-plane boundary or bimaterial interface," *Philosophical Magazine Letters,* vol. 84, no. 12, pp. 771-779, 2004.

[8] Q. H. Qin and J.-Q. Ye, "Thermoelectroelastic solutions for internal bone remodeling under axial and transverse loads," *International Journal of Solids and Structures,* vol. 41, no. 9, pp. 2447-2460, 2004.

[9] Q. H. Qin, C. Qu, and J. Ye, "Thermoelectroelastic solutions for surface bone remodeling under axial and transverse loads," *Biomaterials,* vol. 26, no. 33, pp. 6798-6810, 2005.

[10] C. Qu, Q. H. Qin, and Y. Kang, "A hypothetical mechanism of bone remodeling and modeling under electromagnetic loads," *Biomaterials,* vol. 27, no. 21, pp. 4050-4057, 2006.

[11] J.-S. Wang and Q. H. Qin, "Symplectic model for piezoelectric wedges and its application in analysis of electroelastic singularities," *Philosophical Magazine,* vol. 87, no. 2, pp. 225-251, 2007.

[12] S. W. Yu and Q. H. Qin, "Damage analysis of thermopiezoelectric properties: Part I—crack tip singularities," *Theoretical and Applied Fracture Mechanics,* vol. 25, no. 3, pp. 263-277, 1996.

[13] Q. H. Qin and Y.-W. Mai, "A closed crack tip model for interface cracks in thermopiezoelectric materials," *International Journal of Solids and Structures,* vol. 36, no. 16, pp. 2463-2479, 1999.

[14] H. Wang and Q.H. Qin, "Voronoi Polygonal Hybrid Finite Elements with Boundary Integrals for Plane Isotropic Elastic Problems," *International Journal of Applied Mechanics,* vol. 9, no. 3, p. 1750031, Apr 2017, Art. no. 1750031.

[15] C. Cao and Q.H. Qin, "Hybrid fundamental solution based finite element method: theory and applications," *Advances in Mathematical Physics,* vol. 2015, p. 916029, 2015.

[16] H. Wang and Q.H. Qin, "Voronoi Polygonal Hybrid Finite Elements and Their Applications," in *Current Trends in Mathematical Analysis and Its Interdisciplinary Applications*: Springer, 2019, pp. 521-563.

[17] H. Wang, Q.H. Qin, and X.-P. Liang, "Solving the nonlinear Poisson-type problems with F-Trefftz hybrid finite element model," *Engineering Analysis with Boundary Elements,* vol. 36, no. 1, pp. 39-46, 2012.

[18] Y.-T. Gao, H. Wang, and Q.H. Qin, "Orthotropic Seepage Analysis using Hybrid Finite Element Method," *Journal of Advanced Mechanical Engineering,* vol. 2, no. 1, pp. 1-13, 2015.

[19] H. Wang and Q. H. Qin, "Fundamental solution-based hybrid finite element analysis for non-linear minimal surface problems," in *Recent Developments in Boundary Element Methods: A Volume to Honour Professor John T. Katsikadelis*, E. J. Sapountzakis, Ed. Southampton: WIT Press, 2010, pp. 309-321.

[20] H. Wang, Y.-T. Gao, and Q.H. Qin, "Green's function based finite element formulations for isotropic seepage analysis with free surface," *Latin American Journal of Solids and Structures,* vol. 12, no. 10, pp. 1991-2005, 2015.

[21] Z.J. Fu, W. Chen, and Q. H. Qin, "Hybrid Finite Element Method Based on Novel General Solutions for Helmholtz-Type Problems," *Computers Materials and Continua,* vol. 21, no. 3, pp. 187-208, 2011.

[22] H. Wang and Q.H. Qin, "Numerical implementation of local effects due to two-dimensional discontinuous loads using special elements based on boundary integrals," *Engineering Analysis with Boundary Elements,* vol. 36, no. 12, pp. 1733-1745, 2012.

[23] H. Wang and Q.H. Qin, "Fundamental-solution-based hybrid FEM for plane elasticity with special elements," *Computational Mechanics,* vol. 48, no. 5, pp. 515-528, 2011.

[24] H. Wang, Q. H. Qin, and W. Yao, "Improving accuracy of opening-mode stress intensity factor in two-dimensional media using fundamental solution based finite element model," *Australian Journal of Mechanical Engineering,* vol. 10, no. 1, pp. 41-51, 2012.

[25] N. Nanda, "Spectral finite element method for wave propagation analysis in smart composite beams containing delamination," *Aircraft Engineering and Aerospace Technology,* vol. 92, no. 3, pp. 440-451, 2020.

[26] Q. H. Qin and H. Wang, "Special Elements for Composites Containing Hexagonal and Circular Fibers," *International Journal of Computational Methods,* vol. 12, no. 04, p. 1540012, 2015.

[27] H. Wang and Q. H. Qin, "Special fiber elements for thermal analysis of fiber-reinforced composites," *Engineering Computations,* vol. 28, no. 8, pp. 1079-1097, 2011.

[28] H. Wang and Q. H. Qin, "A fundamental solution based FE model for thermal analysis of nanocomposites," in *Boundary elements and other mesh Reduction methods XXXIII', 33rd International Conference on Boundary Elements and other Mesh Reduction Methods*, UK, 2011, pp. 191-202: WIT Press.

[29] H. Wang and Q. H. Qin, "Implementation of fundamental-solution based hybrid finite element model for elastic circular inclusions," in *Proceedings of the Asia-Pacific Congress for Computational Mechanics*, Singapore 2013, pp. 1-8.

[30] C. Cao, Q. H. Qin, and A. Yu, "Modelling of Anisotropic Composites by Newly Developed HFS-FEM," in *Proceedings of the 23rd International Congress of Theoretical and Applied Mechanics, Yilong Bai , Jianxiang Wang , Daining Fang (eds), SM08-016*, Beijing, China 2012: The International Union of Theoretical and Applied Mechanics (IUTAM).

[31] H. Wang, Q.H. Qin, and Y. Xiao, "Special n-sided Voronoi fiber/matrix elements for clustering thermal effect in natural-hemp-fiber-filled cement composites," *International Journal of Heat and Mass Transfer,* vol. 92, pp. 228-235, Jan 2016.

[32] H. Wang and Q.H. Qin, "A new special coating/fiber element for analyzing effect of interface on thermal conductivity of composites," *Applied Mathematics and Computation,* vol. 268, pp. 311-321, Oct 1 2015.

[33] C. Cao, Q. H. Qin, and A. Yu, "Hybrid fundamental-solution-based FEM for piezoelectric materials," *Computational Mechanics,* vol. 50, no. 4, pp. 397-412, 2012.

[34] C. Cao, A. Yu, and Q.H. Qin, "A new hybrid finite element approach for plane piezoelectricity with defects," *Acta Mechanica,* vol. 224, no. 1, pp. 41-61, 2013.

[35] H. Wang and Q. H. Qin, "Fracture analysis in plane piezoelectric media using hybrid finite element model," in *13th International Conference of fracture*, Beijing, 2013: China Science Literature Publishing House.

[36] C. Cao, Q.H. Qin, and A. Yu, "A new hybrid finite element approach for three-dimensional elastic problems," *Archives of Mechanics,* vol. 64, no. 3, pp. 261-292, 2012.

[37] C. Y. Lee, H. Wang, and Q.H. Qin, "Method of fundamental solutions for 3D elasticity with body forces by coupling compactly supported radial basis functions," *Engineering Analysis with Boundary Elements,* vol. 60, pp. 123-136, Nov 2015.

[38] L. Cao, H. Wang, and Q.H. Qin, "Fundamental solution based graded element model for steady-state heat transfer in FGM," *Acta Mechanica Solida Sinica,* vol. 25, no. 4, pp. 377-392, 2012.

[39] L.-L. Cao, Q.H. Qin, and N. Zhao, "Hybrid graded element model for transient heat conduction in functionally graded materials," *Acta Mechanica Sinica,* vol. 28, no. 1, pp. 128-139, 2012.

[40] H. Wang and Q.H. Qin, "Boundary integral based graded element for elastic analysis of 2D functionally graded plates," *European Journal of Mechanics-A/Solids,* vol. 33, pp. 12-23, 2012.

[41] H. Wang, Q.H. Qin, and C.-Y. Lee, "n-sided polygonal hybrid finite elements with unified fundamental solution kernels for topology optimization," *Applied Mathematical Modelling,* vol. 66, pp. 97-117, Feb 2019.

[42] H. Wang and Q. H. Qin, "FE approach with Green's function as internal trial function for simulating bioheat transfer in the human eye," *Archives of Mechanics,* vol. 62, no. 6, pp. 493-510, 2010.

[43] H. Wang and Q. H. Qin, "Computational bioheat modeling in human eye with local blood perfusion effect," in *Human Eye Imaging and Modeling*, E. Y. K. Ng, J. H. Tan, U. R. Acharya, and J. S. Suri, Eds. Boca Raton: CRC Press, 2012, pp. 311-328.

[44] H. Wang and Q.H. Qin, "A fundamental solution-based finite element model for analyzing multi-layer skin burn injury," *Journal of Mechanics in Medicine and Biology,* vol. 12, no. 05, p. 1250027, 2012.

[45] Z.-W. Zhang, H. Wang, and Q.H. Qin, "Transient bioheat simulation of the laser-tissue interaction in human skin using hybrid finite element formulation," *Molecular & Cellular Biomechanics,* vol. 9, no. 1, pp. 31-53, 2012.

[46] Z. W. Zhang, H. Wang, and Q. H. Qin, "Analysis of transient bioheat transfer in the human eye using hybrid finite element model," in *Applied Mechanics and Materials*, 2014, vol. 553, pp. 356-361: Trans Tech Publ.

[47] J. Tao, Q. H. Qin, and L. Cao, "A Combination of Laplace Transform and Meshless Method for Analysing Thermal Behaviour of Skin Tissues," *Universal Journal of Mechanical Engineering,* vol. 1, no. 2, pp. 32-42, 2013.

[48] Z. W. Zhang, H. Wang, and Q. H. Qin, "Method of fundamental solutions for nonlinear skin bioheat model," *Journal of Mechanics in Medicine and Biology,* vol. 14, no. 4, p. 1450060, 2014.

[49] C. Cao, Q.H. Qin, and A. Yu, "A novel boundary-integral based finite element method for 2D and 3D thermo-elasticity problems," *Journal of Thermal Stresses,* vol. 35, no. 10, pp. 849-876, 2012.

[50] Q. H. Qin and H. Wang, "Special circular hole elements for thermal analysis in cellular solids with multiple circular holes," *International Journal of Computational Methods,* vol. 10, no. 04, p. 1350008, 2013.

[51] H. Wang and Q.H. Qin, "A new special element for stress concentration analysis of a plate with elliptical holes," *Acta Mechanica,* vol. 223, no. 6, pp. 1323-1340, 2012.

[52] C. F. Meng, C. Gu, and B. Hager, "An Eshelby Solution-Based Finite-Element Approach to Heterogeneous Fault-Zone Modeling," *Seismological Research Letters,* vol. 91, no. 1, pp. 465-474, Jan 2020.

[53] Q. H. Qin and H. Wang, "Fundamental solution based FEM for nonlinear thermal radiation problem," in *12th International Conference on Boundary Element and Meshless Techniques (BeTeq 2011), ed. EL Albuquerque, MH Aliabadi, EC Ltd, Eastleigh, UK,* 2011, pp. 113-118.

[54] H. Wang, W. Lin, and Q.H. Qin, "Fundamental-solution-based hybrid finite element with singularity control for two-dimensional mixed-mode crack problems," *Engineering Analysis with Boundary Elements,* vol. 108, pp. 267-278, Nov 2019.

[55] J. C. Zhou, K. Y. Wang, and P. C. Li, "Hybrid fundamental solution based finite element method for axisymmetric potential problems with arbitrary boundary conditions," *Computers & Structures,* vol. 212, pp. 72-85, Feb 2019.

[56] C. Cao, Q. H. Qin, and A. Yu, "Micromechanical Analysis of Heterogeneous Composites using Hybrid Trefftz FEM and Hybrid Fundamental Solution Based FEM," *Journal of Mechanics,* vol. 29, no. 4, pp. 661-674, 2013.

[57] C. Cao, A. Yu, and Q. H. Qin, "Evaluation of effective thermal conductivity of fiber-reinforced composites by boundary integral based finite element method," *International Journal of Architecture, Engineering and Construction,* vol. 1, no. 1, pp. 14-29, 2012.

[58] C. Cao, A. Yu, and Q.H. Qin, "A novel hybrid finite element model for modeling anisotropic composites," *Finite Elements in Analysis and Design,* vol. 64, pp. 36-47, 2013.

[59] C. Cao, A. Yu, and Q. H. Qin, "Mesh reduction strategy: Special element for modelling anisotropic materials with defects," *Boundary Elements and Other Mesh Reduction Methods XXXVI,* vol. 56, p. 61, 2013.

[60] H. Wang and Q.H. Qin, "Fundamental-solution-based finite element model for plane orthotropic elastic bodies," *European Journal of Mechanics-A/Solids,* vol. 29, no. 5, pp. 801-809, 2010.

[61] H. Wang and Q. H. Qin, "Meshless approach for thermo-mechanical analysis of functionally graded materials," *Engineering Analysis with Boundary Elements,* vol. 32, no. 9, pp. 704-712, 2008.

[62] H. Wang and Q. H. Qin, "Some problems with the method of fundamental solution using radial basis functions," *Acta Mechanica Solida Sinica,* vol. 20, no. 1, pp. 21-29, 2007.

[63] Q. H. Qin, *The Trefftz finite and boundary element method.* Southampton, WIT Press, 2000.

[64] L. Cao, Q. H. Qin, and N. Zhao, "An RBF–MFS model for analysing thermal behaviour of skin tissues," *International Journal of Heat and Mass Transfer,* vol. 53, no. 7, pp. 1298-1307, 2010.

[65] Y. Sun, X. He, and Q. H. Qin, "A new procedure for the nonlinear analysis of reissner plate by boundary element method," *Computers & Structures,* vol. 53, no. 3, pp. 649-652, 1994.

[66] Q. H. Qin, "Micromechanics-BEM Analysis for Piezoelectric Composites," *Tsinghua Science & Technology,* vol. 10, no. 1, pp. 30-34, 2005.

[67] Q. H. Qin, "Boundary Element Method," in *Special Topics in the Theory of Piezoelectricity,* J. S. Yang, Ed.: Springer, Cambridge Massachusetts, 2009, pp. 137-168.

[68] A. Sutradhar, G. H. Paulino, and L. Gray, "Transient heat conduction in homogeneous and non-homogeneous materials by the Laplace transform Galerkin boundary element method," *Engineering Analysis with Boundary Elements,* vol. 26, no. 2, pp. 119-132, 2002.

[69] H. Wang, Q. H. Qin, and Y. Kang, "A meshless model for transient heat conduction in functionally graded materials," *Computational Mechanics,* vol. 38, no. 1, pp. 51-60, 2006.

[70] A. Sutradhar and G. H. Paulino, "The simple boundary element method for transient heat conduction in functionally graded materials," *Computer Methods in Applied Mechanics and Engineering,* vol. 193, no. 42, pp. 4511-4539, 2004.

[71] Q. H. Qin, "Variational formulations for TFEM of piezoelectricity," *International Journal of Solids and Structures,* vol. 40, no. 23, pp. 6335-6346, 2003.

[72] B. Davies and B. Martin, "Numerical inversion of the Laplace transform: a survey and comparison of methods," *Journal of computational physics,* vol. 33, no. 1, pp. 1-32, 1979.

[73] S. Zhu, P. Satravaha, and X. Lu, "Solving linear diffusion equations with the dual reciprocity method in Laplace space," *Engineering analysis with boundary elements,* vol. 13, no. 1, pp. 1-10, 1994.

[74] C. Mao, "Seepage computation analysis & control," *China Hydraulic and Hydropower, Beijing,* 2003.

[75] Q. H. Qin, "Nonlinear analysis of Reissner plates on an elastic foundation by the BEM," *International Journal of Solids and Structures,* vol. 30, no. 22, pp. 3101-3111, 1993.

[76] Q. H. Qin, "Green's function and boundary elements of multifield materials," ed: Elsevier, Oxford, 2007.

[77] C. Brebbia and O. Chang, "Boundary elements applied to seepage problems in zoned anisotropic soils," *Advances in Engineering Software (1978),* vol. 1, no. 3, pp. 95-105, 1979.

[78] Q. H. Qin, "Trefftz finite element method and its applications," *Applied Mechanics Reviews,* vol. 58, no. 5, pp. 316-337, 2005.

CHAPTER 6

Hybrid Fundamental Solution based Finite Element Method for Axisymmetric Potential Problems

Keyong Wang[1] and **Junchen Zhou**[1*]

[1] *School of Mechanical and Automotive Engineering, Shanghai University of Engineering Science, Shanghai, China*

Abstract: A hybrid fundamental solution based finite element method is proposed for analyzing axisymmetric potential problems with axisymmetric or arbitrary boundary conditions. In the proposed method, the intra-element potential field is established by using a linear combination of fundamental solutions as interior trial functions. The frame potential field is independently defined on the element boundary. To save computational time and memory, the original three-dimensional problem needs to be reduced to a two-dimensional one. Both the axisymmetric geometry and the boundary conditions are expressed in the form of the Fourier series for the solution of those problems where arbitrary boundary conditions are concerned. In doing so, the two assumed fields of an element are expanded into a series of symmetric and asymmetric components. Based on the axisymmetric form of Hellinger-Reissner functional, the resultant element stiffness equation for each Fourier term involves integrals along the element boundary only. The principle of superposition is finally exploited for the final solution. Several numerical examples demonstrate high-efficiency and insensitivity to mesh distortion of the proposed method.

Keywords: Axisymmetric potential problem, Arbitrary boundary condition, Hellinger-Reissner functional, Finite element method, Fundamental solution, Fourier series expansion.

INTRODUCTION

Potential problems are frequently encountered in different branches of engineering. Applications include heat conduction, mass transfer, seepage, chemical reaction, acoustic fluid behavior, and electrostatic field [1]. When the geometrical body of the problem and related boundary conditions are both axisymmetric, the solution will depend on the radial and axial coordinates only in the meridian plane [2-5].

[*]**Corresponding author Keyong Wang:** School of Mechanical and Automotive Engineering, Shanghai University of Engineering Science, Shanghai, China; E-mail: keyong.wang@hotmail.com

When arbitrary boundary conditions are applied on the axisymmetric body, the solution will depend not only on the radial and axial coordinates, but also on the circumferential coordinate. At present, the potential problem involving complex boundaries is usually solved by the finite element method (FEM), the boundary element method (BEM) and other numerical approaches [6-10]. It is remarkable, however, that the numerical analysis of the three-dimensional (3D) model is very labor-intensive and time-consuming. To overcome this shortcoming, a good choice is to reduce the original 3D model to a three-two-dimensional (2D) case in the meridian plane. One of the earliest studies was reported by Karageorghis and Fairweather [3] who applied the method of fundamental solutions (MFS) to the heat conduction and torsion problems. Smyrlis and Karageorghis [11] proposed a meshless boundary collocation method for analyzing steady-state heat conduction in an isotropic bi-material.

It is noteworthy that the hybrid Trefftz finite element method (HT-FEM) [12-15] is frequently reported in the literature. In the HT-FEM model, the intra-element field and auxiliary frame field are independently assumed within the element domain and the elemental boundary. The two-field interpolation mode enhances the computational accuracy and lowers the sensitivity to mesh distortion. However, the drawback is obvious. The intra-element trial functions come from truncated T-complete solutions of the problem under consideration. Unfortunately, T-complete solutions are not available for some physical problems. As an alternative, the hybrid fundamental solution based finite element method (HFS-FEM) employs the fundamental solutions to construct intra-element trial functions, which has attracted more attention in recent years [16-19].

Previous investigations have focused on 2D potential problems using HT-FEM/HFS-FEM, with a few being devoted to axisymmetric or 3D ones. Wang and Qin [20] studied the thermal behavior of carbon fiber or nanotube composites using HFS-FEM. By virtue of quasi-harmonic polynomials, Wang *et al.* [2] developed a HT-FEM for solving axisymmetric potential problems.

This chapter is devoted to providing a novel HFS-FEM for analyzing axisymmetric potential problems. To investigate the problem with arbitrary boundary conditions, the Fourier series is used to expand the two assumed potential fields and their derivatives into a series of symmetric and asymmetric components [21-24]. The modified variational functional (an axisymmetric form of Hellinger-Reissner functional) [2] is utilized to link the two assumed fields for the enforcement of inter-element continuity. The thermal behavior is finally evaluated by the principle of superposition.

GOVERNING EQUATIONS

In the cylindrical coordinate system (r, θ, z), the governing equation for 3D potential problems can be written by the Laplace equation

$$\frac{\partial^2 u^*}{\partial r^2}+\frac{1}{r}\frac{\partial u^*}{\partial r}+\frac{1}{r^2}\frac{\partial^2 u^*}{\partial \theta^2}+\frac{\partial^2 u^*}{\partial z^2}=0 \quad \text{in } \Omega^* \tag{1}$$

with the Dirichlet boundary condition

$$u^*(r,\theta,z)=\bar{\alpha}^*(r,\theta,z) \quad \text{on } \Gamma_u^* \tag{2}$$

and the Neumann boundary condition

$$q^*(r,\theta,z)=\lambda\frac{\partial u^*}{\partial r}n_r+\lambda\frac{\partial u^*}{\partial \theta}n_\theta+\lambda\frac{\partial u^*}{\partial z}n_z=\bar{\beta}^*(r,\theta,z) \quad \text{on } \Gamma_q^* \tag{3}$$

where u^* and q^* are the potential and its normal derivative on the boundary $\Gamma^*(=\Gamma_u^*\cup\Gamma_q^*)$, $\bar{\alpha}^*$ and $\bar{\beta}^*$ are the prescribed potential and normal derivative, n_r, n_θ and n_z are direction cosines of the outward unit normal vector **n** to the boundary Γ^*, λ is the property coefficient. It is worth noting that we have $n_\theta \equiv 0$ for the axisymmetric geometry.

Axisymmetric Bodies with Axisymmetric Boundary Conditions

When the geometry and boundary conditions are both axisymmetric, the original 3D potential problem can be simply reduced to an axisymmetric one. The simplified problem becomes independent of the azimuthal angle θ. Thus, the governing equation can be rewritten in the cylindrical coordinates r and z only

$$\frac{\partial^2 u}{\partial r^2}+\frac{1}{r}\frac{\partial u}{\partial r}+\frac{\partial^2 u}{\partial z^2}=0 \quad \text{in } \Omega \tag{4}$$

with boundary conditions

$$u(r,z)=\bar{\alpha}(r,z) \text{ on } \Gamma_u \tag{5}$$

$$q(r,z)=\lambda\frac{\partial u}{\partial r}n_r+\lambda\frac{\partial u}{\partial z}n_z=\bar{\beta}(r,z) \text{ on } \Gamma_q \tag{6}$$

where Ω is the working meridian plane which is usually placed in the $\theta = 0^{\circ}$ plane (later called the origin meridian plane), $\Gamma(=\Gamma_u \cup \Gamma_q)$.

Axisymmetric Bodies with Arbitrary Boundary Conditions

If the boundary conditions are non-axisymmetric, the solution *u* needs to be written in terms of Fourier series expansion [9,22]

$$u=\sum_{l=0}^{\infty}\left[u_l(r,z)\cos(l\theta)+\widehat{u}_l(r,z)\sin(l\theta)\right] \tag{7}$$

where *l* stands for the harmonic number. The terms with u_l and $\widehat{u}_l$ represent the symmetric and asymmetric potential amplitudes in the origin meridian plane. Each of u_l and $\widehat{u}_l$ satisfies the following differential equation

$$\frac{\partial^2 u}{\partial r^2}+\frac{1}{r}\frac{\partial u}{\partial r}+\frac{\partial^2 u}{\partial z^2}-\left(\frac{l}{r}\right)^2 u=0 \text{ in } \Omega \tag{8}$$

It is worth noting that the symmetric term with $l = 0$ is related to the axisymmetric component. The non-axisymmetric boundary conditions can be approximately expanded in terms of Fourier series by selecting a sufficient number of harmonics *l*.

HFS FINITE ELEMENT FORMULATION

The establishment of a hybrid finite element model based on fundamental solutions is identical to HT-FEM [2]. The entire domain Ω of interest requires a mesh devision including a number of elements. Over each element, two potential fields are independently assumed within the element domain and along the element boundary [2,13,25,26]. Fig. (**1**) demonstrates an eight-node annular element.

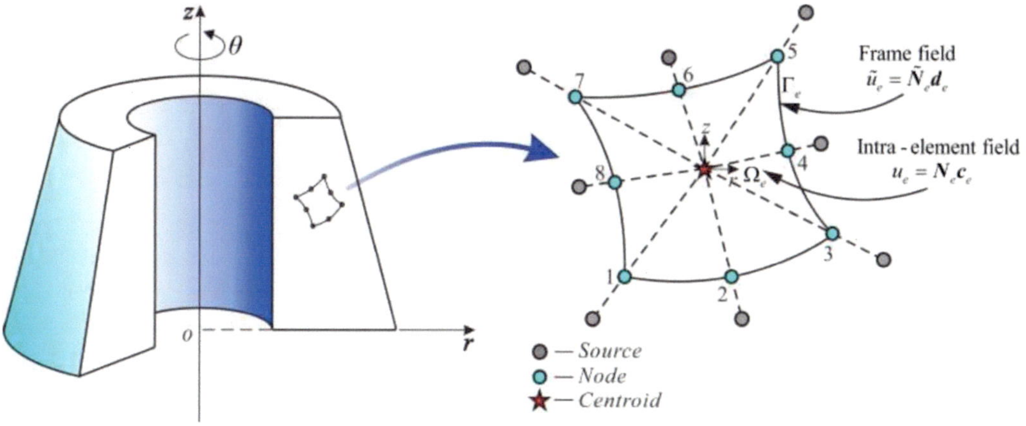

Fig. (1). An eight-node annular element for axisymmetric problems [26].

Axisymmetric Boundary Conditions Case

Two potential fields are assumed for each element e occupying the subdomain Ω_e. To weaken the singularity of fundamental solutions, the potential field defined inside the element domain is constructed by a linear combination of fundamental solutions at several source points positioned outside the element [4] (Fig. **1**). In doing so, the non-conforming intra-element potential field is approximated as:

$$u_e(\mathbf{Q}) = \sum_{j=1}^{n_s} N_e(\mathbf{P}_j, \mathbf{Q}) c_{ej} = \mathbf{N}_e(\mathbf{Q})\mathbf{c}_e \qquad \forall \mathbf{Q} \in \Omega_e,\ \mathbf{P}_j \notin \Omega_e \tag{9}$$

where n_s is the number of source points and c_{ej} are undetermined coefficients. The fundamental solution $N_e(\mathbf{P}_j, \mathbf{Q})$ of Eq. (9) satisfies

$$\nabla^2 N_e(\mathbf{P}, \mathbf{Q}) + \delta(\mathbf{P}, \mathbf{Q}) = 0 \qquad \forall \mathbf{P}, \mathbf{Q} \in \Re^2 \tag{10}$$

which gives

$$N_e(\mathbf{P}, \mathbf{Q}) = \frac{4K(k)}{R} \tag{11}$$

where $\mathbf{P} = [r_P,\ z_P]^{\mathrm{T}}$ is the source point outside the element, $\mathbf{Q} = [r_Q,\ z_Q]^{\mathrm{T}}$ is the field point inside the element domain, $R^2 = (r_Q + r_P)^2 + (z_Q - z_P)^2$, $K(k)$ denotes the complete elliptic integral of the first kind defined as

$$K(k) = \int_0^{\pi/2} \left[1 - k^2 \sin^2 \phi(\mathbf{P})\right]^{-1/2} d\phi(\mathbf{P}) \tag{12}$$

with

$$k^2 = 4 r_Q r_P / R^2 \tag{13}$$

In the numerical implementation, the sources are usually located with the help of the following relation used in the MFS [27,28]:

$$\mathbf{P} = \mathbf{Q}_b + \gamma(\mathbf{Q}_b - \mathbf{Q}_c) \tag{14}$$

where γ denotes a dimensionless parameter, $\mathbf{Q}_b$ and $\mathbf{Q}_c$ are the elemental boundary point and centroid, respectively. The source points need to maintain a distance away from the elemental boundary. Generally, a small distance leads to the inaccurate solution while a large distance results in ill-conditioned discretization matrix. In order to diminish this singularity to the greatest extent, numerical experiments are usually conducted for the optimal value of γ. Fig. (**1**) depicts the source points generated by the elemental nodes.

The corresponding outward normal derivative of u_e on the elemental boundary Γ_e can be written as

$$q_e = \sum_{j=1}^{n_s} \Theta_e(\mathbf{P}_j, \mathbf{Q}) c_{ej} = \mathbf{\Theta}_e \mathbf{c}_e \tag{15}$$

where

$$\mathbf{\Theta}_e = \mathbf{A}\mathbf{T}_e \tag{16}$$

With

$$\mathbf{A}=\left[\lambda n_r \quad \lambda n_z\right] \text{ and } \mathbf{T}_e=\left[\frac{\partial \mathbf{N}_e}{\partial r} \quad \frac{\partial \mathbf{N}_e}{\partial z}\right]^{\mathrm{T}} \tag{17}$$

in which

$$\frac{\partial \mathbf{N}_e}{\partial r}=\frac{2\left\{R^2\left[E(k)-K(k)\left(1-k^2\right)\right]-2r_Q\left(r_Q+r_P\right)E(k)\right\}}{r_Q R^3\left(1-k^2\right)} \tag{18}$$

$$\frac{\partial \mathbf{N}_e}{\partial z}=-\frac{4\left(z_Q-z_P\right)E(k)}{R^3\left(1-k^2\right)} \tag{19}$$

Here, the complete elliptic integral of the second kind $E(k)$ is defined by

$$E(k)=\int_0^{\pi/2}\left[1-k^2\sin^2\phi(\mathbf{P})\right]^{1/2}d\phi(\mathbf{P}) \tag{20}$$

For a multiply connected domain, the solid (denoted by a plane region Ω) is placed on one side of the axis of rotation, and no elemental nodes are located on the axis of rotation. To avoid the singularity involved in Eqs. (11), (18) and (19), the source points cannot be located beyond the axis of rotation nor into the interior of the element of interest.

The continuity between adjacent elements is ensured by the auxiliary frame field $\tilde{u}_e$, which is a function in terms of degrees of freedom (DOFs) of the element. As for the vector of frame functions $\tilde{\mathbf{N}}_e(\mathbf{Q})$, it can be easily constructed in a conventional way. It should be noted that, $\tilde{u}_e$ is confined to the entire elemental boundary as follows

$$\tilde{u}_e(\mathbf{Q})=\tilde{\mathbf{N}}_e(\mathbf{Q})\mathbf{d}_e \tag{21}$$

where $\tilde{\mathbf{N}}_e$ is a row vector including one-dimensional (1D) shape functions as defined in the conventional FEM and $\mathbf{d}_e$ is a column vector including nodal DOFs of the element. As shown in Fig. (**1**), the potential distribution on Element side 1-2-3 can be simply written as

$$\tilde{u}_{1-2-3}=\tilde{N}_1u_1+\tilde{N}_2u_2+\tilde{N}_3u_3 \tag{22}$$

where $\tilde{N}_i (i = 1, 2, 3)$ are the shape functions expressed in the natural coordinate ξ [25,26,29].

The modified variational functional Π_m of the axisymmetric body can be expressed as a sum of all elements Π_{me} [4,29]

$$\Pi_m = \sum_e \Pi_{me} \tag{23}$$

where

$$\Pi_{me} = \pi \iint_{\Omega_e} \lambda \left(q_{er}^2 + q_{ez}^2 \right) r d\Omega - 2\pi \int_{\Gamma_{eu}} q_e \bar{u}_e r d\Gamma + 2\pi \int_{\Gamma_{eq}} \left(\bar{q}_e - q_e \right) \tilde{u}_e r d\Gamma - 2\pi \int_{\Gamma_{el}} q_e \tilde{u}_e r d\Gamma \tag{24}$$

in which $q_{er} = \partial u_e / \partial r$, $q_{ez} = \partial u_e / \partial z$, Γ_{el} is the inter-element boundary, $\Gamma_{eu} = \Gamma_e \bigcup \Gamma_u$, $\Gamma_{eq} = \Gamma_e \bigcup \Gamma_q$ and $\Gamma_e = \Gamma_{eu} \bigcup \Gamma_{eq} \bigcup \Gamma_{el}$.

Integrating the domain integral appeared in Eq. (24) by parts produces a boundary integral functional:

$$\Pi_{me} = \pi \int_{\Gamma_e} q_e u_e r d\Gamma - 2\pi \int_{\Gamma_{eu}} q_e \bar{u}_e r d\Gamma + 2\pi \int_{\Gamma_{eq}} \left(\bar{q}_e - q_e \right) \tilde{u}_e r d\Gamma - 2\pi \int_{\Gamma_{el}} q_e \tilde{u}_e r d\Gamma \tag{25}$$

Noting that on the elemental boundary Γ_{eu}, we have the relation $\tilde{u}_e = \bar{u}_e$. Thus, Eq. (25) can be further written in a concise form:

$$\Pi_{me} = \pi \int_{\Gamma_e} q_e u_e r d\Gamma - 2\pi \int_{\Gamma_{eq}} \bar{q}_e \tilde{u}_e r d\Gamma + 2\pi \int_{\Gamma_e} q_e \tilde{u}_e r d\Gamma \tag{26}$$

Substituting Eqs. (9), (15) and (21) into Eq. (25) produces the formulation:

$$\Pi_{me} = \frac{1}{2} \mathbf{c}_e^{\mathrm{T}} \mathbf{H}_e \mathbf{c}_e - \mathbf{c}_e^{\mathrm{T}} \mathbf{G}_e \mathbf{d}_e + \mathbf{d}_e^{\mathrm{T}} \mathbf{\Phi}_e + \text{terms without } \mathbf{c}_e \text{ and / or } \mathbf{d}_e \tag{27}$$

where

$$\mathbf{H}_e = \pi\int_{\Gamma_e} \left(\mathbf{Q}_e^{\mathrm{T}}\mathbf{N}_e + \mathbf{N}_e^{\mathrm{T}}\mathbf{Q}_e\right) r d\Gamma$$
$$\mathbf{G}_e = 2\pi\int_{\Gamma_e} \mathbf{Q}_e^{\mathrm{T}}\tilde{\mathbf{N}}_e r d\Gamma \tag{28}$$
$$\mathbf{\Phi}_e = 2\pi\int_{\Gamma_{eq}} \tilde{\mathbf{N}}_e \bar{q}_e r d\Gamma$$

To enforce the inter-element continuity, we should establish a relationship between the unknown vector $\mathbf{c}_e$ and the nodal DOFs vector $\mathbf{d}_e$. For this purpose, we minimizate Eq. (27) with respect to $\mathbf{c}_e$ and obtain:

$$\frac{\partial \Pi_{me}}{\partial \mathbf{c}_e} = 0 \Rightarrow \mathbf{c}_e = \mathbf{H}_e^{-1}\mathbf{G}_e\mathbf{d}_e \tag{29}$$

Then, substituting Eq. (29) into Eq. (27) yields

$$\Pi_{me} = -\frac{1}{2}\mathbf{d}_e^{\mathrm{T}}\left(\mathbf{G}_e^{\mathrm{T}}\mathbf{H}_e^{-1}\mathbf{G}_e\right)\mathbf{d}_e + \mathbf{d}_e^{\mathrm{T}}\mathbf{\Phi}_e + \text{terms without } \mathbf{d}_e \tag{30}$$

Finally, the minimization of Eq. (30) with respect to $\mathbf{d}_e$ leads to the following standard element stiffness equation

$$\frac{\partial \Pi_{me}}{\partial \mathbf{d}_e} = 0 \Rightarrow \mathbf{K}_e\mathbf{d}_e - \mathbf{\Phi}_e = 0 \tag{31}$$

where $\mathbf{K}_e = \mathbf{G}_e^{\mathrm{T}}\mathbf{H}_e^{-1}\mathbf{G}_e$ represents the element stiffness matrix and $\mathbf{\Phi}_e$ the equivalent nodal load vector. In a conventional way, a global system of equations can be obtained by assembling Eq. (31) for every element. To seek the unique solution of nodal potential $\tilde{u}_e$, the Dirichlet boundary condition given by Eq. (5) needs to be introduced into the global system of equations. As is known, the rigid-body motion modes of fundamental solutions are discarded to construct the intra-element trial functions. However, the missing rigid-body motion modes must be recovered for the true potential field u_e inside the element of interest [29].

Arbitrary Boundary Conditions Case

To seek the solution of the axisymmetric potential problem with arbitrary or non-axisymmetric boudnary conditions, we must devide the intra-element potential field defined by Eq. (9) into axisymmetric and non-axisymmetric components as follows [27]

$$
\begin{aligned}
u_e(\mathbf{Q}) &= \sum_{l=0}^{L}\left[u_{le}\cos(l\theta)+\widehat{u}_{le}\sin(l\theta)\right] \\
&= \sum_{l=0}^{L}\left[\sum_{j=1}^{n_s} N_{lj}\left(\mathbf{P}_j,\mathbf{Q}\right)\cos(l\theta)c_{lj}+\sum_{j=1}^{n_s} N_{lj}\left(\mathbf{P}_j,\mathbf{Q}\right)\widehat{\sin}(l\theta)c_{lj}\right] \\
&= \sum_{l=0}^{L}\mathbf{N}_{le}\left[\cos(l\theta)+\sin(l\theta)\right]\mathbf{c}_{le}
\end{aligned}
\tag{32}
$$

where L stands for the truncation number of harmonics, $\mathbf{c}_{le}$ is the column vector collecting undetermined coefficients c_{lj}, $\mathbf{N}_{le}$ is the row vector collecting a series of fundamental solutions N_{lj} which fullfil the Laplace equation (33)

$$
\nabla^2 N_{lj}\left(\mathbf{P},\mathbf{Q}\right)+\delta\left(\mathbf{P},\mathbf{Q}\right)=0 \quad \forall\mathbf{P},\mathbf{Q}\in\Re^2
\tag{33}
$$

which yields

$$
N_{lj}\left(\mathbf{P},\mathbf{Q}\right)=\frac{4K_l\left(k\right)}{R}
\tag{34}
$$

$$
K_0\left(k\right)=K\left(k\right)
\tag{35}
$$

$$
K_l\left(k\right)=\left(-1\right)^l l\sum_{i=0}^{L}H_i^l 2^{2(l-i)}C_{l-i},\quad l\geq 1
\tag{36}
$$

The basic approach is still the same. The source points $\mathbf{P}_{sj}$ $(j=1,2,...,n_s)$ are generated based on Eq. (14) . H_i^l and $C_0, C_1,\cdots,C_n$ in Eq. (36) are explicitly espressed as [27]

$$H_i^l = (-1)^i \frac{(2l-i-1)!}{i!(2l-2i)!}, \quad i = 0, 1, ..., l \tag{37}$$

$$C_0 = K(k), \quad C_1 = \frac{1}{k^2}\left[E(k) - (1-k^2)K(k)\right] \tag{38}$$

$$C_n = \frac{1}{(2n-1)k^2}\left[(2n-2)(2k^2-1)C_{n-1} + (2n-3)(1-k^2)C_{n-2}\right], \quad n = 2, 3, ..., l \tag{39}$$

By virtue of Eq. (32), the corresponding outward normal derivative q_e on the elemental boundary Γ_e gives

$$\begin{aligned} q_e(\mathbf{Q}) &= \sum_{l=0}^{L}\left[q_{le}\cos(l\theta) + \hat{q}_{le}\sin(l\theta)\right] \\ &= \sum_{l=0}^{L}\left[\sum_{j=1}^{n_s}\Theta_{lj}(\mathbf{P}_j, \mathbf{Q})\cos(l\theta)c_{lj} + \sum_{j=1}^{n_s}\Theta_{lj}(\mathbf{P}_j, \hat{\mathbf{Q}})\sin(l\theta)c_{lj}\right] \\ &= \sum_{l=0}^{L}\mathbf{\Theta}_{le}\left[\cos(l\theta) + \sin(l\theta)\right]\mathbf{c}_{le} \end{aligned} \tag{40}$$

where

$$\mathbf{\Theta}_{le} = \mathbf{A}\mathbf{\Phi}_{le} \tag{41}$$

with

$$\mathbf{\Phi}_{le} = \left[\frac{\partial \mathbf{N}_{le}}{\partial r}, \frac{\partial \mathbf{N}_{le}}{\partial z}\right]^{\mathrm{T}} \tag{42}$$

in which [3]

$$\frac{\partial N_{lj}}{\partial r} = \frac{4}{R^2}\left(R\frac{\partial K_l}{\partial k}\frac{\partial k}{\partial r_Q} - K_l\frac{\partial R}{\partial r_Q}\right) \tag{43}$$

$$\frac{\partial N_{lj}}{\partial z}=\frac{4}{R^2}\left(R\frac{\partial K_l}{\partial k}\frac{\partial k}{\partial z_Q}-K_l\frac{\partial R}{\partial z_Q}\right) \tag{44}$$

$$\frac{\partial R}{\partial r_Q}=\frac{r_Q+r_P}{R},\quad \frac{\partial R}{\partial z_Q}=\frac{z_Q-z_P}{R} \tag{45}$$

$$\frac{\partial k}{\partial r_Q}=\frac{2r_P-k^2\left(r_Q+r_P\right)}{kR^2},\quad \frac{\partial k}{\partial z_Q}=\frac{k\left(z_P-z_Q\right)}{R^2} \tag{46}$$

$$\frac{\partial K_0(k)}{\partial k}=\frac{\partial K(k)}{\partial k} \tag{47}$$

$$\frac{\partial K_l}{\partial k}=(-1)^l\, l\sum_{i=0}^{L}H_i^l\, 2^{2(l-i)}\frac{\partial C_{l-i}}{\partial k},\quad l\geq 1 \tag{48}$$

and the partial derivatives of C_i with respect to k in Eq. (48) are expressed as

$$\frac{\partial C_0}{\partial k}=\frac{E(k)-\left(1-k^2\right)K(k)}{k\left(1-k^2\right)},\quad \frac{\partial C_1}{\partial k}=\frac{\left(2-k^2\right)K(k)-2E(k)}{k^3} \tag{49}$$

$$\begin{aligned}\frac{\partial C_n}{\partial k}&=\frac{2n-2}{(2n-1)k^2}\left[\left(2k^2-1\right)\frac{\partial C_{n-1}}{\partial k}+\frac{2}{k}C_{n-1}\right]\\&+\frac{2n-3}{(2n-1)k^2}\left[\left(1-k^2\right)\frac{\partial C_{n-2}}{\partial k}-\frac{2}{k}C_{n-2}\right],\quad n\geq 2\end{aligned} \tag{50}$$

Like the axisymmetric boundary conditions case, a frame potential field $\tilde{u}_e$ is independently defined on the elemental boundary for enforcing the continuity across the inter-element boundary:

$$\begin{aligned}\tilde{u}_e(\mathbf{Q})&=\sum_{l=0}^{L}\tilde{u}_{le}\cos(l\theta)+\hat{\tilde{u}}_{le}\sin(l\theta)\\&=\sum_{l=0}^{L}\left[\sum_{j=1}^{n_s}\tilde{N}_{lj}\left(\mathbf{P}_j,\mathbf{Q}\right)\cos(l\theta)d_{lj}+\sum_{j=1}^{n_s}\tilde{N}_{lj}\left(\mathbf{P}_j,\mathbf{Q}\right)\sin(l\theta)d_{lj}\right]\\&=\sum_{l=0}^{L}\tilde{\mathbf{N}}_{le}\left[\cos(l\theta)+\sin(l\theta)\right]\mathbf{d}_{le}\end{aligned} \tag{51}$$

where $\tilde{\mathbf{N}}_{le}$ and $\mathbf{d}_{le}$ have the same forms as $\tilde{\mathbf{N}}_e$ and $\mathbf{d}_e$ in the axisymmetric boundary conditions case.

Next, the procedure is described for the establishment of hybrid finite element model with fundamental solutions as interior trial functions. An axisymmetric form of the Hellinger-Reissner functional Π_m [27]:

$$\Pi_m = \sum_e \left(\sum_l \Pi_{me}^l + \sum_l \widehat{\Pi}_{me}^l \right) \tag{52}$$

where Π_{me}^l and $\widehat{\Pi}_{me}^l$ are the symmetric and asymmetric components for each harmonic l:

$$\Pi_{me}^l = -\frac{1}{2}\int_{\Omega_e^*} \lambda\left(q_{rle}^{*2} + q_{\theta le}^{*2} + q_{zle}^{*2}\right) \mathrm{d}\Omega^* - \int_{\Gamma_{qe}^*} \bar{q}_{le}\tilde{u}_{le} \mathrm{d}\Gamma^* + \int_{\Gamma_e^*} q_{le}\left(\tilde{u}_{le} - T_{le}\right)\mathrm{d}\Gamma^* \tag{53}$$

$$\widehat{\Pi}_{me}^l = -\frac{1}{2}\int_{\Omega_e^*} \lambda\left(\widehat{q}_{rle}^{*2} + \widehat{q}_{\theta le}^{*2} + \widehat{q}_{zle}^{*2}\right) \mathrm{d}\Omega^* - \int_{\Gamma_{qe}^*} \bar{q}_{le}\widehat{\tilde{u}}_{le} \mathrm{d}\Gamma^* + \int_{\Gamma_e^*} \widehat{q}_{le}\left(\widehat{\tilde{u}}_{\mathbf{le}} - \widehat{u}_{le}\right)\mathrm{d}\Gamma^* \tag{54}$$

with $q_{rle}^* = \left(\partial u_{le} / \partial r\right)\cos(l\theta)$, $q_{\theta le}^* = u_{le}\sin(l\theta)/r$, $q_{zle}^* = \left(\partial u_{le} / \partial z\right)\cos(l\theta)$, $\widehat{q}_{rle}^* = \left(\partial \widehat{u}_{le} / \partial r\right)\cos(l\theta)$, $\widehat{q}_{\theta le}^* = \widehat{u}_{le}\sin(l\theta)/r$, $\widehat{q}_{zle}^* = \left(\partial \widehat{u}_{le} / \partial z\right)\cos(l\theta)$ and $\Gamma_{ue}^* = \Gamma_u^* \cap \Gamma_e^*$ and $\Gamma_{qe}^* = \Gamma_q^* \cap \Gamma_e^*$.

It is worth noting that the intra-element potential field guarantees Eq. (1) being automatically satisfied. With the help of the divergence theorem over the entire domain Ω, the subfunctionals Π_{me}^l and $\widehat{\Pi}_{me}^l$ are finally simplified to

$$\Pi_{me}^l = -\frac{1}{2}\pi\int_{\Gamma_e} q_{le}u_{le} r\mathrm{d}\Gamma - \pi\int_{\Gamma_{qe}} \bar{q}_{le}\tilde{u}_{le} r\mathrm{d}\Gamma + \pi\int_{\Gamma_e} q_{le}\tilde{u}_{le} r\mathrm{d}\Gamma \tag{55}$$

$$\widehat{\Pi}_{me}^l = -\frac{1}{2}\pi\int_{\Gamma_e} \widehat{q}_{le}\widehat{u}_{le} r\mathrm{d}\Gamma - \pi\int_{\Gamma_{qe}} \widehat{\bar{q}}_{le}\widehat{\tilde{u}}_{le} r\mathrm{d}\Gamma + \pi\int_{\Gamma_e} \widehat{q}_{le}\widehat{\tilde{u}}_{le} r\mathrm{d}\Gamma \tag{56}$$

The derivation process of Eqs. (55) and (56) are provided in Appendix A of Ref. [27]. Considering $q_{le} = \widehat{q}_{le}$, $u_{le} = \widehat{u}_{le}$ and $\tilde{u}_{le} = \widehat{\tilde{u}}_{le}$, we have $\Pi_{me}^l = \widehat{\Pi}_{me}^l$ for each harmonic term. Therefore, substituting Eqs. (32), (40) and (51) into Eqs. (55) and (56) gives rise to a united form for both symmetric and asymmetric components

$$\Pi_{me}^{l}=-\frac{1}{2}\mathbf{c}_{le}^{\mathrm{T}}\mathbf{H}_{le}\mathbf{c}_{le}+\mathbf{c}_{le}^{\mathrm{T}}\mathbf{G}_{le}\mathbf{d}_{le}-\mathbf{d}_{le}^{\mathrm{T}}\mathbf{P}_{le}+\text{terms without }\mathbf{c}_{le}\text{ or }\mathbf{d}_{le} \quad \textbf{(57)}$$

where

$$\mathbf{H}_{0e}=\pi\int_{\Gamma_e}\mathbf{Q}_{0e}^{\mathrm{T}}\mathbf{N}_{0e}r\mathrm{d}\Gamma,\ \mathbf{G}_{0e}=2\pi\int_{\Gamma_e}\mathbf{Q}_{0e}^{\mathrm{T}}\tilde{\mathbf{N}}_{0e}r\mathrm{d}\Gamma,\ \mathbf{P}_{0e}=2\pi\int_{\Gamma_{qe}}\tilde{\mathbf{N}}_{0e}^{\mathrm{T}}\bar{q}_{0e}r\mathrm{d}\Gamma \quad \textbf{(58)}$$

for the axisymmetric component when $l=0$, and

$$\mathbf{H}_{le}=\frac{\pi}{2}\int_{\Gamma_e}\mathbf{Q}_{le}^{\mathrm{T}}\mathbf{N}_{le}r\mathrm{d}\Gamma,\ \mathbf{G}_{le}=\pi\int_{\Gamma_e}\mathbf{Q}_{le}^{\mathrm{T}}\tilde{\mathbf{N}}_{le}r\mathrm{d}\Gamma,\ \mathbf{P}_{le}=\pi\int_{\Gamma_{qe}}\tilde{\mathbf{N}}_{le}^{\mathrm{T}}\bar{q}_{le}r\mathrm{d}\Gamma \quad \textbf{(59)}$$

for the non-axisymmetric component when $l=1,2,\cdots,L$.

Minimization of the subfunctional Π_{me}^{l} with respect to $\mathbf{c}_{le}$ yields

$$\frac{\partial\Pi_{me}^{l}}{\partial\mathbf{c}_{le}}=\mathbf{H}_{le}\mathbf{c}_{le}-\mathbf{G}_{le}\mathbf{d}_{le}=\mathbf{0} \quad \textbf{(60)}$$

which yields the relationship between $\mathbf{c}_{le}$ and $\mathbf{d}_{le}$:

$$\mathbf{c}_{le}=\mathbf{H}_{le}^{-1}\mathbf{G}_{le}\mathbf{d}_{le} \quad \textbf{(61)}$$

Back substitution of Eq. (61) into Eq. (57) produces

$$\Pi_{me}^{l}=\frac{1}{2}\mathbf{d}_{le}^{\mathrm{T}}\mathbf{G}_{le}^{\mathrm{T}}\mathbf{H}_{le}^{-1}\mathbf{G}_{le}\mathbf{d}_{le}-\mathbf{d}_{le}^{\mathrm{T}}\mathbf{P}_{le}+\text{terms without }\mathbf{d}_{le} \quad \textbf{(62)}$$

Similarly, minimization of Eq. (62) with respect to $\mathbf{d}_{le}$ leads to

$$\frac{\partial\Pi_{me}^{l}}{\partial\mathbf{d}_{le}}=\mathbf{K}_{le}\mathbf{d}_{le}-\mathbf{P}_{le}=\mathbf{0} \quad \textbf{(63)}$$

where $\mathbf{K}_{le} = \mathbf{G}_{le}^{\mathrm{T}}\mathbf{H}_{le}^{-1}\mathbf{G}_{le}$ and $\mathbf{P}_{le}$ are the element stiffness matrix and the equivalent nodal loading vector respectively for each harmonic of symmetric/asymmetric component. Matrix operations and Gaussian quadrature are employed to evaluate the boundary integrals involved in $\mathbf{K}_{le}$ and $\mathbf{P}_{le}$. For each harmonic, a global system of equations can be obtianed by assembling Eq. (63) for every element. Introducing the Dirichlet boundary conditions given by Eq. (5), we can solve for the nodal unknowns of potential $\tilde{u}_{le}$. Thus, the final solutions for $\tilde{u}_e$ can be calculated by a superposition of truncated harmonics. The discarded rigid-body motion modes are recovered by means of the least square technique [27].

NUMERICAL EXAMPLES

To validate the proposed method, several numerical examples are investigated. In all the examples, the eight-node quadrilateral annular element is utilized for mesh division due to its great popularity in boundary matching and most preprocessing algorithms.

Example 1: Steady-state Heat Conduction through a Hollow Cylinder

As the first example, Fig. (**2**) demonstrates the steady-state heat conduction through a hollow cylinder with insulated ends [26]. It is worth noting that the MFS-based methods require a parametric study to determine the optimal location of source points.

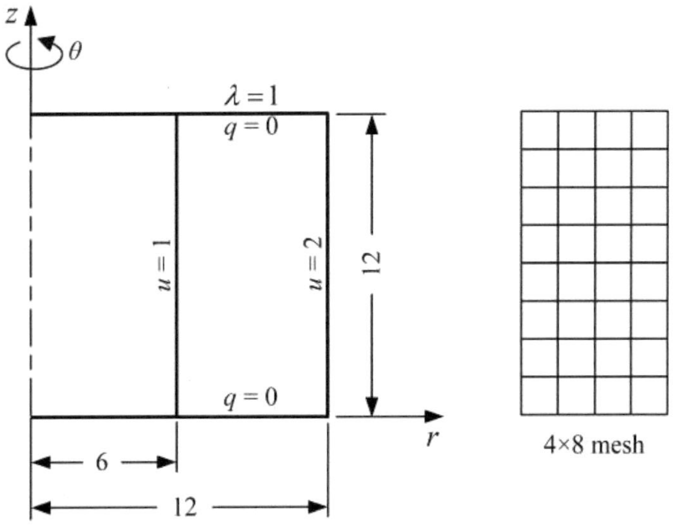

Fig. (2). Steady-state heat conduction through a hollow cylinder.

Fig. (**3**) depicts four schemes with 4, 8, 12 or 16 source points respecitvely. The average relative error (*ARE*) of u at all elemental nodes and centroids is plotted in Fig. (**4**) based on the 4×8 mesh (Fig. **2**). All results of *ARE* are less than 0.1% when γ ranges from 2.2 to 3.5 with the number of source points n_s being 12. This is acceptable. When n_s is larger, say $n_s = 16$, the matrix $\mathbf{H}_e$ becomes larger. This is not beneficial to the inverse operation of $\mathbf{H}_e$. Also, this may result in numerical overflow. To balance the computational effort and accuracy, the optimal combination of $n_s = 12$ and $\gamma = 3.0$ are selected in all the following calculations.

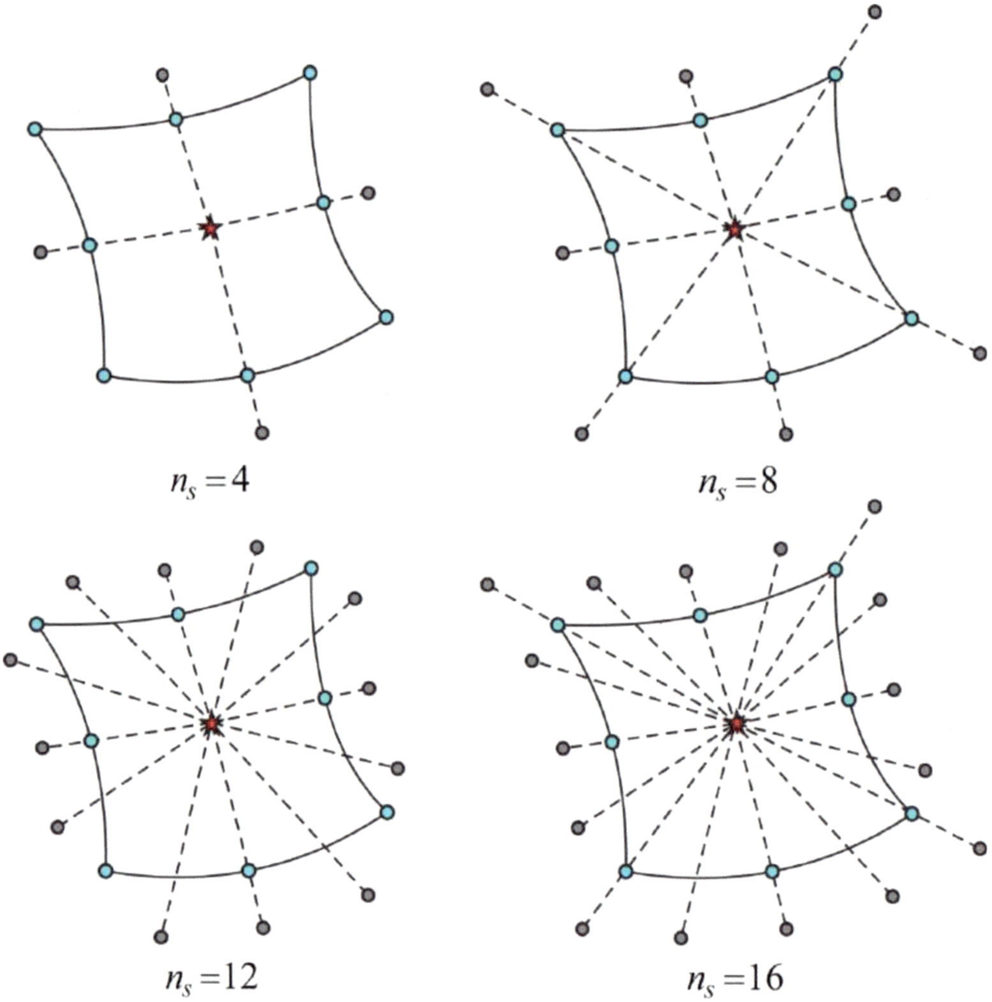

Fig. (3). Configuration of source points for the eight-node annular element.

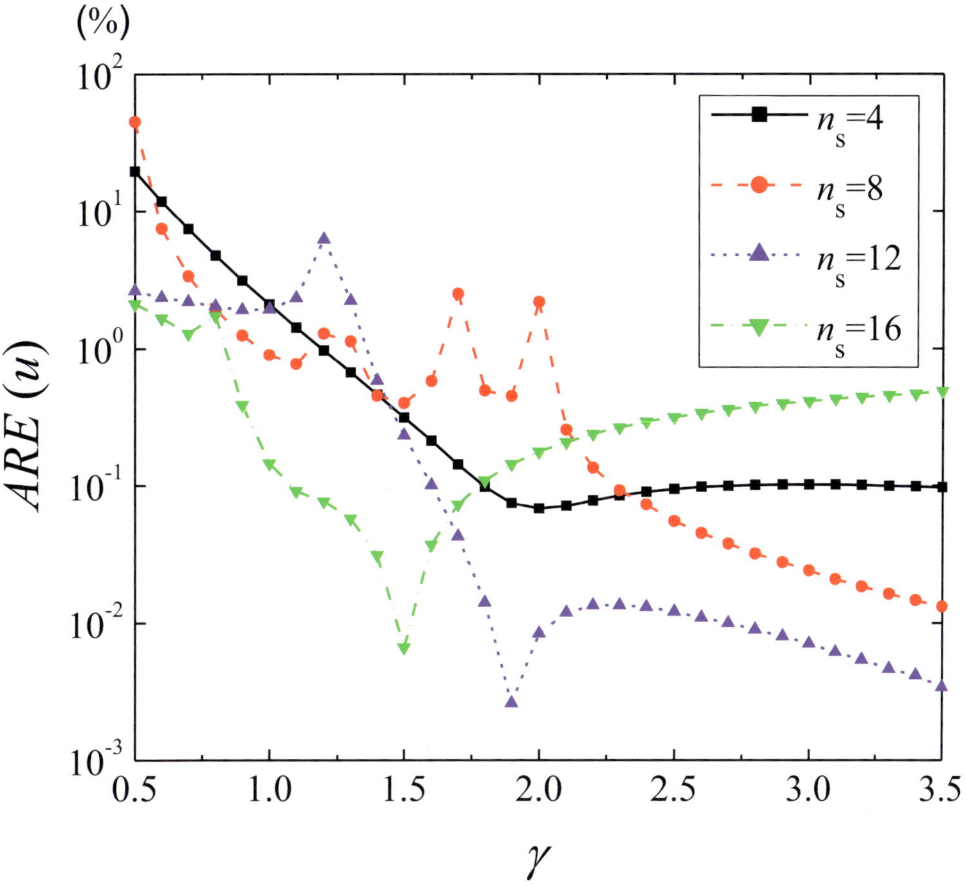

Fig. (4). Effect of the dimensionless parameter γ on the computational precision.

Next, the sensitivity to mesh distortion is performed using the 3×6 mesh scheme as shown in Fig. (**5**). The distortion degree is described by a dimensionless parameter $\psi = d/l$ [2,29]. Table **1** lists the relative errors for the temperature gradient $\partial u / \partial r$ based on the distorted meshes with ψ=0.1, 0.167 and 0.2. The maximum relative error is only 0.41%. When ψ=0.167, elemental sides 3-4-5 and 5-6-7 will be collinear, which leads to element ① becoming an eight-node triangle. When ψ is greater than 0.167 (here, 0.2), element ① collapses to a concave quadrilateral. It cannot be dealt with in the conventional FEM. Surprisingly, HFS-FEM can go beyond this limitation easily. In other words, HFS-FEM is of insensitivity to mesh distortion.

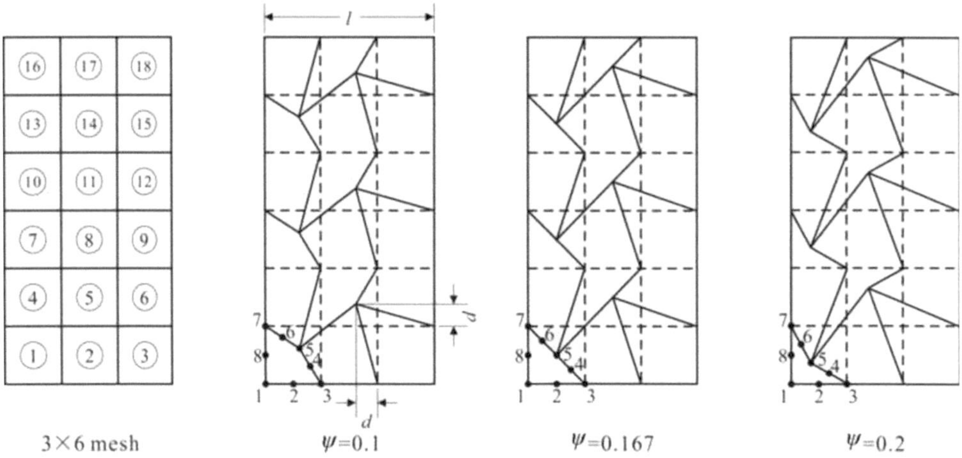

Fig. (5). Mesh distortion scheme.

Table 1. Relative errors for the temperature gradient $\partial u / \partial r$ at the selected points.

Coordinates		Relative error (%)		
r	z	ψ=0.1	ψ=0.167	ψ=0.2
7.0	1.0	-0.179125	-0.072618	-0.048412
7.0	5.0	-0.164601	-0.072618	-0.043571
7.0	9.0	-0.164601	-0.067777	-0.024206
9.0	1.0	-0.354831	-0.180528	-0.385956
9.0	5.0	-0.404631	-0.192978	-0.404631
9.0	9.0	-0.410857	-0.192978	-0.404631
11.0	1.0	0.076086	-0.205433	-0.213041
11.0	5.0	0.114129	-0.197824	-0.190215
11.0	9.0	0.114129	-0.197824	-0.182607

Example 2: Steady-state Heat Conduction through a Hollow Ring

To assess the performance of HSF-FEM for the problem with curved geometry, a hollow ring is considered [26]. As shown in Fig. (**6**), zero temperature is applied on the outer surface whereas a constant uniform heat flux $q=5$ is specified on the inner surface. The entire domain is discretized with 88 eight-node quadrilateral annular elements. Fig. (**7**) depicts the contour maps of the temperature u. It is obvious that the results obtained by HFS-FEM agree well with those done by ABAQUS under the same mesh in the meridian plane.

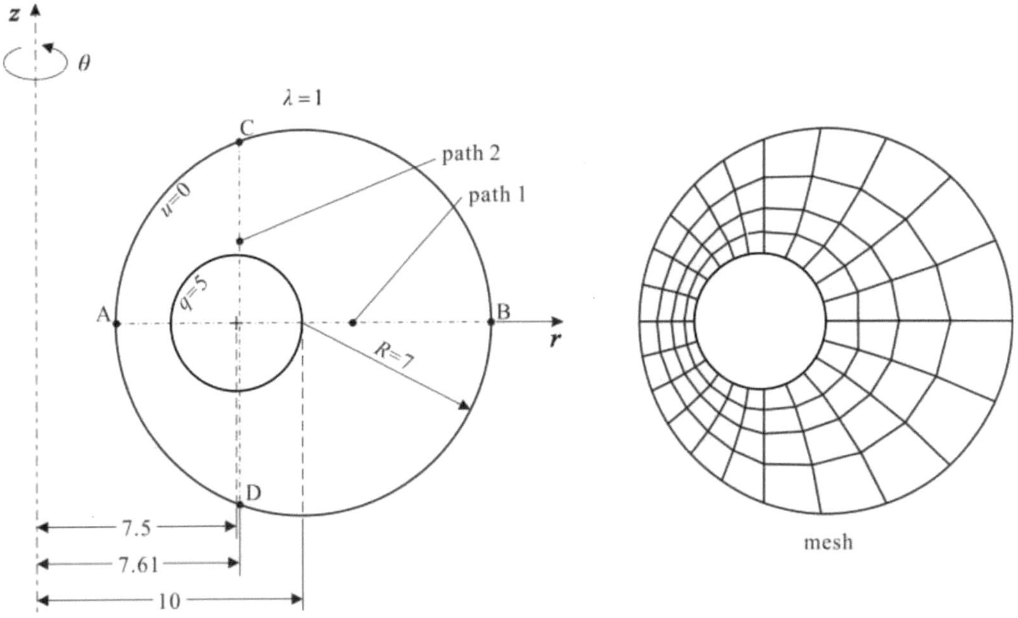

Fig. (6). Steady-state heat conduction through a hollow ring.

Fig. (**8**) shows the variation of temperature gradients $\partial u/\partial r$ and $\partial u/\partial z$ along Path 1 (dot dash line A-B) and Path 2 (dot dash line C-D), respectively. It is seen that the temperature gradient $\partial u/\partial r$ around the inner surface is of somewhat deviation. This deviation gradually decreases as more elements are employed in the ABAQUS model. This clearly demonstrates the efficiency of HFS-FEM.

Example 3: Uniaxial Steady-state Heat Conduction through a Hollow Cylinder

In this example, the application of the proposed method to the problem with arbitrary bondary conditions is illustrated [27]. As shown in Fig. (**9**), the harmonic number is chosen as l=1. The heat flow direction is not along the axis of the hollow cylindrical body, which leads to non-axisymmetric boundary conditions:

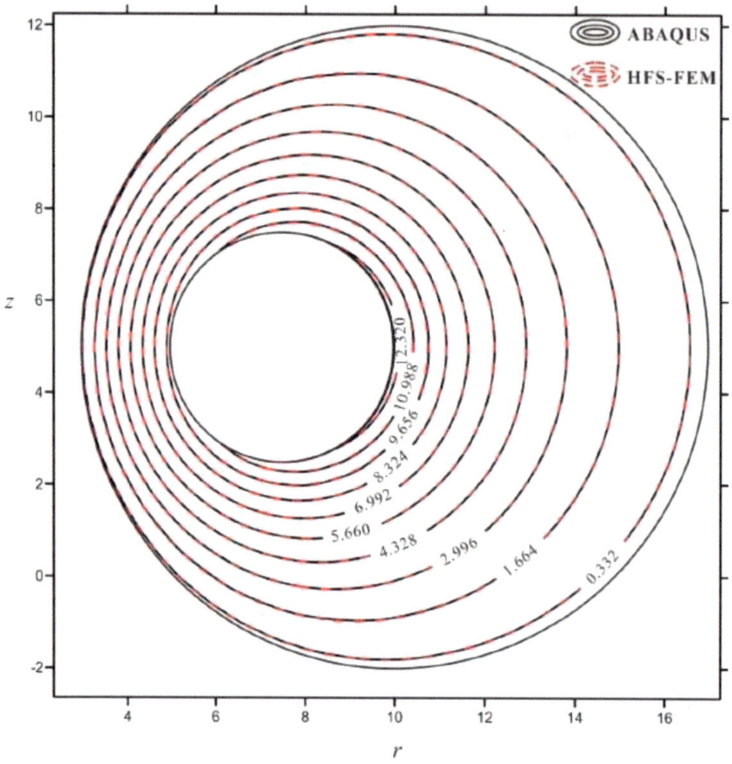

Fig. (7). Contour maps of the temperature *u* in the hollow ring.

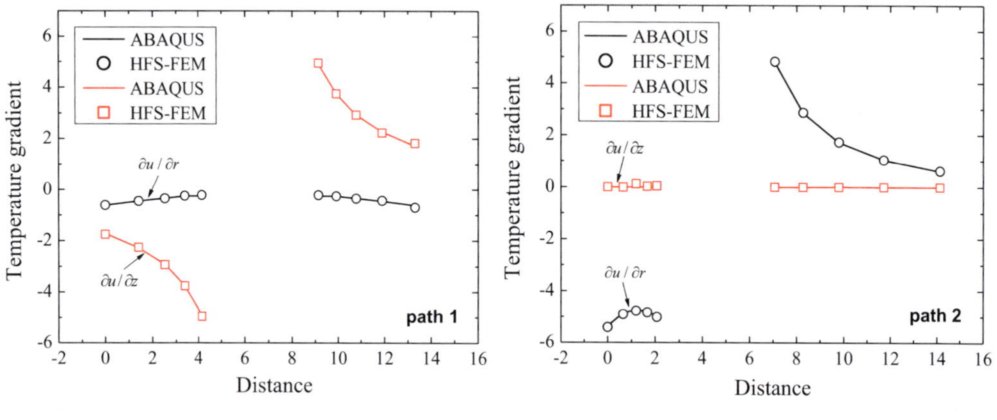

Fig. (8). Temperature gradients $\partial u / \partial r$ and $\partial u / \partial z$ along the two selected paths.

$$\begin{cases} u_1 = u_2 = 2r\cos\theta \\ q_1 = -q_2 = -2\cos\theta \end{cases} \quad (64)$$

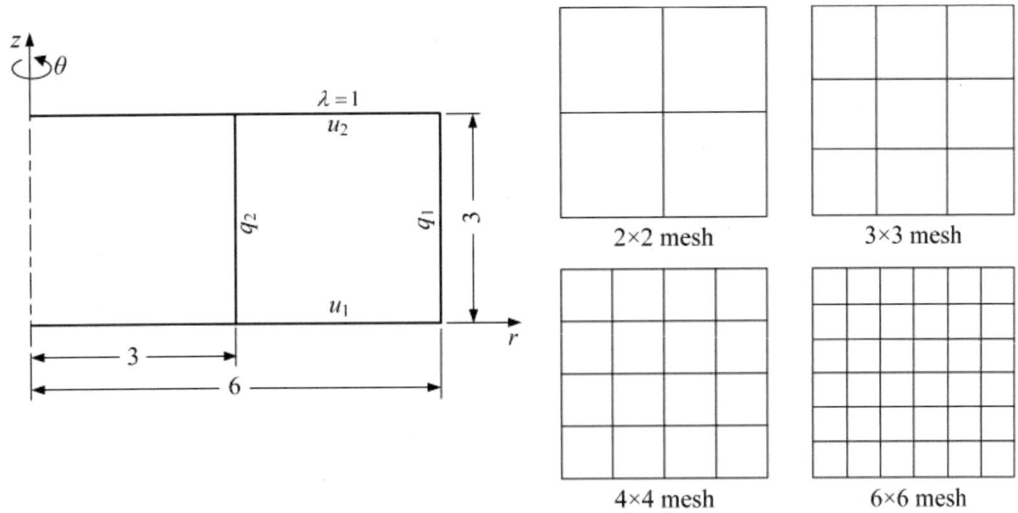

Fig. (9). Uniaxial steady-state heat conduction through a hollow cylinder.

To know the convergence, the total potential capacity dissipation $PCD = \frac{1}{2}\int_V \lambda q^2 dV$ [27] is employed. Here, V is the geometrical volume of the hollow cylinder. Fig. (**10**) depicts the error norm $\eta = \left|PCD - PCD_{reference}\right|^{1/2}$ against the length of element side h. It is seen that HFS-FEM exhibits a faster convergence rate than ABAQUS under four mesh schemes in the origin medirian plane as shown in Fig. **9**. It is noticed that the PCD results obtained by ABAQUS are based on the 30×30 mesh (totally 252,000 DC3D20 elements).

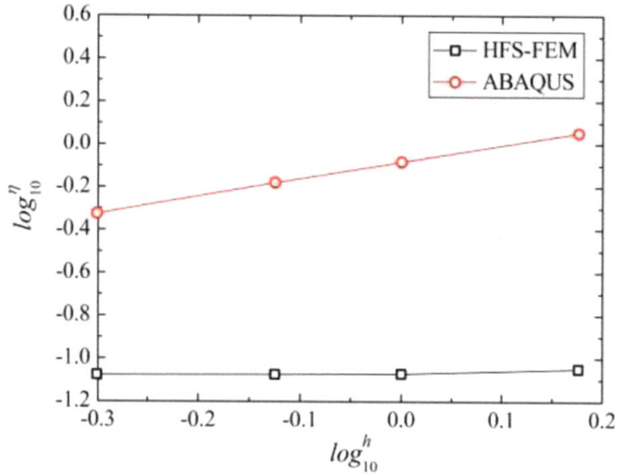

Fig. (10). Error norm of the hollow cylinder with regular elements.

Example 4: Steady-state Heat Conduction through a Hollow Ring under a Non-axisymmetric Heat Flux Boundary Condition

Different from previous examples, for general axisymmetic potential problems this example begins to consider more complex boundary conditions. Fig. **(11)** shows a hollow ring whose inner surface is assumed to be zero temperature [27]. But the outer surface is applied by a heat flux including 4 harmonics:

$$\begin{cases} u_1 = 0 \\ q_1 = 2 + \sin\theta + 2\sin 2\theta + 3\sin 3\theta \end{cases} \quad \textbf{(65)}$$

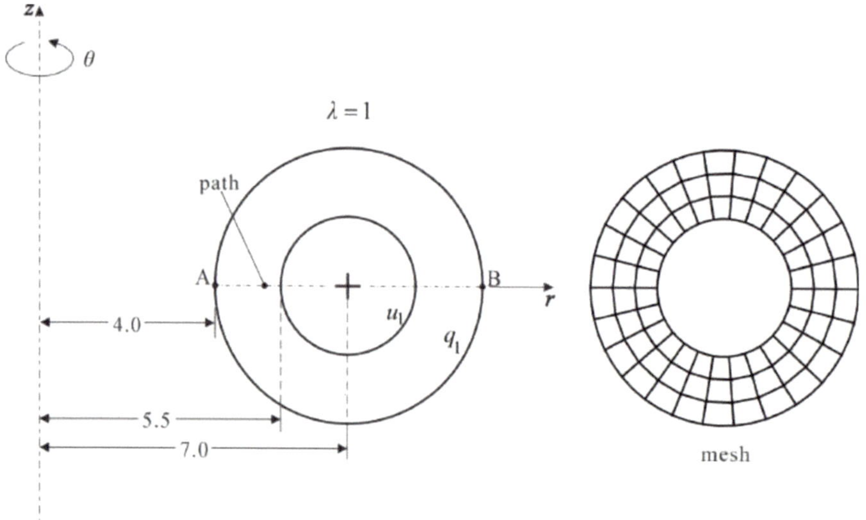

Fig. (11). Steady-state heat conduction through a hollow ring.

Fig. **(12)** depicts the contour maps of temperature u. HFS-FEM uses 78 annular elements while ABAQUS uses 5,772 DC3D20 elements in the θ=0°, 45° and 90° meridian planes. For convenience of comparison, the mesh configurations in the origin meridian plane of ABAQUS are identical to those of HFS-FEM. The results obtained by both numerical methods are in good agreement. Besides, Fig. **(13)** plots the variation of temperature gradients $\partial u / \partial r$ and $\partial u / r\partial\theta$ along the selected path (dash dot line denoted by A-B) in the θ=0°, 45° and 90° meridian planes. The agreement is obvious.

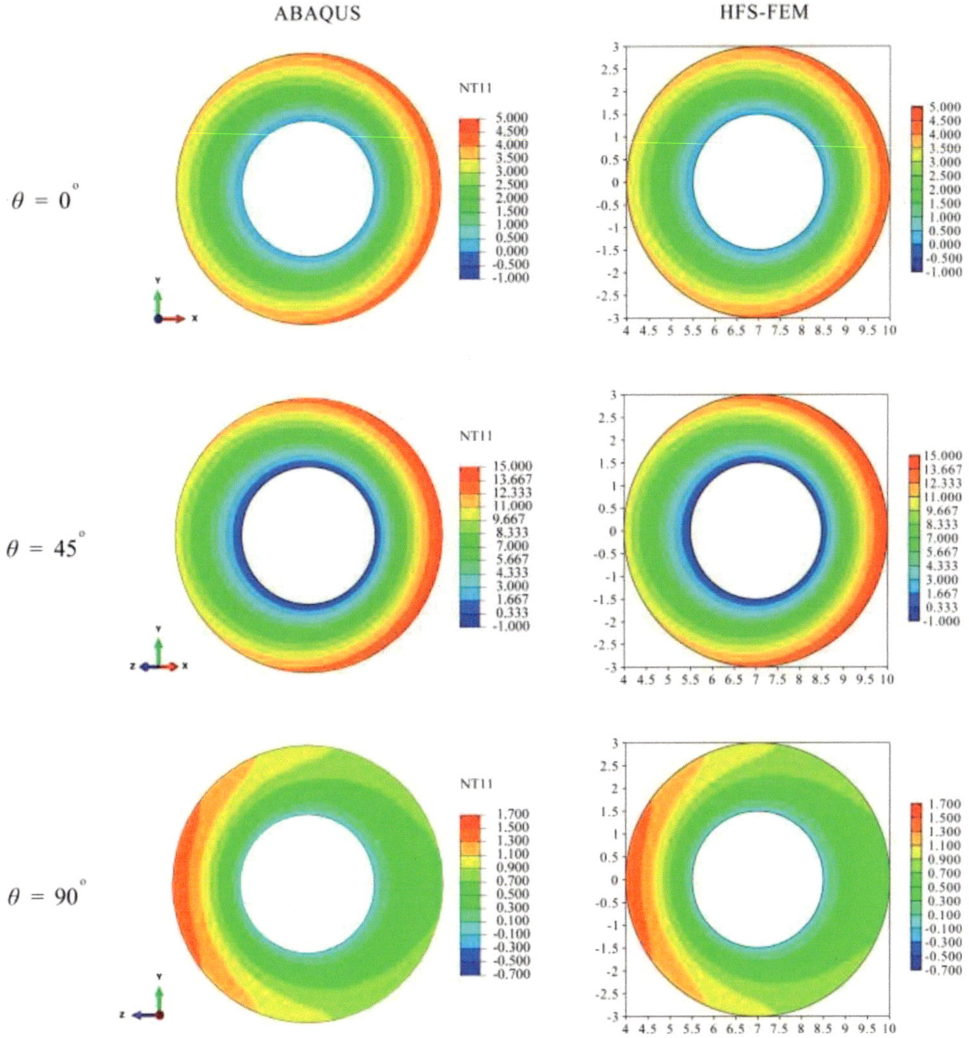

Fig. (12). Contour maps of the temperature u for h = 0, 45 and 90 meridian planes.

Example 5: Steady-state Heat Conduction through a Tower under a Non-axisymmetric Heat Flux Boundary Condition

As the final example, we consider a tower with non-axisymmetric boundary conditions. As shown in Fig. (**14**), a non-axisymmetric heat flux is assumed on the outer surface of the tower while zero temperature is specified on the remaining surfaces [27]. The related boundary conditions are given by

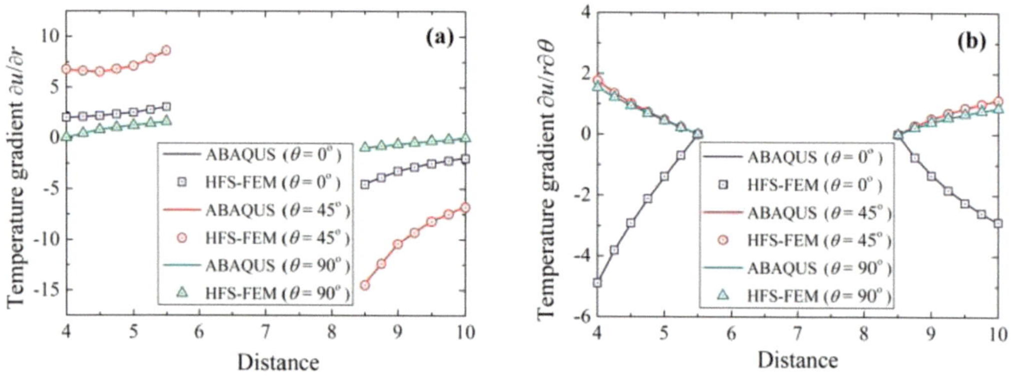

Fig. (13). Temperature gradients $\partial u/\partial r$ and $\partial u/r\partial\theta$ along the selected path.

$$u_1 = u_2 = u_3 = 0$$
$$q_1 = \begin{cases} 10\cos\theta & \theta \in (0, \pi/2] \\ 0 & \theta \in (\pi/2, 3\pi/2] \\ 10\cos\theta & \theta \in (3\pi/2, 2\pi] \end{cases} \tag{66}$$

By virtue of the Fourier series, the axisymmetric part of heat flux boundary condition is readily fitted by the series:

$$q_1 = \frac{10}{\pi} + 5\cos\theta + \frac{20}{\pi}\sum_{k=1}^{\infty}\frac{(-1)^{k-1}}{4k^2-1}\cos(2k\theta) \tag{67}$$

To know the effect of Fourier series expansion on the numerical accuracy, the variation of number of truncated harmonics (l=2, 4, 6, 8 and 10) is investigated. Fig. (**15**) depicts the average relative errors of temperature u and its gradients $\partial u/\partial r$, $\partial u/r\partial\theta$ and $\partial u/\partial z$ against the number of harmonics l. The results are compared to ABAQUS which uses 12,960 DC3D20 elements. As expected, the accuracy is improved by increasing the number of harmonics. It is found that the results change slightly when the number of harmonics l is greater than or equal to 8. Therefore, $l = 2k = 8$, which corresponds to the first six terms of Eq. (67), is selected in the following analysis.

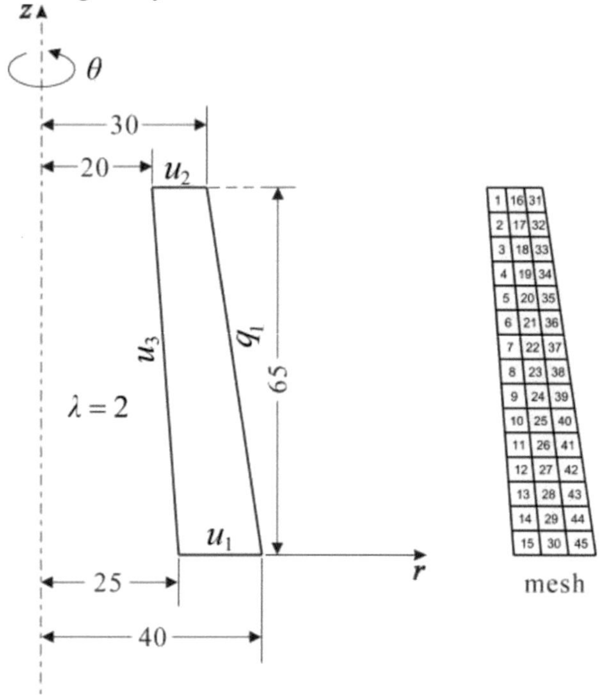

Fig.(14). Steady-state heat conduction through a tower.

Fig. (**16**) depicts the results for the temperature gradients $\partial u / \partial r$ and $\partial u / \partial z$ at 45 elemental centroids in the origin meridian plane. A good agreement is again observed between HFS-FEM (45 annular elements) and ABAQUS (12,960 DC3D20 elements). We can conclude that the proposed method can yield accurate results under a coarse 2D mesh.

Fig. (**17**) displays the contour maps of temperature u and its gradients $\partial u / \partial r$, $\partial u / r\partial \theta$ and $\partial u / \partial z$ at elevation z=32.5. As is seen, the results are exactly similar for both numerical methods except for some corners in $\partial u / r\partial \theta$ obtained by ABAQUS. This demonstrates that HFS-FEM can predict more reliable information of field variable distribution. Fig. (**18**) depicts the variation of temperature gradients $\partial u / \partial r$, $\partial u / r\partial \theta$ and $\partial u / \partial z$ at centroids of Elements 8, 23 and 38. The results obtianed by HFS-FEM coincide with those done by ABAQUS along the circumferential direction.

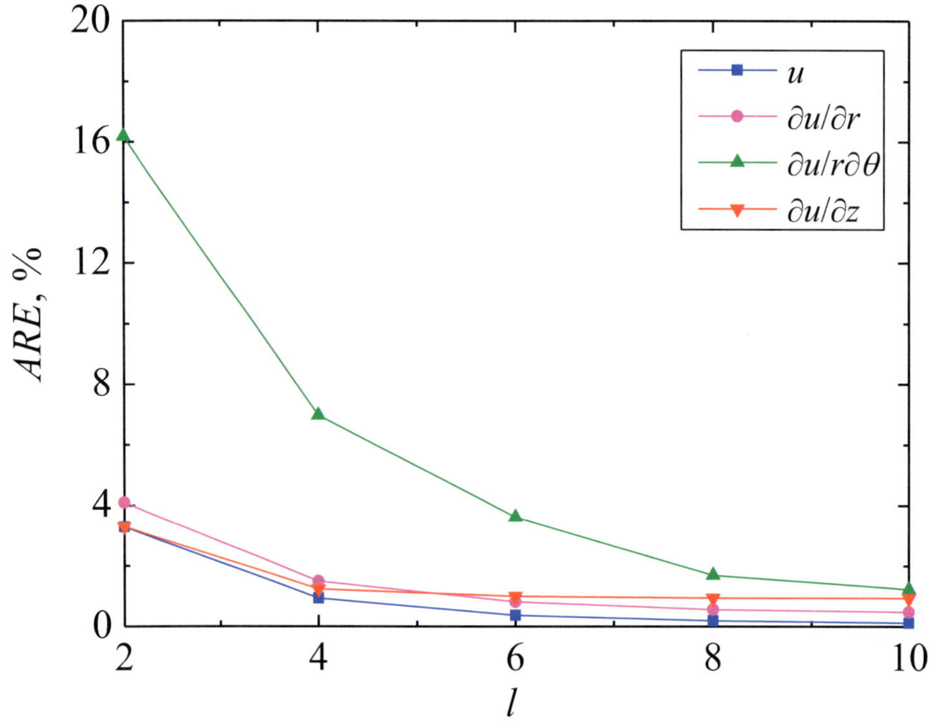

Fig. (15): Effect of the harmonic number on the computational accuracy.

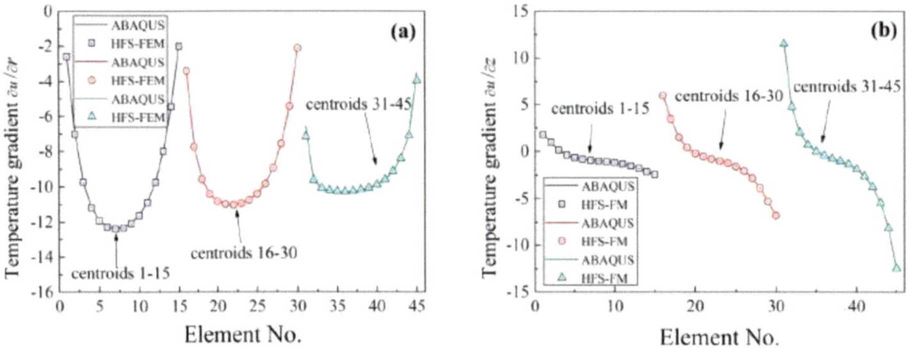

Fig. (16). Temperature gradients $\partial u / \partial r$ and $\partial u / \partial z$ at the selected elemental centroids.

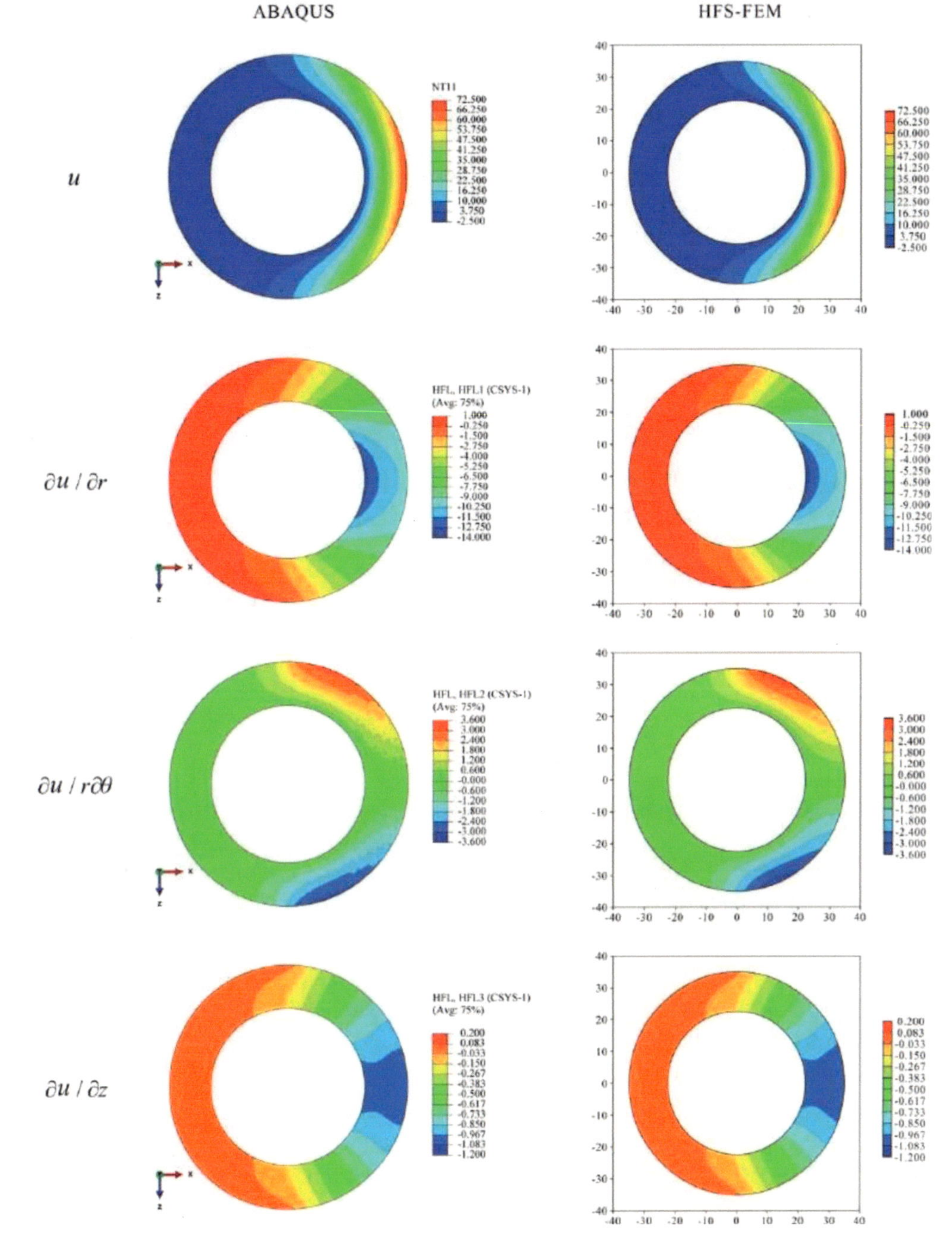

Fig. (17). Contour maps of the temperature *u* and its gradients $\partial u / \partial r$, $\partial u / r\partial\theta$ and $\partial u / \partial z$ at elevation *z*=32.5.

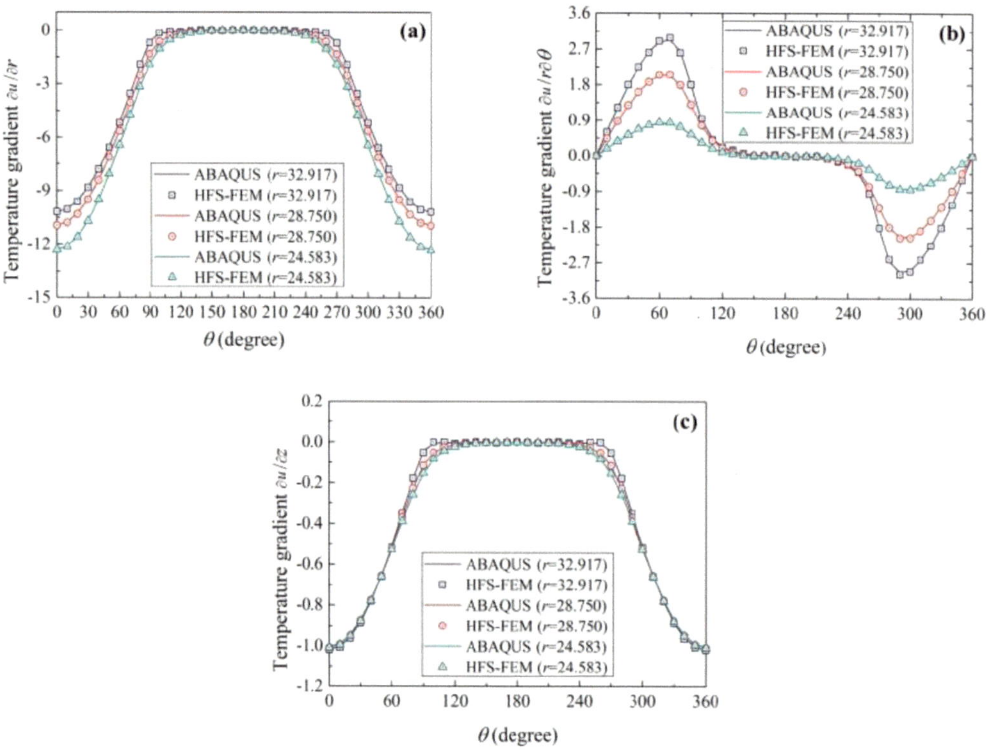

Fig. (18). Temperature gradients $\partial u / \partial r$, $\partial u / r\partial \theta$ and $\partial u / \partial z$ at elevation z=32.5 along the circumferential direction.

CONCLUDING REMARKS

An axisymmetic HFS-FEM is proposed for potential problems with axisymmetric or non-axisymmetric boundary conditions. In this method, all the calculations are performed in the origin meridian plane by reducing the 3D geometry to 2D. For the axisymmetric boudnary conditions case, the linear combination of fundamental solutions at source points is used to construct the trial functions for the intra-element potential field. The Fourier series is employed to expand the arbitrary boundary conditions into the summation of Fourier series. As a result, the assumed potential fields are both written in the form of Fourier series expansion. To maintain the efficience, a finite number of harmonics is generally truncated in the analysis. For each harmonic, both potential fields are identical with those for the axisymmetric boudnary conditions case. The axisymmetric functional consists of symmetric and asymmetric components, which brings about the linkage of both assumed fields. The element stiffness equation for each harmonic can be obtained for subsequent superposition of the solution.

Numerical examples illustrate that the proposed method exhibits insensitivity to mesh distortion and high efficiency compared with ABAQUS. Most importantly, the idea of HFS-FEM is not only applicable for axisymmetric potential problems, but also may be possible for general problems if the fundamental solutions are available.

CONSENT FOR PUBLICATION

Not applicable.

CONFLICT OF INTEREST

The author declares no conflict of interest, financial or otherwise.

ACKNOWLEDGEMENTS

Declared none.

REFERENCES

[1] M. Kamiński, "Potential problems with random parameters by the generalized perturbation-based stochastic finite element method", *Comput. Struct.*, vol. 88, pp. 437-445, 2010.

[2] K.Y. Wang, L.Q. Zhang, and P.C. Li, "A four-node hybrid-Trefftz annular element for analysis of axisymmetric potential problems", *Finite Elem. Anal. Des.*, vol. 60, pp. 49-56, 2012.

[3] A. Karageorghis, and G. Fairweather, "The method of fundamental solutions for axisymmetric potential problems", *Int. J. Numer. Methods Eng.*, vol. 44, pp. 1653-1669, 1999.

[4] A. Karageorghis, and G. Fairweather, "The method of fundamental solutions for axisymmetric elasticity problems", *Comput. Mech.*, vol. 25, pp. 524-532, 2000.

[5] A. Karageorghis, and G. Fairweather, "The method of fundamental solutions for axisymmetric acoustic scattering and radiation problems", *J. Acoust. Soc. Am.*, vol. 104, pp. 3212-3218, 1998.

[6] C. Provatidis, and A. Kanarachos, "Performance of a macro-FEM approach using global interpolation (Coons') functions in axisymmetric potential problems", *Comput. Struct.*, vol. 79, pp. 1769-1779, 2001.

[7] C.S. Ling, and K.S. Surana, "*p*-version least squares finite element formulation for axisymmetric heat conduction with temperature-dependent thermal conductivities", *Comput. Struct.*, vol. 52, pp. 353-364, 1994.

[8] H. Zhou, Z. Niu, C. Cheng, and Z. Guan, "Analytical integral algorithm applied to boundary layer effect and thin body effect in BEM for anisotropic potential problems", *Comput. Struct.*, vol. 86, pp. 1656-1671, 2008.

[9] C. Provatidis, "A fast fourier-boundary element method for axisymmetric potential and elasticity problems with arbitrary boundary conditions", *Comput. Mech.*, vol. 23, pp. 258-270, 1999.

[10] H. Wang and Q.H. Qin, "Meshless approach for thermo-mechanical analysis of functionally graded materials", *Eng. Anal. Bound. Elem.*, vol. 32, pp. 704-712, 2008.

[11] Y. Smyrlis, and A. Karageorghis, "The method of fundamental solutions for stationary heat conduction problems in rotationally symmetric domains", *SIAM J. Sci. Comput.*, vol. 27, pp. 1493-1512, 2006.

[12] K.Y. Wang, P.C. Li, and D.Z. Wang, "Trefftz-type FEM for solving orthotropic potential problems", *Lat. Am. J. Solids Struct.*, vol. 11, pp. 2537-2554, 2014.

[13] Z.N Xia, K.Y Wang, and F.Y. Ge, "Special hole elements for simulating the heat conduction in two-dimensional cellular materials", *Compos. Struct.* vol. 246, pp. 112383, 2020.

[14] Q.H. Qin, "Hybrid Trefftz finite-element approach for plate bending on an elastic foundation", *Appl. Mathe. Model.*, vol. 18, pp.334-339, 1994.

[15] J. Jirousek and Q.H. Qin, "Application of hybrid-Trefftz element approach to transient heat conduction analysis", *Comput. Struct.*, vol. 58, pp.195-201, 1996.

[16] C.Y. Cao, and Q.H. Qin, "Hybrid fundamental solution based finite element method: theory and applications", *Adv. Math. Phys.*, vol. 2015, pp. 1-38, 2015.

[17] C.Y. Cao, Q.H. Qin, and A.B. Yu, "Micromechanical analysis of heterogeneous composites using hybrid Trefftz FEM and hybrid fundamental solution based FEM", *J. Mech.*, vol. 29, pp. 661-674, 2013.

[18] H. Wang, and Q.H. Qin, "Hybrid FEM with fundamental solutions as trial functions for heat conduction simulation", *Acta Mech. Solida Sin.*, vol. 22, pp. 487-498, 2009.

[19] H. Wang, Q.H. Qin, and Y. Xiao, "Special n-sided Voronoi fiber/matrix elements for clustering thermal effect in natural-hemp-fiber-filled cement composites", *Int. J. Heat Mass Tran.*, vol. 92, pp. 228-235, 2016.

[20] A.K. Gupta, "The boundary integral equation method for potential problems involving axisymmetric geometry and arbitrary boundary conditions", M.S. thesis, University of Kentucky, KY, USA,1979.

[21] F.J. Rizzo, and D.J. Shippy, "A boundary integral approach to potential and elasticity problems for axisymmetric bodies with arbitrary boundary conditions", *Mech. Res. Commun.*, vol. 6, pp. 99-103, 1979.

[22] Q.H. Qin, and H. Wang, MATLAB and C Programming for Trefftz Finite Element Methods, Boca Raton: CRC, 2008.

[23] Q.H. Qin, The Trefftz Finite and Boundary Element Method, Southampton: WIT, 2000.

[24] H. Wang, Q.H. Qin, and Y.L. Kang, "A meshless model for transient heat conduction in functionally graded materials", *Comput. Mech.*, vol. 38, pp. 51-60, 2006.

[25] H. Wang, and Q.H. Qin, "Some problems with the method of fundamental solution using radial basis functions", *Acta Mech. Solida Sin.*, vol. 20, pp. 21-29, 2007.

[26] J.C. Zhou, K.Y. Wang, P.C. Li, and X.D. Miao, "Hybrid fundamental solution based finite element method for axisymmetric potential problems", *Eng. Anal. Bound. Elem.*, vol. 91, pp. 82-91, 2018.

[27] J.C. Zhou, K.Y Wang, and P.C. Li, "Hybrid fundamental solution based finite element method for axisymmetric potential problems with arbitrary boundary conditions", *Comput. Struct.*, vol. 212, pp. 72-85, 2019.

[28] Y. Gu, W. Chen, and C.Z. Zhang, "Singular boundary method for solving plane strain elastostatic problems", *Int. J. Solids Struct.*, vol. 48, pp. 2549-2556, 2011.

[29] C. Gáspár, "Some variants of the method of fundamental solutions: regularization using radial and nearly radial basis functions", *Cent. Eur. J. Math.*, vol. 11, pp. 1429-1440, 2013.

CHAPTER 7

Rotordynamic Response of Tapered Composite Driveshafts based on Hierarchical Shaft Finite Element

Rajamohan Ganesan[1*]

[1]*Concordia Centre for Composites (CONCOM), Department of Mechanical, Industrial and Aerospace Engineering, Concordia University, Montreal, Canada*

Abstract: The use of conventional finite element method for the determination of flexural vibrations and rotordynamic response of rotating tapered composite shafts requires a fine mesh of Lagrangian as well as Hermitian finite elements. Such a fine mesh leads to discontinuities in the shear force and bending moments across the interfaces of the elements. In addition, the strains in individual plies are discontinuous across the element interfaces. As a consequence, the natural frequencies, critical speeds and other dynamic response variables could not be calculated with the required accuracy and efficiency. The Hierarchical finite element method combines the features and advantages of the conventional finite element method and the Rayleigh-Ritz method. It can be used in such a way that the above-mentioned limitations associated with the use of conventional finite element method can be eliminated. In the present chapter, an efficient hierarchical finite element formulation is developed for the vibration and rotordynamic analysis of tapered composite driveshafts. The accuracy and efficiency of this formulation are demonstrated. Using the developed formulation, the rotordynamic response including the natural frequencies and critical speeds of tapered composite shafts are determined. A design parametric study is also presented.

Keywords: Composite shaft, Critical speeds, Driveshaft, Free vibration, Natural frequencies, Rotordynamics, Taper angle, Tapered shaft.

INTRODUCTION

Polymer-matrix fiber-reinforced composite materials have widely been used in aerospace, mechanical and automobile engineering structures and components, due to their high strength- and stiffness-to-weight ratios, good damping capacity, and high resistance to fatigue and corrosion damage. In such engineering applications, the driveshafts made with metallic materials are increasingly being replaced by

[*]**Corresponding author Rajamohan Ganesan:** Concordia Centre for Composites (CONCOM), Department of Mechanical, Industrial and Aerospace Engineering, Concordia University, Montreal, Canada; E-mail: r.ganesan@concordia.ca

driveshafts that are made using polymer-matrix fiber-reinforced composite materials. The advantages of composite shaft over aluminum shaft have been described in the work of Zinberg and Symonds [1], in which the natural frequencies and critical speed of a boron-epoxy composite shaft have experimentally been determined and compared with that of an aluminum shaft. However, due to the lack of adequate understanding of the dynamic response of such shafts, potential applications and high-volume use of composite driveshafts are hindered.

During the operation of the driveshaft, the resonance condition needs to be avoided. This avoidance can be achieved by designing the driveshaft to have high natural frequencies in bending and also high first critical speed value. In this regard, in the case of metallic driveshaft, it has been found that tapering the shaft, so as to have variable cross-sectional area along the longitudinal axis of the shaft, results in such desired natural frequencies and first critical speed. Similar tapering of the shaft can more effectively be achieved in composite driveshaft using traditional manufacturing processes such as filament winding process as well as modern processes such as automated fiber placement process.

The Finite Element Method (FEM) has been used over the past few decades as a reliable and efficient method in the analysis of driveshafts. In this regard, the conventional finite element formulation has extensively been developed and used. A Lagrangian finite element formulation has been developed by Chang *et al.* [2] to perform the rotordynamic analysis of uniform-cross-section composite shafts. Almuslmani and Ganesan [3] developed a Hermitian finite element for the rotordynamic analysis of uniform composite shafts. Conventional finite elements that are suitable and efficient for the rotordynamic analysis of shafts have been described in the work of Gmür and Rodrigues [4]. In order to obtain accurate solutions, in the conventional FEM, the finite element mesh that models the structure or component has to be refined into finer meshes. This refinement, known as the h-version of the FEM, introduces correspondingly more discontinuities in the stress distributions across the nodes and across element interfaces. In addition, the finite element model becomes stiffer than the actual structure. For composite structures, these issues cause significant errors in the solutions. On the other hand, the use of p-version finite element formulation leads to computational issues especially in advanced analyses such as the rotordynamic analysis, in addition to discontinuities in the solutions across the element interfaces.

Hierarchical Finite Element Method (HFEM) is an advanced finite element method that offers the combined advantages of both the conventional finite element method as well as the Rayleigh-Ritz method. The HFEM is versatile and powerful and it offers different options such as the formulations based on polynomial and trigonometric hierarchical basis functions. Using a hierarchical finite element, a mechanical component or a structure can adequately be modelled using very few elements, sometimes even using only one element or two elements, with variable number of hierarchical terms. The desired accuracy in the solution can be achieved not necessarily by increasing the number of finite elements in the mesh as in the case of conventional h-version finite element method but by increasing the number of hierarchical terms yet keeping the number of elements in the mesh at the minimum. As a result, the issues of solution discontinuities across the nodes and element interfaces are avoided. More accurate stress distributions within the component or structure and correspondingly, more accurate solutions to response variables of the component or structure such as displacements and natural frequencies can be obtained, with much lesser computational effort than that required by the conventional FEM. Compared to the conventional finite element method, the HFEM yields more accurate results specifically in eigenvalue problems [5]. The HFEM has been developed through many works [5 - 7] in the literature. The hierarchical terms for a finite element can be constructed based on either trigonometric functions or Legendre orthogonal polynomials. Legendre orthogonal polynomials generally have been used as hierarchical terms in static analyses. However, their use in dynamic analyses such as modal and high-frequency wave propagation analyses is rather limited [5]. Each version is suitable for a specific analysis of mechanical components. Efficient hierarchical finite element formulations for uniform and tapered composite beams have been developed in the works of Ganesan and Chen [8] and Fazili and Ganesan [9]. Using the hierarchical finite element method, Boukhalfa and Hadjoui [10] studied the free vibration response of uniform composite shafts.

In the present chapter, the development of the hierarchical finite element formulation for the composite driveshaft that has uniform or variable cross-section is described. The hierarchical interpolation functions are developed based on trigonometric functions, while the polynomial shape functions are associated with the physical nodal degrees of freedom of the shaft element.

The finite element formulation takes into account the effects of the axial load, the rotary inertia, the gyroscopic forces, the taper angle, and the coupling effects due to the lamination of composite plies. In addition, the effect of shear deformation has been incorporated in the formulation through the use of Timoshenko beam theory. For the purposes of validation and comparison, the development of conventional finite element formulation based on Lagrangian shape functions is also described. The rotordynamic analysis of tapered composite driveshafts using the developed hierarchical finite element and the Lagrangian conventional finite element formulations is then conducted. Application for a design parametric study is also demonstrated.

TAPERED COMPOSITE DRIVESHAFT

A rotating shaft of tapered longitudinal profile made of laminated polymer-matrix fiber-reinforced composite material, as shown in Fig. (**1**) is considered. The uniform-diameter composite shaft is a special case of this tapered shaft when the taper angle is set to zero.

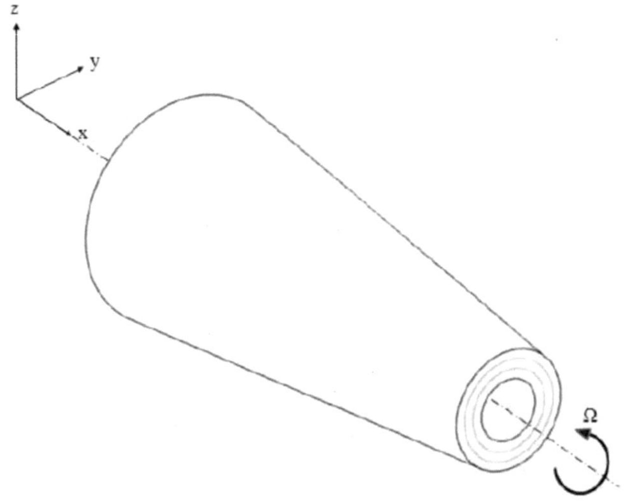

Fig. (1). Tapered composite driveshaft.

The mass per unit length, the diametral mass moment of inertia, and the polar mass moment of inertia of the shaft are expressed as

$$m_c(x) = \pi \sum_{s=1}^{n} \rho_s (r_{os}^2(x) - r_{is}^2(x)) \quad \textbf{(1)}$$

$$I_{dc}(x) = \frac{\pi}{4} \sum_{s=1}^{n} \rho_s (r_{os}^4(x) - r_{is}^4(x)) \quad \textbf{(2)}$$

$$I_{pc}(x) = \frac{\pi}{2}\sum_{s=1}^{n} \rho_s (r_{os}^4(x) - r_{is}^4(x)) \quad \textbf{(3)}$$

where n is the number of the composite material laminae (plies) in the laminate, ρ_s is the density of the ply, and further, r_{os} and r_{is} denote respectively the outer radius and the inner radius of sth ply of the composite shaft as shown in Fig. (**1**). For the composite shaft with a linear taper profile, the outer radius and the inner radius for each composite ply of the tapered shaft are given using linear Lagrange interpolation functions by:

$$r_{os}(x) = \left(1 - \frac{x}{L}\right) r_{o1} + \frac{x}{L} r_{o2} \quad \textbf{(4)}$$

$$r_{is}(x) = \left(1 - \frac{x}{L}\right) r_{i1} + \frac{x}{L} r_{i2} \quad \textbf{(5)}$$

where, as shown in Fig. (**2**), r_{i1} and r_{o1} denote the inner and outer radii of the hollow cross-section of the tapered shaft at the end with the smallest cross-sectional area, and r_{i2} and r_{o2} denote the inner and outer radii of the hollow cross-section of the sth ply in the tapered shaft at the end with the largest cross-sectional area.

Fig. (2). A composite material ply in the tapered shaft.

Fig. (**3**) shows a single composite material ply in the tapered composite shaft with the taper angle α. The principal material directions are denoted by 1, 2, and 3. The axis 1' extends along the plane that is tangent to the surface of the tapered tube, while axis 3' is perpendicular to the same surface. Axis 2' is perpendicular to both the axes 1' and 3'. The fiber orientation angle η is the angle between the axes 1 and 1'. The angle between the axes 2 and 2' is also η.

The stress-strain relations corresponding to the cylindrical coordinate system (x, θ, r) of a material point in the composite material ply can be obtained by considering the stress-strain relations for a composite material ply in the principal material coordinate system (1, 2, 3) and then performing a set of transformations, one from the coordinate system (1,2,3) to the coordinate system (1', 2', 3') and another from the coordinate system (1', 2', 3') to the coordinate system (x, θ, r). These transformations in turn take into account the influences of the fiber orientation angle and the taper angle.

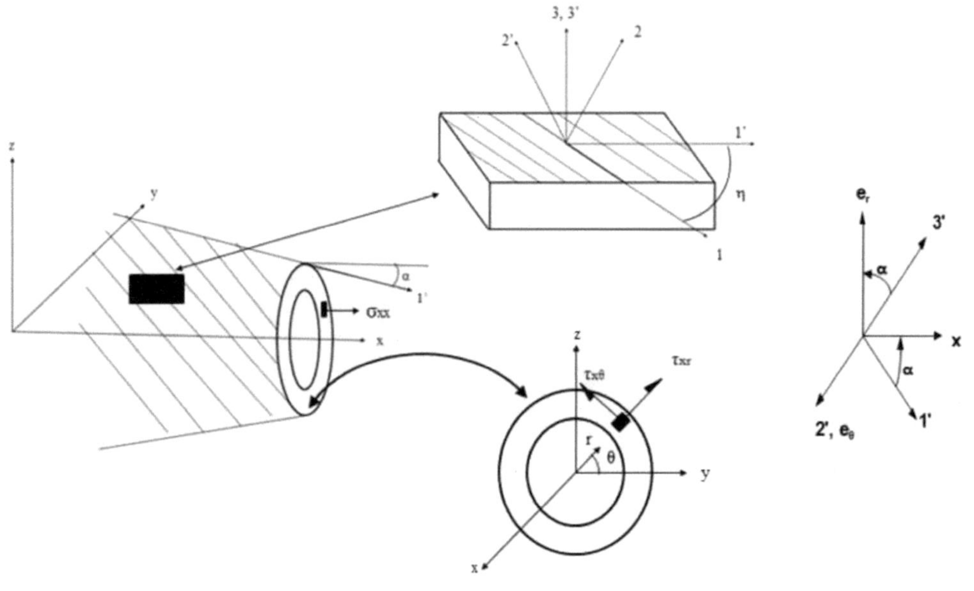

Fig. (3). Coordinate systems for the tapered shaft and the composite material ply.

Upon performing these two transformations, the stress-strain relations in the cylindrical coordinate system (x, θ, r) are obtained as

$$[\sigma_{x\theta r}] = [\bar{\bar{Q}}][\varepsilon_{x\theta r}] = [T_3][T_1][Q]\,[T_2]^{-1}[T_4]^{-1}[\varepsilon_{x\theta r}] \quad \textbf{(6)}$$

where [Q] is the so-called reduced stiffness matrix of a composite material ply of composite material. The matrix $[\bar{\bar{Q}}]$ is the transformed stiffness of the composite material ply that is obtained after performing the transformation corresponding to the rotation η about axis 3 in the principal material coordinate system (1, 2, 3) and then performing the transformation corresponding to the rotation α about axis 2' in the coordinate system (1', 2', 3'). The corresponding transformation matrices are given below.

$$[T_1] = \begin{bmatrix} m^2 & n^2 & 0 & 0 & 0 & 2mn \\ n^2 & m^2 & 0 & 0 & 0 & -2mn \\ 0 & 0 & 1 & 0 & 0 & 0 \\ 0 & 0 & 0 & m & -n & 0 \\ 0 & 0 & 0 & n & m & 0 \\ -mn & mn & 0 & 0 & 0 & m^2 - n^2 \end{bmatrix} \tag{7}$$

$$[T_2] = \begin{bmatrix} m^2 & n^2 & 0 & 0 & 0 & mn \\ n^2 & m^2 & 0 & 0 & 0 & -mn \\ 0 & 0 & 1 & 0 & 0 & 0 \\ 0 & 0 & 0 & m & -n & 0 \\ 0 & 0 & 0 & n & m & 0 \\ -2mn & 2mn & 0 & 0 & 0 & m^2 - n^2 \end{bmatrix} \tag{8}$$

$$[T_3] = \begin{bmatrix} c^2 & 0 & s^2 & 0 & 2sc & 0 \\ 0 & 1 & 0 & 0 & 0 & 0 \\ s^2 & 0 & c^2 & 0 & -2sc & 0 \\ 0 & 0 & 0 & c & 0 & -s \\ -sc & 0 & sc & 0 & c^2 - s^2 & 0 \\ 0 & 0 & 0 & s & 0 & c \end{bmatrix} \tag{9}$$

$$[T_4] = \begin{bmatrix} c^2 & 0 & s^2 & 0 & sc & 0 \\ 0 & 1 & 0 & 0 & 0 & 0 \\ s^2 & 0 & c^2 & 0 & -sc & 0 \\ 0 & 0 & 0 & c & 0 & -s \\ -2sc & 0 & 2sc & 0 & c^2 - s^2 & 0 \\ 0 & 0 & 0 & s & 0 & c \end{bmatrix} \tag{10}$$

where

$$m = \cos\eta \; ; \; n = \sin\eta$$

$$c = \cos\alpha \; ; \; s = \sin\alpha$$

HIERARCHICAL COMPOSITE SHAFT FINITE ELEMENT

In Fig. (**4**), a hierarchical finite element that can represent both the x-y plane and x-z plane motions of the tapered composite shaft is shown. The element is of length L and has two end nodes and each of the two nodes has four degrees of freedom, that are, two translational (in y and z directions) and two rotational (about z and y axes) degrees of freedom. The local coordinate x and the non-dimensional coordinate ξ, where $\xi = \frac{x}{L}$, are shown in Fig. (**4**).

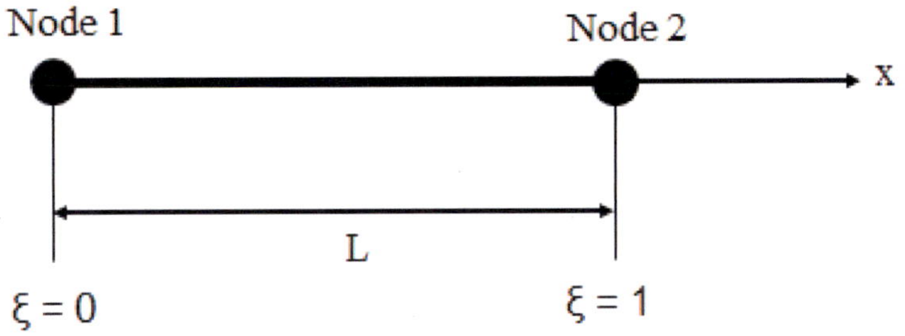

Fig. (4): The 2-node finite element.

The interpolation functions for the hierarchical finite element are developed using polynomial or trigonometric functions. For the shaft hierarchical element, the trigonometric functions are more efficient and therefore, are considered. In the hierarchical finite element formulation, the transverse displacement in the y-direction of the shaft element, denoted by *v*, is expressed as follows:

$$v = c_1 + c_2\xi + \sum_{n=1}^{N} c_{n+2}\,\mathrm{Sin}(\,n\pi\xi) \tag{11}$$

where N represents the total number of the hierarchical terms used for the translational displacement field. Recasting the above equation in the form :

$$v = [1 \quad \xi \quad \sin\pi\xi \quad \dots\dots \quad \mathrm{Sin}\,N\pi\xi]\begin{bmatrix} c_1 \\ c_2 \\ c_3 \\ c_4 \\ \cdot \\ \cdot \\ c_{N+2} \end{bmatrix} \tag{12}$$

one can obtain:

$$\boldsymbol{v} = [\boldsymbol{A^v}]\{\boldsymbol{c}\} \tag{13}$$

where $[\boldsymbol{A^v}]$ is the generalized co-ordinate matrix for the hierarchical finite element. The "generalized" element nodal displacement matrix, denoted as $\{\boldsymbol{q_v}\}$, is now formed by including the nodal displacements in y-direction at the two ends of the element, $\boldsymbol{v_1}$ and $\boldsymbol{v_2}$, and the hierarchical terms.

$$\{q_v\} = \begin{bmatrix} v_1 \\ v_2 \\ v_3 \\ v_4 \\ \cdot \\ \cdot \\ v_{N+2} \end{bmatrix} = \begin{bmatrix} 1 & 0 & 0 & 0 & & & 0 \\ 1 & L & 0 & 0 & & & 0 \\ 0 & 0 & 1 & 0 & & & 0 \\ 0 & 0 & 0 & 1 & & & 0 \\ \vdots & \vdots & \vdots & \vdots & \vdots & \vdots & \vdots \\ 0 & 0 & 0 & 0 & & & 1 \end{bmatrix} \begin{bmatrix} c_1 \\ c_2 \\ c_3 \\ c_4 \\ \cdot \\ \cdot \\ c_{N+2} \end{bmatrix} \tag{14}$$

$$\{q_v\} = [K^{q_v}]\{c\} \tag{15}$$

The translational displacement field corresponding to the y-direction can be expressed in terms of the "generalized" element nodal displacement matrix using a matrix of interpolation functions, as in the following.

$$v = [f_1 \quad f_2 \quad f_3 \quad f_4 \quad \cdots\cdots \quad f_{N+2}] \begin{bmatrix} v_1 \\ v_2 \\ v_3 \\ v_4 \\ \cdot \\ \cdot \\ v_{N+2} \end{bmatrix} \tag{16}$$

$$v = [N_v]\{q_v\} \tag{17}$$

Substituting for $\{q_v\}$ from Equation (15) into Equation (17), one can get:

$$v = [N_v][K^{q_v}]\{c\} = [K^v]\{c\} \tag{18}$$

One can obtain the shape functions of the displacement v as:

$$[N_v] = [1 - \xi \quad \xi \quad \sin \pi\xi \quad \sin 2\pi\xi \quad \sin 3\pi\xi \quad \cdots\cdots \quad \sin N\pi\xi] \tag{19}$$

It can now be observed that: (a) the interpolation functions f_1 and f_2 are linear polynomial functions and they correspond to the end nodal translational displacements of the hierarchical element, (b) the functions f_{n+2}, where $n = 1, 2,$

3,…., N are trigonometric functions and they correspond to the hierarchical terms, and (c) the functions f_{n+2} contribute only to the internal displacement field of the element.

By applying the above procedure, one can obtain the matrix of interpolation functions $[N_w]$ for the z-directional displacement function w, and the matrices of interpolation functions $\left[N_{\beta_y}\right]$ and $\left[N_{\beta_z}\right]$ for the rotational motions about y and z axes β_y and β_z. Using these matrices of interpolation functions, the displacement vector consisting of v, w, β_y and β_z can be formed as:

$$\begin{bmatrix} v \\ w \\ \beta_y \\ \beta_z \end{bmatrix} = \begin{bmatrix} [N_v] & [0] & [0] & [0] \\ [0] & [N_w] & [0] & [0] \\ [0] & [0] & \left[N_{\beta_y}\right] & [0] \\ [0] & [0] & [0] & \left[N_{\beta_z}\right] \end{bmatrix} \begin{bmatrix} \{q_v\} \\ \{q_w\} \\ \left\{q_{\beta_y}\right\} \\ \left\{q_{\beta_z}\right\} \end{bmatrix} \tag{20}$$

where

$$[N_v] = [N_w] = \left[N_{\beta_y}\right] = \left[N_{\beta_z}\right] = [f_1 \;\; f_2 \;\; f_3 \; \dots \dots \dots f_{N+2}] \tag{21}$$

$$\{q_v\} = \{v_1, v_2, v_3, \dots \dots \dots, v_{N+2}\}^T \tag{22}$$

$$\{q_w\} = \{w_1, w_2, w_3, \dots \dots \dots, w_{N+2}\}^T \tag{23}$$

$$\left\{q_{\beta_y}\right\} = \{\beta_{y1}, \beta_{y2}, \beta_{y3}, \dots \dots \dots, \beta_{y(N+2)}\}^T \tag{24}$$

$$\left\{q_{\beta_z}\right\} = \{\beta_{z1}, \beta_{z2}, \beta_{z3}, \dots \dots \dots, \beta_{z(N+2)}\}^T \tag{25}$$

STRESS-STRAIN RELATIONS

Based on Timoshenko beam theory, the displacement field of the cross-section of the tapered composite shaft can be described by the following equations:

$$\begin{aligned} u_x(x,y,z,t) &= z\beta_y(x,t) - y\beta_z(x,t) \\ &= (r\sin\theta)\beta_y(x,t) - (r\cos\theta)\beta_z(x,t) \end{aligned} \tag{26}$$

$$u_y(x, y, z, t) = v(x, t) \quad (27)$$

$$u_z(x, y, z, t) = w(x, t) \quad (28)$$

where $y = r\cos\theta$ and $z = r\sin\theta$. Further, u_x, u_y and u_z are the displacements of any point of the composite shaft in x, y and z directions, and v *and* w are the displacements of a point on the reference axis of the shaft in y and z directions and β_y and β_z are the rotation angles of the cross-section about y-axis and z-axis. Using Equations (26) - (28), the strains can be obtained as

$$\epsilon_{xx} = \frac{\partial u_x}{\partial x} = z\frac{\partial \beta_y}{\partial x} - y\frac{\partial \beta_z}{\partial x} \quad (29)$$

$$\epsilon_{yy} = \frac{\partial u_y}{\partial y} = 0 \quad (30)$$

$$\epsilon_{zz} = \frac{\partial u_z}{\partial z} = 0 \quad (31)$$

$$\gamma_{yz} = \frac{\partial u_z}{\partial y} + \frac{\partial u_y}{\partial z} = 0 \quad (32)$$

$$\gamma_{xz} = \frac{\partial u_x}{\partial z} + \frac{\partial u_z}{\partial x} = \beta_y + \frac{\partial w}{\partial x} \quad (33)$$

$$\gamma_{xy} = \frac{\partial u_x}{\partial y} + \frac{\partial u_y}{\partial x} = -\beta_z + \frac{\partial v}{\partial x} \quad (34)$$

The strain components in the cylindrical coordinate system (x, θ, r) can be written in terms of the strains in the Cartesian coordinate system (x, y, z) as

$$\begin{bmatrix} \epsilon_{xx} \\ \epsilon_{\theta\theta} \\ \epsilon_{rr} \\ \gamma_{\theta r} \\ \gamma_{xr} \\ \gamma_{x\theta} \end{bmatrix} = \begin{bmatrix} 1 & 0 & 0 & 0 & 0 & 0 \\ 0 & h^2 & g^2 & -2gh & 0 & 0 \\ 0 & g^2 & h^2 & 2gh & 0 & 0 \\ 0 & -gh & gh & g^2 - h^2 & 0 & 0 \\ 0 & 0 & 0 & 0 & h & g \\ 0 & 0 & 0 & 0 & g & -h \end{bmatrix} \begin{bmatrix} \epsilon_{xx} \\ \epsilon_{yy} \\ \epsilon_{zz} \\ \gamma_{yz} \\ \gamma_{xz} \\ \gamma_{xy} \end{bmatrix} \quad (35)$$

where $g = \cos\theta$ and $h = \sin\theta$. Substituting Equations (29) – (34) into Equation (35), the strain components can be obtained:

$$\epsilon_{xx} = r\,(\sin\theta)\frac{\partial\beta_y}{\partial x} - r(\cos\theta)\frac{\partial\beta_z}{\partial x} \tag{36}$$

$$\epsilon_{\theta\theta} = 0 \tag{37}$$

$$\epsilon_{rr} = 0 \tag{38}$$

$$\gamma_{\theta r} = 0 \tag{39}$$

$$\gamma_{xr} = (\sin\theta)\left(\frac{\partial w}{\partial x} + \beta_y\right) + (\cos\theta)\left(-\beta_z + \frac{\partial v}{\partial x}\right) \tag{40}$$

$$\gamma_{x\theta} = (\cos\theta)\left(\frac{\partial w}{\partial x} + \beta_y\right) - (\sin\theta)\left(-\beta_z + \frac{\partial v}{\partial x}\right) \tag{41}$$

By considering Equations (36)-(41) the stress – strain relations in Equation (6) can be written as

$$\begin{bmatrix}\sigma_{xx}\\ \sigma_{\theta\theta}\\ \sigma_{rr}\\ \tau_{\theta r}\\ \tau_{xr}\\ \tau_{x\theta}\end{bmatrix} = \begin{bmatrix}\bar{\bar{Q}}_{11} & \bar{\bar{Q}}_{15} & \bar{\bar{Q}}_{16}\\ \bar{\bar{Q}}_{21} & \bar{\bar{Q}}_{25} & \bar{\bar{Q}}_{26}\\ \bar{\bar{Q}}_{31} & \bar{\bar{Q}}_{35} & \bar{\bar{Q}}_{36}\\ \bar{\bar{Q}}_{41} & \bar{\bar{Q}}_{45} & \bar{\bar{Q}}_{46}\\ \bar{\bar{Q}}_{51} & \bar{\bar{Q}}_{55} & \bar{\bar{Q}}_{56}\\ \bar{\bar{Q}}_{61} & \bar{\bar{Q}}_{65} & \bar{\bar{Q}}_{66}\end{bmatrix}\begin{bmatrix}\varepsilon_{xx}\\ \gamma_{xr}\\ \gamma_{x\theta}\end{bmatrix} \tag{42}$$

In Timoshenko beam theory [2], the shear correction factor k_s is used to more accurately determine the shear stresses at a material point. By introducing the shear correction factor , the stress-strain relations of Equation (42) can be modified as

$$\begin{bmatrix}\sigma_{xx}\\ \sigma_{\theta\theta}\\ \sigma_{rr}\\ \tau_{\theta r}\\ \tau_{xr}\\ \tau_{x\theta}\end{bmatrix} = \begin{bmatrix}\bar{\bar{Q}}_{11} & k_s\bar{\bar{Q}}_{15} & k_s\bar{\bar{Q}}_{16}\\ \bar{\bar{Q}}_{21} & k_s\bar{\bar{Q}}_{25} & k_s\bar{\bar{Q}}_{26}\\ \bar{\bar{Q}}_{31} & k_s\bar{\bar{Q}}_{35} & k_s\bar{\bar{Q}}_{36}\\ k_s\bar{\bar{Q}}_{41} & k_s\bar{\bar{Q}}_{45} & k_s\bar{\bar{Q}}_{46}\\ k_s\bar{\bar{Q}}_{51} & k_s\bar{\bar{Q}}_{55} & k_s\bar{\bar{Q}}_{56}\\ k_s\bar{\bar{Q}}_{61} & k_s\bar{\bar{Q}}_{65} & k_s\bar{\bar{Q}}_{66}\end{bmatrix}\begin{bmatrix}\varepsilon_{xx}\\ \gamma_{xr}\\ \gamma_{x\theta}\end{bmatrix} \tag{43}$$

ENERGY FUNCTIONALS OF THE COMPOSITE SHAFT

The kinetic energy of the composite shaft is given by the following expression:

$$T_{comp} = \frac{1}{2}\int_0^L m_c(x)\,(\dot{v}^2 + \dot{w}^2)dx + \frac{1}{2}\int_0^L I_{dc}(x)\big(\dot{\beta}_y^2 + \dot{\beta}_z^2\big)dx - \int_0^L I_{pc}(\mathrm{x})\Omega\dot{\beta}_z\beta_y dx \tag{44}$$

In the above expression, m_c, I_{dc}, I_{pc}, $\dot{v}$, $\dot{w}$, $\dot{\beta}_y$, $\dot{\beta}_z$, and Ω are, respectively, the mass per unit length of the shaft, the diametral mass moment of inertia of the shaft, the polar mass moment of inertia of the shaft, the velocity of the shaft cross-section in the y-direction, the velocity of the shaft cross-section in the z-direction, the rotational velocity of the cross-section of the shaft around the y-direction, the rotational velocity of the cross-section of the shaft around the z-direction, and the rotational speed of the shaft (in radians/second).

The strain energy of the tapered composite shaft due to its bending and shear deformations is given by:

$$U_{BS} = \frac{1}{2}\iiint_V [\sigma_{x\theta r}]^T[\varepsilon_{x\theta r}]\,dV \tag{45}$$

Considering Equations (36) – (41), the strain energy U_{BS} can be expressed in terms of the appropriate force and moment resultants of the laminate as:

$$U_{BS} = \frac{1}{2}\int_0^L \left[M_y\frac{\partial\beta_y}{\partial x} + M_z\frac{\partial\beta_z}{\partial x} + Q_y\left(\frac{\partial v}{\partial x} - \beta_z\right) + Q_z\left(\beta_y + \frac{\partial w}{\partial x}\right)\right]dx \tag{46}$$

wherein the expressions for the moment and force resultants are given by

$$M_y = \int_A \sigma_{xx}\, r\, sin\theta\, dA \tag{47}$$

$$M_z = -\int_A \sigma_{xx} r \cos\theta \, dA \tag{48}$$

$$Q_{xr}^{(1)} = \int_A \tau_{xr} \sin\theta \, dA \tag{49}$$

$$Q_{xr}^{(2)} = \int_A \tau_{xr} \cos\theta \, dA \tag{50}$$

$$Q_{x\theta}^{(1)} = \int_A \tau_{x\theta} \sin\theta \, dA \tag{51}$$

$$Q_{x\theta}^{(2)} = \int_A \tau_{x\theta} \cos\theta \, dA \tag{52}$$

have been used. After applying the integrations in Equations (47) – (52), the force and moment resultants of the tapered composite shaft can be shown to be:

$$M_y = \bar{D}_{11}\frac{\partial \beta_y}{\partial x} + \bar{B}_{15}k_s\left(\beta_y + \frac{\partial w}{\partial x}\right) + \frac{1}{2}\bar{B}_{16}k_s\left(\beta_z - \frac{\partial v}{\partial x}\right) \tag{53}$$

$$M_z = \bar{D}_{11}\frac{\partial \beta_z}{\partial x} - \bar{B}_{15}k_s\left(\frac{\partial v}{\partial x} - \beta_z\right) - \frac{1}{2}\bar{B}_{16}k_s(\beta_y + \frac{\partial w}{\partial x}) \tag{54}$$

$$Q_{xr}^{(1)} = \bar{B}_{51}\, k_s \frac{\partial \beta_y}{\partial x} + k_s\, \bar{A}_{55}\left(\beta_y + \frac{\partial w}{\partial x}\right) + k_s\, \bar{A}_{56}\left(\beta_z - \frac{\partial v}{\partial x}\right) \tag{55}$$

$$Q_{xr}^{(2)} = -\bar{B}_{51}\, k_s \frac{\partial \beta_z}{\partial x} + k_s\, \bar{A}_{55}\left(\frac{\partial v}{\partial x} - \beta_z\right) + k_s\, \bar{A}_{56}\left(\beta_y + \frac{\partial w}{\partial x}\right) \tag{56}$$

$$Q_{x\theta}^{(1)} = \frac{1}{2}\bar{B}_{61}\, k_s \frac{\partial \beta_y}{\partial x} + k_s\, \bar{A}_{65}\left(\beta_y + \frac{\partial w}{\partial x}\right) + k_s\, \bar{A}_{66}\left(\beta_z - \frac{\partial v}{\partial x}\right) \tag{57}$$

$$Q_{x\theta}^{(2)} = -\frac{1}{2}\bar{B}_{61}\, k_s \frac{\partial \beta_z}{\partial x} + k_s\, \bar{A}_{65}\left(\frac{\partial v}{\partial x} - \beta_z\right) + k_s\, \bar{A}_{66}\left(\beta_y + \frac{\partial w}{\partial x}\right) \tag{58}$$

where

$$\bar{A}_{66}(x) = \frac{\pi}{2}\sum_{s=1}^{n} \bar{\bar{Q}}_{66}\left(r_{os}^2(x) - r_{is}^2(x) \right) \tag{59}$$

$$\bar{A}_{65}(x) = \frac{\pi}{2}\sum_{s=1}^{n} \bar{\bar{Q}}_{65}\left(r_{os}^2(x) - r_{is}^2(x) \right) \tag{60}$$

$$\bar{A}_{55}(x) = \frac{\pi}{2}\sum_{s=1}^{n} \bar{\bar{Q}}_{55}\left(r_{os}^2(x) - r_{is}^2(x) \right) \tag{61}$$

$$\bar{A}_{56}(x) = \frac{\pi}{2}\sum_{s=1}^{n} \bar{\bar{Q}}_{56}\left(r_{os}^2(x) - r_{is}^2(x) \right) \tag{62}$$

$$\bar{B}_{16}(x) = \frac{2}{3}\pi\sum_{s=1}^{n} \bar{\bar{Q}}_{16}\left(r_{os}^3(x) - r_{is}^3(x) \right) \tag{63}$$

$$\bar{B}_{15}(x) = \frac{\pi}{3}\sum_{s=1}^{n} \bar{\bar{Q}}_{15}\left(r_{os}^3(x) - r_{is}^3(x) \right) \tag{64}$$

$$\bar{B}_{51}(x) = \frac{\pi}{3}\sum_{s=1}^{n} \bar{\bar{Q}}_{51}\left(r_{os}^3(x) - r_{is}^3(x) \right) \tag{65}$$

$$\bar{B}_{61}(x) = \frac{2}{3}\pi\sum_{s=1}^{n} \bar{\bar{Q}}_{61}\left(r_{os}^3(x) - r_{is}^3(x) \right) \tag{66}$$

$$\bar{D}_{11}(x) = \frac{\pi}{4}\sum_{s=1}^{n} \bar{\bar{Q}}_{11}\left(r_{os}^4(x) - r_{is}^4(x)\right) \tag{67}$$

If the composite shaft is subjected to a constant axial force, the external work done by the constant axial force P in going through the shaft deformations, denoted as U_F can be written as:

$$U_F = \frac{1}{2}\int_0^L P[\left(\frac{\partial v}{\partial x}\right)^2 + \left(\frac{\partial w}{\partial x}\right)^2]dx \tag{68}$$

The total strain energy of the composite shaft can now be obtained as the summation of the internal strain energy of the shaft and the external work done by the axial force, and it is given as:

$$U_{comp} = U_{BS} + U_F \tag{69}$$

Lagrange's equations [11] will be used to obtain the equations of motion for the composite shaft element. In this regard, the generalized coordinates are assigned as:

$$\begin{array}{lllll} q_1 = v_1 & q_2 = v_2 & q_3 = v_3 & \cdots\cdots & q_b = v_{N+2} \\ q_{b+1} = w_1 & q_{b+2} = w_2 & q_{b+3} = w_3 & \cdots\cdots & q_{2b} = w_{N+2} \\ q_{2b+1} = \beta_{y1} & q_{2b+2} = \beta_{y2} & q_{2b+3} = \beta_{y3} & \cdots\cdots & q_{3b} = \beta_{y(N+2)} \\ q_{3b+1} = \beta_{z1} & q_{3b+2} = \beta_{z2} & q_{3b+3} = \beta_{z3} & \cdots\cdots & q_{4b} = \beta_{z(N+2)} \end{array} \tag{70}$$

where $b = N + 2$.

It can be noted from the above that, in addition to the number of nodes of the hierarchical finite element, the number of the generalized co-ordinates correspond also to the number of the hierarchical terms N. The generalized co-ordinates given above can be put together into a vector $\{q\}$ as:

$$\{q\} = \begin{bmatrix} \{q_v\} \\ \{q_w\} \\ \{q_{\beta_y}\} \\ \{q_{\beta_z}\} \end{bmatrix} \tag{71}$$

After applying Lagrange's equations, the matrix equation of motion of free vibrations of the rotating tapered composite driveshaft is obtained:

$$[M]\{\ddot{q}\} + [G]\{\dot{q}\} + ([K] + [K_P])\{q\} = \{0\} \tag{72}$$

where

$$[M] = \begin{bmatrix} [M_{11}] & [0] & [0] & [0] \\ [0] & [M_{22}] & [0] & [0] \\ [0] & [0] & [M_{33}] & [0] \\ [0] & [0] & [0] & [M_{44}] \end{bmatrix} \tag{73}$$

$$[G] = \begin{bmatrix} [0] & [0] & [0] & [0] \\ [0] & [0] & [0] & [0] \\ [0] & [0] & [0] & [G_{34}] \\ [0] & [0] & [G_{43}] & [0] \end{bmatrix} \tag{74}$$

$$[K] = \begin{bmatrix} [K_{11}] & 0 & [K_{13}] & [K_{14}] \\ [0] & [K_{22}] & [K_{23}] & [K_{24}] \\ [K_{31}] & [K_{32}] & [K_{33}] & [K_{34}] \\ [K_{41}] & [K_{42}] & [K_{43}] & [K_{44}] \end{bmatrix} \tag{75}$$

$$[K_P] = \begin{bmatrix} [K_{pv}] & [\ 0\ ] & [\ 0\ ] & [\ 0\ ] \\ [\ 0\ ] & [K_{pw}] & [\ 0\ ] & [\ 0\ ] \\ [\ 0\ ] & [\ 0\ ] & [\ 0\ ] & [\ 0\ ] \\ [\ 0\ ] & [\ 0\ ] & [\ 0\ ] & [\ 0\ ] \end{bmatrix} \tag{76}$$

The mass sub-matrices, gyroscopic sub-matrices, and stiffness sub-matrices for the hierarchical tapered composite driveshaft finite element are given in detail in Appendix A.

The equation of motion for the finite element model of the driveshaft system that is modelled using a number of hierarchical finite elements can be obtained by "assembling" the element matrices following the standard finite element method, with the consideration that the hierarchical degrees of freedom will not participate in the inter-element compatibility conditions since they are not physical degrees of freedom.

LAGRANGIAN COMPOSITE SHAFT ELEMENT

A conventional Lagrangian-type finite element formulation for the composite driveshaft has been developed by Al Muslmani and Ganesan [12], based on the Timoshenko beam theory, for the dynamic analysis of rotating composite driveshafts. As in the case of hierarchical finite element formulation, the effects of rotary inertia, transverse shear deformation, gyroscopic force, axial force, and taper angle have been incorporated in the Lagrangian finite element formulation, and further, the equations of motion have been developed using the Lagrange's equation.

In the conventional finite element formulation, the Lagrangian interpolation functions are used to approximate the displacement fields of the tapered composite shaft. Fig. (**5**) shows the finite element with three nodes, wherein two nodes are at the ends of the element and one node is at the center of the element. Each node has four degrees of freedom, that are, two translational motions in y and z directions and two rotational motions about y and z axes. The displacement fields are approximated in each case by a quadratic polynomial function. Hence,

$$v(x) = a + bx + cx^2 \tag{77}$$

$$w(x) = d + ex + fx^2 \tag{78}$$

$$\beta_y(x) = \tilde{a} + \tilde{b}x + \tilde{c}x^2 \tag{79}$$

$$\beta_z(x) = \tilde{d} + \tilde{e}x + \tilde{f}x^2 \tag{80}$$

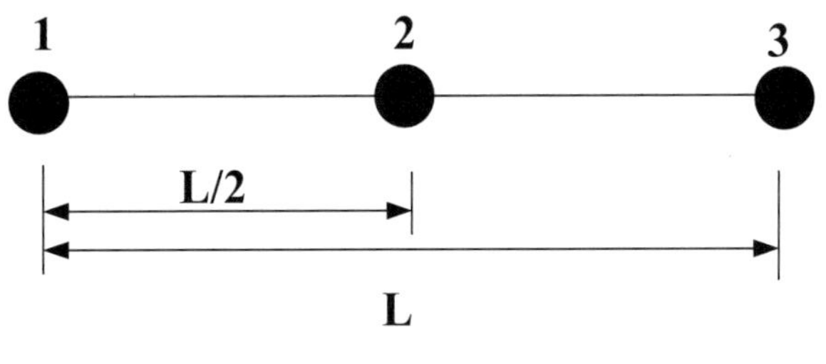

Fig. (5). The 3-node Lagrangian finite element.

For this approximation, one can get the interpolation functions following the conventional finite element formulation as:

$$v(x) = N_{11}v_1 + N_{22}v_2 + N_{33}v_3 \tag{81}$$

where

$$N_{11} = \left(1 - \frac{2x}{L}\right)\left(1 - \frac{x}{L}\right) \tag{82}$$

$$N_{22} = \frac{4x}{L}\left(1 - \frac{x}{L}\right) \tag{83}$$

$$N_{33} = \left(\frac{2x}{L} - 1\right)\left(\frac{x}{L}\right) \tag{84}$$

The interpolation functions N_{11}, N_{22}, and N_{33} are used to obtain an approximate solution for $v(x)$. Following the same procedure, for $w(x)$, $\beta_y(x)$, and $\beta_z(x)$, one can get the same interpolation functions as that of $v(x)$. Hence, the displacement field of the shaft can be established as:

$$v(x) = \sum_{i=1}^{3} v_i(t)N_{ii}(x) \tag{85}$$

$$w(x) = \sum_{i=1}^{3} w_i(t)N_{ii}(x) \tag{86}$$

$$\beta_y(x) = \sum_{i=1}^{3} \beta_{yi}(t)N_{ii}(x) \tag{87}$$

$$\beta_z(x) = \sum_{i=1}^{3} \beta_{zi}(t)N_{ii}(x) \tag{88}$$

Using the kinetic and potential energy functions given in the previous sections, and applying Lagrange's equations [11], one can get the equations of motion for the tapered composite driveshaft finite element as:

$$[M]\{\ddot{q}\} + [G]\{\dot{q}\} + ([K] + [K_P])\{q\} = \{0\} \tag{89}$$

where

$$\{q\} = \{\{q_1\} \quad \{q_2\} \quad \{q_3\} \quad \{q_4\}\}^T \tag{90}$$

$$\{q_1\} = \{v_1 \quad v_2 \quad v_3\} \tag{91}$$

$$\{q_2\} = \{w_1 \quad w_2 \quad w_3\} \tag{92}$$

$$\{q_3\} = \{\beta_{y1} \quad \beta_{y2} \quad \beta_{y3}\} \tag{93}$$

$$\{q_4\} = \{\beta_{z1} \quad \beta_{z2} \quad \beta_{z3}\} \tag{94}$$

and further,

$$[M] = \begin{bmatrix} [M_{11}]_{3\times3} & [\,0\,]_{3\times3} & [\,0\,]_{3\times3} & [\,0\,]_{3\times3} \\ [\,0\,]_{3\times3} & [M_{22}]_{3\times3} & [\,0\,]_{3\times3} & [\,0\,]_{3\times3} \\ [\,0\,]_{3\times3} & [\,0\,]_{3\times3} & [M_{33}]_{3\times3} & [\,0\,]_{3\times3} \\ [\,0\,]_{3\times3} & [\,0\,]_{3\times3} & [\,0\,]_{3\times3} & [M_{44}]_{3\times3} \end{bmatrix}_{12\times12} \tag{95}$$

$$[G] = \begin{bmatrix} [\,0\,]_{3\times3} & [\,0\,]_{3\times3} & [\,0\,]_{3\times3} & [\,0\,]_{3\times3} \\ [\,0\,]_{3\times3} & [\,0\,]_{3\times3} & [\,0\,]_{3\times3} & [\,0\,]_{3\times3} \\ [\,0\,]_{3\times3} & [\,0\,]_{3\times3} & [\,0\,]_{3\times3} & [G_{34}]_{3\times3} \\ [\,0\,]_{3\times3} & [\,0\,]_{3\times3} & [G_{43}]_{3\times3} & [\,0\,]_{3\times3} \end{bmatrix}_{12\times12} \tag{96}$$

$$[K] = \begin{bmatrix} [K_{11}]_{3\times3} & [\,0\,]_{3\times3} & [K_{13}]_{3\times3} & [K_{14}]_{3\times3} \\ [\,0\,]_{3\times3} & [K_{22}]_{3\times3} & [K_{23}]_{3\times3} & [K_{24}]_{3\times3} \\ [K_{31}]_{3\times3} & [K_{32}]_{3\times3} & [K_{33}]_{3\times3} & [K_{34}]_{3\times3} \\ [K_{41}]_{3\times3} & [K_{42}]_{3\times3} & [K_{43}]_{3\times3} & [K_{44}]_{3\times3} \end{bmatrix}_{12\times12} \tag{97}$$

$$[K_P] = \begin{bmatrix} [K_{pv}]_{3\times3} & [\,0\,]_{3\times3} & [\,0\,]_{3\times3} & [\,0\,]_{3\times3} \\ [\,0\,]_{3\times3} & [K_{pw}]_{3\times3} & [\,0\,]_{3\times3} & [\,0\,]_{3\times3} \\ [\,0\,]_{3\times3} & [\,0\,]_{3\times3} & [\,0\,]_{3\times3} & [\,0\,]_{3\times3} \\ [\,0\,]_{3\times3} & [\,0\,]_{3\times3} & [\,0\,]_{3\times3} & [\,0\,]_{3\times3} \end{bmatrix}_{12\times12} \tag{98}$$

$$[N_i] = [N_{11} \quad N_{22} \quad N_{33}] \tag{99}$$

$$[N_i'] = [N_{11}' \quad N_{22}' \quad N_{33}'] \tag{100}$$

The mass sub-matrices, gyroscopic sub-matrices, and stiffness sub-matrices for the Lagrangian tapered composite driveshaft finite element are given in detail in Appendix B.

The equation of motion for the finite element model of the driveshaft system that is modelled using a number of Lagrangian finite elements can be obtained by "assembling" the element matrices following the standard finite element method.

NATURAL FREQUENCIES AND CRITICAL SPEEDS OF THE SHAFT

The critical speeds of the rotating shaft can be determined using three methods [11], that are, the Direct method, the Iteration method and the Campbell diagram method. The Direct method is used here. In this method, the natural frequencies of the shaft that correspond to a specific rotating speed are calculated and the natural frequency that is equal to the forcing frequency is deemed as the critical speed of the shaft. The forcing frequency corresponds to the mass unbalance force of the disk or the skew of the disk. The forcing frequency is written in terms of the rotational speed as:

$$\omega_f = \mathrm{n}\Omega \quad \mathbf{(101)}$$

where n refers to the so-called "level of the lateral force" on the shaft. In the case of mass unbalance n = 1, and in the case of an m-bladed rotor n = m.

The solution to the equation of motion, corresponding to both the hierarchical and Lagrangian finite element formulations is considered in the form: $\{q\} = \{q_0\}e^{j\omega_f t}$ Using this form in the equation of motion results in the following algebraic eigenvalue problem:

$$\left(-\Omega^2\left[\mathrm{n}^2[\mathrm{M}] + \mathrm{jn}[\mathrm{G}]\right] + \mathrm{j}\Omega\mathrm{n}\,[\mathrm{C}] + ([\mathrm{K}] + [\mathrm{K}]_P)\right)\{\mathrm{q}_0\} = \{0\} \quad \mathbf{(102)}$$

Each eigenvalue Ω is a complex number and the real part of Ω is the critical speed of the shaft.

VALIDATION

A hollow tapered shaft made of a graphite-epoxy composite material is considered. The shaft is supported at the ends by two bearings as shown in Fig. (**6**), and these bearing supports are modeled together as the simply supported boundary conditions. The outer diameter of the composite shaft at the left end is taken to be

12.69 cm. The outer diameter of the shaft at the right end can be calculated for the considered value of taper angle. The outer diameter of the composite shaft at the left end is kept constant for all the taper angle values considered in the analysis. The properties of the graphite-epoxy composite material [2] are listed in Table **1**. The composite shaft is considered to be made of ten composite material plies, and the thickness is the same for all plies. Hence, the wall thickness of the composite shaft is 1.321 mm. The laminate configuration of the composite shaft is [90°/45°/-45°/$0°_6$/90°] with the lay-up starting from the inside. The length of the composite shaft is 2.47 m. The value of shear correction factor k_s for the uniform-cross-section composite shaft is 0.503 [2]. The value of the shear correction factor depends only on the type of the composite material of the shaft, the laminate theory that has been used to represent the mechanics of the shaft, and the structural theory that has been used for describing the shaft motion. These are the same for both the uniform and tapered composite shafts. Hence, the same value of shear correction factor, that is 0.503, is considered here for the tapered composite shaft.

Table 1. Properties of the Graphite-epoxy composite material.

Property	**Value**
E_{11}	139 GPa
E_{22}	11 GPa
$G_{12} = G_{13}$	6.05 GPa
G_{23}	3.78 GPa
ν_{12}	0.313
Density	1578 Kg/m^3

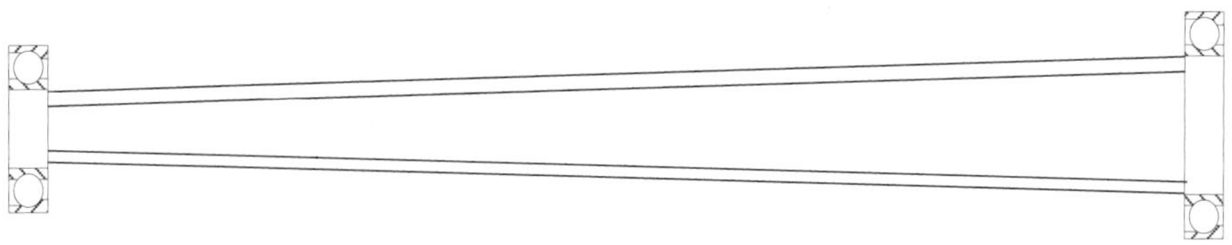

Fig. (6). Longitudinal sectional view of a tapered composite shaft mounted on bearings.

A total of six hierarchical finite elements of equal length and a total of twenty conventional Lagrangian finite elements have been used in the analysis. The first critical speeds of the tapered composite shaft obtained from the dynamic analyses [13, 14] are given in Table **2**. It can be observed from this table that the difference between the values of the first critical speed obtained using the conventional and the hierarchical finite elements is very small.

Table 2. The first critical speed of the tapered composite shaft for different taper angles.

Taper angle in degrees	First critical speed in rpm	
	Hierarchical finite element	Lagrangian finite element
0	5220	5219
1	6647	6645
2	7721	7718
3	8531	8525
4	9126	9121

The natural frequencies of the tapered composite shaft corresponding to the rotational speed of 10,000 rpm have also been determined [13] and are given in Table **3**.

Table 3. The natural frequencies (in Hz) of the tapered composite shaft at 10,000 rpm for different taper angles.

Taper angle, degrees	Mode	Hierarchical finite element model	Lagrangian finite element model
0 (uniform-diameter shaft)	BW1	87	86
	FW1	88	87
	BW2	317	316
	FW2	319	319
1	BW1	111	110
	FW1	112	112
	BW2	387	386
	FW2	390	390
2	BW1	128	128

(Table 3) cont.....

	FW1	131	131
	BW2	437	436
	FW2	441	440
3	BW1	142	142
	FW1	146	145
	BW2	473	472
	FW2	477	476
4	BW1	152	152
	FW1	157	157
	BW2	499	499
	FW2	504	502

It can be observed from the above table that the natural frequencies increase as the taper angle increases and this increase is not linear. Compared to the first forward natural frequency of the uniform-diameter composite shaft (*i.e.* shaft with 0° taper angle), the first backward natural frequency of the composite shaft with 1° taper angle is 28% higher and that with 2° taper angle is 47% higher, whereas that of the shaft with 3° taper angle is 63% higher. When the taper angle is increased, the distances to the mid-planes of the composite plies from the shaft centerline increase correspondingly, and in turn, the flexural stiffness of the composite shaft correspondingly increases, by the order 3. In addition, the volume of the composite material in the shaft increases when the taper angle increases, and due to this, the mass per unit length of the shaft and correspondingly the inertia of the shaft increases by the order 1. As a result of the combined effects of these increases, the natural frequency of the shaft increases in a non-linear manner when the taper angle of the composite shaft is increased.

The solution that has been obtained using the Rayleigh-Ritz method [14] is given in Table **4**. A very good agreement between the results calculated using the hierarchical finite element model and the Rayleigh-Ritz solution can be observed from the results presented in Tables **2** and **4**.

Table 4: First critical speed of the tapered composite shaft based on the Rayleigh-Ritz solution.

Taper angle in degrees	First critical speed in rpm
0	5220
1	6667
2	7820
3	8772
4	9510

TAPERED COMPOSITE SHAFT – A DESIGN PARAMETRIC STUDY

A tapered composite shaft that is supported by a bearing at one end and free at the other end, and made of a Boron-epoxy composite material is considered. The properties of the composite material [2] are given in Table **5**. The tapered composite shaft is made of ten plies, and the thickness of each ply is 0.025 mm. The length of the shaft is 0.5 m and the inner diameter at the free end is 1 cm. The inner diameter at the supported end is changed by changing the taper angle. The stiffnesses of the bearings in the y- and z-directions are taken to be equal to 10 GN/m. A rotational speed of 5,000 rpm is considered to calculate the natural frequencies. Seven hierarchical composite finite elements of equal length are used to model the shaft, with the same number of hierarchical terms per element as in the previous case. The same value of the shear correction factor as before is used here.

Table 5. Properties of the Boron-epoxy composite material.

Property	Value
E_{11}	211 GPa
E_{22}	24 GPa
$G_{12} = G_{13}$	6.9 GPa
G_{23}	6.9 GPa
ν_{12}	0.36
Density	1967 kg/m^3

In the following, the influence of the ply orientation angle on the natural frequencies of the tapered composite shaft is studied. The ten plies of the shaft have the same fiber orientation, and the fiber orientation angle of these plies is varied within the range of 0° to 90°. Different taper angle values are considered. In Table **6** the natural frequencies of the tapered composite shaft for different fiber orientation angles are presented. As can be observed from this table, the natural frequencies of the tapered composite shaft decrease when the increase in the fiber orientation angle of the plies.

Table 6. Natural frequencies in Hz of the tapered composite shaft rotating at 5000 rpm.

Fiber orientation angle	Whirl Mode	Taper angle				
		1°	2°	3°	4°	5°
0°	BW1	810	1078	1281	1431	1542
	FW1	811	1079	1282	1432	1543
45°	BW1	445	640	831	1016	1192
	FW1	446	642	832	1019	1196
90°	BW1	420	599	769	928	1073
	FW1	421	601	771	930	1076

Majed and Ganesan [14] have conducted an extensive rotordynamic analysis of rotating tapered composite shafts using the hierarchical finite element method as well as the Rayleigh-Ritz method. To validate the solution obtained using the hierarchical finite element, the Rayleigh-Ritz method has been used to obtain an approximate solution. It has been shown through numerical examples that the bending natural frequencies and the critical speeds of tapered composite shafts, determined using the Rayleigh-Ritz method are in very good agreement with that obtained using the hierarchical finite element.

A detailed design parametric study has also been conducted using the hierarchical finite element method. It has been determined that increasing the taper angle increases the bending natural frequencies and the critical speeds of the tapered composite shaft. However, it has been shown the first critical speed reaches its maximum value at a certain taper angle and then reduces for increasing taper angles.

It can be observed from the numerical results and the parametric study that the critical speed of the tapered composite shaft increases when the taper angle is increased, because the cross-sectional area as well as the distance to the center of the laminate in the cross-section from the shaft centerline increases throughout the length of the shaft when the taper angle is increased. This makes the tapered composite shaft stiffer than the uniform composite shaft.

The effects of ply stacking sequence, fiber orientation angles in the laminate, taper angle, axial loading, bearing stiffness, the inner diameter of the shaft, and length of the shaft have been studied. The key observations are mentioned in the following:

(a) Increasing the taper angle of the shaft when using low-stiffness bearings leads to decreasing first critical speed of the shaft. On the other hand, for high-stiffness bearings, the first critical speed is increased.

(b) Using composite material plies that have high flexural stiffness at and near the outer surface of the shaft significantly increases the natural frequencies and first critical speed.

(c) By decreasing the length of the tapered composite shaft and by increasing the shaft mean diameter, the natural frequencies and the first critical speed can be increased considerably.

CONCLUSION

In the present chapter, the formulation of a hierarchical finite element for modeling and analyzing the rotating (spinning) composite shaft are described. The hierarchical finite element is developed based on the Timoshenko beam theory. In the formulation, trigonometric functions have been used for the hierarchical terms and the Lagrangian polynomial interpolation (shape) functions have been used for the part of the interpolation that corresponds to the physical nodal degrees of freedom of the composite shaft element. The effects of rotary inertia, transverse shear deformation, gyroscopic force, axial force, and taper angle are incorporated in the hierarchical finite element formulation for the rotating composite shaft. The application of the developed hierarchical composite shaft finite element to the dynamic analysis of composite driveshafts is demonstrated.

CONSENT FOR PUBLICATION

Not applicable.

CONFLICT OF INTEREST

The author declares no conflict of interest, financial or otherwise.

ACKNOWLEDGEMENTS

Declared none.

APPENDIX A

$$[M_{11}] = L\int_0^1 m(\xi)\,[N_v]^T[N_v]d\xi \tag{A.1}$$

$$[M_{22}] = L\int_0^1 m(\xi)\,[N_w]^T[N_w]d\xi \tag{A.2}$$

$$[M_{33}] = L\int_0^1 m(\xi)\,[N_{\beta_y}]^T\left[N_{\beta_y}\right]d\xi \tag{A.3}$$

$$[M_{44}] = L\int_0^1 m(\xi)\,[N_{\beta_z}]^T\left[N_{\beta_z}\right]d\xi \tag{A.4}$$

$$[G_{34}] = -L\Omega\int_0^1 I_p(\xi)\,[N_{\beta_y}]^T\left[N_{\beta_y}\right]d\xi \tag{A.5}$$

$$[G_{43}] = L\Omega\int_0^1 I_p(\xi)\,[N_{\beta_z}]^T\left[N_{\beta_z}\right]d\xi \tag{A.6}$$

$$[K_{11}] = \frac{1}{L}\int_0^1 k_s(\bar{A}_{55}(\xi) + \bar{A}_{66}(\xi))\,[N_v']^T[N_v']d\xi \tag{A.7}$$

$$[K_{13}] = -\frac{1}{2L}\int_0^1 k_s\bar{B}_{16}(\xi)\,[N_v']^T\left[N'_{\beta_y}\right]d\xi \tag{A.8}$$

$$\begin{aligned}[K_{14}] = \int_0^1 \Bigg[&-\frac{1}{L}k_s\bar{B}_{15}(\xi)\,[N_v']^T\left[N'_{\beta_z}\right] - k_s(\bar{A}_{55}(\xi)\\ &+ \bar{A}_{66}(\xi))\,[N_v']^T\left[N_{\beta_z}\right]\Bigg]d\xi\end{aligned} \tag{A.9}$$

$$[K_{22}] = \frac{1}{L}\int_0^1 k_s(\bar{A}_{55}(\xi) + \bar{A}_{66}(\xi))\,[N_w']^T[N_w']d\xi \tag{A.10}$$

$$\begin{aligned}[K_{23}] = \int_0^1 \Bigg[&\frac{1}{L}k_s\bar{B}_{15}(\xi)\,[N_w']^T\left[N'_{\beta_y}\right] + k_s(\bar{A}_{55}(\xi)\\ &+ \bar{A}_{66}(\xi))\,[N_w']^T\left[N_{\beta_y}\right]\Bigg]d\xi\end{aligned} \tag{A.11}$$

$$[K_{24}] = -\frac{1}{2L}\int_0^1 k_s\bar{B}_{16}(\xi)\,[N_w']^T\left[N'_{\beta_z}\right]d\xi \tag{A.12}$$

$$[K_{31}] = -\frac{1}{2L}\int_0^1 k_s\bar{B}_{16}(\xi)\,[N'_{\beta_y}]^T[N_v']\,d\xi \tag{A.13}$$

$$[K_{32}] = \int_0^1 \left[\frac{1}{L} k_s \bar{B}_{15}(\xi)[N'_{\beta_y}]^T[N'_w] + k_s(\bar{A}_{55}(\xi) + A_{66}(\bar{\xi}))\,[N_{\beta_y}]^T[N'_v]\right] d\xi \quad \textbf{(A.14)}$$

$$[K_{33}] = \int_0^1 \left[\frac{1}{L}\bar{D}_{11}(\xi)\left[N'_{\beta_y}\right]^T\left[N'_{\beta_y}\right] + k_s\bar{B}_{15}(\xi)\left(\left[N'_{\beta_y}\right]^T\left[N_{\beta_y}\right] + \left[N_{\beta_y}\right]^T\left[N'_{\beta_y}\right]\right) + k_s L(A_{55}(\bar{\xi}) + A_{66}(\bar{\xi}))\left[N_{\beta_y}\right]\left[N_{\beta_y}\right]\right] d\xi \quad \textbf{(A.15)}$$

$$[K_{34}] = \int_0^1 \left[\frac{1}{2} k_s \bar{B}_{16}(\xi)\,[N'_{\beta_y}]^T\left[N_{\beta_z}\right] - \frac{1}{2} k_s \bar{B}_{16}(\xi)\,[N_{\beta_y}]^T\left[N'_{\beta_z}\right]\right] d\xi \quad \textbf{(A.16)}$$

$$[K_{41}] = \int_0^1 \left[-\frac{1}{L} k_s \bar{B}_{15}(\xi)\,[N'_{\beta_z}]^T[N'_v] - k_s(\bar{A}_{55}(\xi) + \bar{A}_{66}(\xi))\,[N'_{\beta_z}]^T[N'_v]\right] d\xi \quad \textbf{(A.17)}$$

$$[K_{42}] = -\frac{1}{2L}\int_0^1 \left[k_s \bar{B}_{16}(\xi)\,[N'_{\beta_z}]^T[N'_w]\right] d\xi \quad \textbf{(A.18)}$$

$$[K_{43}] = \int_0^1 \left[\frac{1}{2} k_s \bar{B}_{16}(\xi)\,[N_{\beta_z}]^T\left[N'_{\beta_y}\right] - \frac{1}{2} k_s \bar{B}_{16}(\xi)\,[N'_{\beta_z}]^T\left[N_{\beta_y}\right]\right] d\xi \quad \textbf{(A.19)}$$

$$[K_{44}] = \int_0^1 \left[\frac{1}{L}\bar{D}_{11}(\xi)\,[N'_{\beta_z}]^T\left[N'_{\beta_z}\right]\right.$$

$$+\,k_s\bar{B}_{15}(\xi)\left([N'_{\beta_z}]^T\left[N_{\beta_z}\right] + [N_{\beta_z}]^T\left[N'_{\beta_z}\right]\right)$$

$$\left. +\;k_sL(A_{55}(\bar{\xi}) + A_{66}(\bar{\xi}))\left[N_{\beta_z}\right]\left[N_{\beta_z}\right]\right]d\xi \qquad \textbf{(A.20)}$$

$$[K_{pv}] = \int_0^1 P\;\;[N_v']^T[N_v']d\xi \qquad \textbf{(A.21)}$$

$$[K_{pw}] = \int_0^1 P\;\;[N_w']^T[N_w']d\xi \qquad \textbf{(A.22)}$$

APPENDIX B

$$[M_{11}]_{3\times3} = \int_0^L m(x)\,[N_i]^T[N_i]dx \qquad \textbf{(B.1)}$$

$$[M_{22}]_{3\times3} = \int_0^L m(x)\,[N_i]^T[N_i]dx \qquad \textbf{(B.2)}$$

$$[M_{33}]_{3\times3} = \int_0^L m(x)\,[N_i]^T[N_i]dx \qquad \textbf{(B.3)}$$

$$[M_{44}]_{3\times3} = \int_0^L m(x)\,[N_i]^T[N_i]dx \qquad \textbf{(B.4)}$$

$$[G_{34}]_{3\times3} = -\Omega\int_0^L I_p(x)\,[N_i]^T[N_i]dx \tag{B.5}$$

$$[G_{43}]_{3\times3} = \Omega\int_0^L I_p(x)\,[N_i]^T[N_i]dx \tag{B.6}$$

$$[K_{11}]_{3\times3} = \int_0^L k_s(\bar{A}_{55}(x) + \bar{A}_{66}(x))\,[N_i']^T[N_i']dx \tag{B.7}$$

$$[K_{13}]_{3\times3} = -\frac{1}{2}\int_0^L k_s\bar{B}_{16}(x)\,[N_i']^T[N_i']dx \tag{B.8}$$

$$\begin{aligned}[K_{14}]_{3\times3} = \int_0^L \Big[-k_s\bar{B}_{15}(x)\,[N_i']^T[N_i'] - k_s(\bar{A}_{55}(x)\\ + \bar{A}_{66}(x))\,[N_i']^T[N_i]\Big]\,dx\end{aligned} \tag{B.9}$$

$$[K_{22}]_{3\times3} = \int_0^L k_s(\bar{A}_{55}(x) + \bar{A}_{66}(x))\;[N_i']^T[N_i']\,dx \tag{B.10}$$

$$\begin{aligned}[K_{23}]_{3\times3} = \int_0^L \Big[k_s\bar{B}_{15}(x)\;[N_i']^T[N_i'] + k_s(\bar{A}_{55}(x)\\ + \bar{A}_{66}(x))\;[N_i']^T[N_i]\Big]\,dx\end{aligned} \tag{B.11}$$

$$[K_{24}]_{3\times3} = -\frac{1}{2}\int_0^L \Big[k_s\bar{B}_{16}(x)\,[N_i']^T[N_i']\Big]\,dx \tag{B.12}$$

$$[K_{31}]_{3\times3} = -\frac{1}{2}\int_0^L \Big[k_s\bar{B}_{16}(x)\,[N_i']^T[N_i']\Big]\,dx \tag{B.13}$$

$$[K_{32}]_{3\times3} = \int_0^L \Big[k_s\bar{B}_{15}(x)\ [N_i']^T[N_i'] + k_s(\bar{A}_{55}(x) + A_{66}\bar{(x)})[N_i]^T[N_i']\Big]\,dx \tag{B.14}$$

$$[K_{33}]_{3\times3} = \int_0^L \Big[\bar{D}_{11}(x)\ [N_i']^T[N_i'] + k_s\bar{B}_{15}(x)\ ([N_i']^T[N_i] + [N_i]^T[N_i']) + k_s\big(A_{55}(\bar{x}) + A_{66}(\bar{x})\big)\ [N_i]^T[N_i]\Big]\,dx \tag{B.15}$$

$$[K_{34}]_{3\times3} = \int_0^L \left[\frac{1}{2}k_s\bar{B}_{16}(x)\ [N_i']^T[N_i] - \frac{1}{2}k_s\bar{B}_{16}(x)[N_i]^T[N_i']\right]dx \tag{B.16}$$

$$[K_{41}]_{3\times3} = \int_0^L \Big[-k_s\bar{B}_{15}(x)\ [N_i']^T[N_i'] - k_s(\bar{A}_{55}(x) + A_{66}\bar{(\xi)})\ [N_i]^T[N_i']\Big]\,dx \tag{B.17}$$

$$[K_{42}]_{3\times3} = -\frac{1}{2}\int_0^L \Big[k_s\bar{B}_{16}(x)\ [N_i']^T[N_i']\Big]\,dx \tag{B.18}$$

$$[K_{43}]_{3\times3} = \frac{1}{2}\int_0^L \Big[k_s\bar{B}_{16}(x)\ [N_i]^T[N_i'] - k_s\bar{B}_{16}(x)[N_i']^T[N_i]\Big]\,dx \tag{B.19}$$

$$[K_{44}]_{3\times3} = \int_0^L \Big[\bar{D}_{11}(x)\ [N_i']^T[N_i'] + k_s\bar{B}_{15}(x)\ ([N_i']^T[N_i] + [N_i]^T[N_i']) + k_s(\bar{A}_{55}(x) + \bar{A}_{66}(x))\ [N_i]^T[N_i]\Big]\,dx \tag{B.20}$$

$$[K_{pv}]_{3\times3} = \int_0^L P\,[N_i']^T[N_i']dx \tag{B.21}$$

$$[K_{pw}]_{3\times3} = \int_0^L P\,[N_i']^T[N_i']dx \tag{B.22}$$

REFERENCES

[1] H. Zinberg and M.F. Symonds, "The Development of an Advanced Composite Tail Rotor Driveshaft," In: *proceedings of the 26th Annual Forum of the American Helicopter Society.*, Washington, D.C.: United States, June 1970.

[2] M. Y. Chang, C. Y. Chang and J. K. Chen, "A simple spinning laminated composite shaft model," *International Journal of Solids and Structures.*, vol. 41, no. 3, pp. 637–662, Feb. 2004.

[3] M. Al Muslmani and R. Ganesan, "Rotor-dynamics of Stepped Composite Shaft-Disk Systems Based on a Conventional Composite Finite Element," In: *Joint American Society for Composites 27th Technical Conference, the 15th US-Japan Conference on Composite Materials and the ASTM-D30 Meeting.*, Texas: USA, October 2012.

[4] T.C. Gmür and J.D. Rodrigues, "Shaft finite elements for rotor dynamics analysis," *Transactions of ASME: Vibration and Acoustics.*, vol. 113, pp. 482-493, October 1991.

[5] L. Meirovitch, and H. Baruh, "On the Inclusion Principle for the Hierarchical Finite Element Method," *International Journal of Numerical Methods in Engineering.*, vol. 19, no. 2, pp. 281–291, 1983.

[6] L. West, N. Bardell, J. Dunsdon and P. Loasby, "Some limitations associated with the use of K-orthogonal polynomials in hierarchical versions of the finite element method," In: *The 6th International Conference on Recent Advances in Structural Dynamics*, Southampton, 1997.

[7] A. Houmat, "A sector Fourier p-element applied to free vibration analysis of sectorial plates," *Journal of Sound and Vibration.*, vol. 243, no. 2, pp. 269-282, May 2001.

[8] R. Ganesan and L. Chen, "Vibration analysis of tapered composite beams using Hierarchical Finite Element and Three-Dimensional ply stiffness analysis", In: *International Conference on Resource Utilization and Intelligent Systems.* (INCRUIS-2006), Erode: India, Jan. 2006.

[9] A. Fazili and R. Ganesan, "Hierarchical Finite Element for Vibration Analysis of Thickness- and Width-Tapered Laminated Composite Beams," In: *24th International Congress of Theoretical and Applied Mechanics.* (ICTAM), Montreal: Canada, August. 2016.

[10] A. Boukhalfa, A. Hadjoui and S.M. Hamza Cherif, "Free vibration analysis of a rotating composite shaft using the p-version of the finite element method," *International Journal of Rotating Machinery*., 2008.

[11] M. I. Friswell, J. Penny, S. D. Garvey and A. W. Lees, *Dynamics of Rotating Machines*, 1st ed. Cambridge University Press: New York, 2010.

[12] M. Al Muslmani and R. Ganesan, "Rotordynamic analysis of tapered composite shaft based on Lagrangian finite element analysis," In: *19th International Conference on Composite Materials* (ICCM19), 2013.

[13] M. Al Muslmani and R. Ganesan, "Rotordynamic Response of Tapered Composite Driveshaft based on a Conventional Composite Timoshenko Finite Element," In: *American Society for Composites 28th Annual Technical Conference*, 2013.

[14] M. Al Muslmani and R. Ganesan, "Vibration of tapered composite driveshaft based on the hierarchical finite element analysis," *Composite Structures*., vol. 209, pp. 905-927, 2019.

CHAPTER 8

n-sided Polygonal Hybrid Finite Elements with Unified Fundamental Solution Kernels

Wan-Qing Lin[1] and **Hui Wang**[2*]

[1]*College of Civil Engineering, Henan University of Technology, Zhengzhou, China*

[2]*School of Civil Engineering and Architecture, Hainan University, Haikou, China*

Abstract: Heat conduction in anisotropic composite materials has received wide attention in material engineering. In this chapter, the *n*-sided polygonal hybrid finite element method with fundamental solution kernels, named as HFS-FEM, is thoroughly studied for two-dimensional heat conduction in fully anisotropic media, which can be regarded as a combination of the traditional finite element method and boundary element method. In this approach, the unknown temperature field within an *n*-sided polygon is represented by the linear combination of anisotropic fundamental solutions of problem to achieve the local satisfaction of the related governing equations, but not the specific boundary conditions and the continuity conditions across the element boundary. To tackle such shortcoming, a frame temperature field is independently defined over the entire boundary of polygonal element by means of the conventional one-dimensional shape function interpolation. Subsequently, by the hybrid functional with the assumed intra- and inter-element temperature fields, the stiffness equation can be obtained including the line integrals along the element boundary only, whose dimension is reduced by one compared to the domain integrals in the traditional finite elements. This means that the higher computing efficiency is expected. Moreover, any shaped polygonal elements can be constructed in a unified form with same fundamental solution kernels, including convex and non-convex polygonal elements, to provide greater flexibility in meshing effort for complex geometries. Besides, the element boundary integration strategy endows the method with higher versatility and with non-conforming mesh in the pre-processing stage of the analysis over the traditional finite elements. No modification to the HFS-FEM formulation is needed for the non-conforming mesh and the element containing hanging nodes is treated normally as the one with more nodes. Finally, the accuracy, convergence, computing efficiency, stability of non-convex element, and straightforward treatment of non-conforming discretization are discussed for the present *n*-sided polygonal hybrid finite elements by a few applications in the context of anisotropic heat conduction.

Keywords: Hybrid finite element, Fundamental solution, *n*-sided polygon, Non-conforming mesh, Anisotropic material, Heat conduction.

[*]**Corresponding author Hui Wang:** School of Civil Engineering and Architecture, Hainan University, Haikou 570228, China; E-mail: huiwang@haut.edu.cn

INTRODUCTION

Heat conduction in fully anisotropic materials has received wide attention in engineering, especially for the extensive applications of composite materials, which generally consist of fillers and matrix materials and involve complex microstructures [1-6]. Although many attempts have been made in solving anisotropic heat conduction by the theoretical methods [7-9], it is difficult to deal with anisotropic structures with complex material definitions, boundary shapes, and/or boundary conditions. Accordingly, various numerical methods have been resorted to for serving this purpose, including finite element method (FEM) [10, 11], boundary element method (BEM) [12-15], meshless method [16-22], hybrid Trefftz element method (HT-FEM) [23-30], scaled boundary finite element method [31], *etc.*

Among various numerical methods, the FEM and BEM are the ubiquitous choice to solve such boundary-value problems in anisotropic solids. For the FEM, the unitary polynomial interpolation of temperature variable is introduced over the entire element level, instead of the whole solution domain, so that it can be applied for multi-material issues flexibly. However, to meet the conforming requirements, the construction of shape functions becomes tedious and generally different types of elements require different forms of shape functions. Besides, the time-consuming domain integration is unavoidable. Moreover, the model having local holes, cracks, or concentrated heat sources needs mesh refinement, which leads to a significant increase in the amount of calculation of the model. Such limitations are encountered with the recently developed polygonal finite elements as well [32-34]. For the BEM, the complex theory background such as the derivation of boundary integral equation from Green's second identity is required. Moreover, it easily encounters difficulties in dealing with singular and supersingular integrals. Besides, the final solution matrix is full, so the computational efficiency issue becomes considerable for problems with a large amount of local defects or multiple material constituents.

Typically, as a hybrid method combining the advantages of FEM and BEM, the hybrid finite element formulation with approximating kernels of fundamental solutions of problem, named as HFS-FEM in literature [35-39], has been extensively developed for isotropic thermal and elastic analysis [40-50]. In the HFS-FEM, the fundamental solutions at different source points are employed as kernel functions for approximating the element interior field, which exactly satisfies the corresponding governing equations, but not the specific boundary conditions and the interfacial continuity conditions across the element boundary.

To enforce them, a conforming approximation with conventional shape functions is independently defined along the element boundary. The substitution of them into the hybrid functional in a weak form leads to element's stiffness equation, which includes element boundary integrals only in computation. Obviously, the integration dimension is reduced by one compared to the domain integrals in the traditional finite elements. More importantly, the element boundary integration strategy makes that any hybrid element with any numbers of sides (n-sided polygonal element) can be constructed in a unified form with same fundamental solution kernels, including convex and non-convex polygonal elements, to provide great flexibility in automatic meshing manipulation in complex geometries. Besides, the element boundary integrals endow the method higher versatility with non-conforming mesh in the pre-processing stage of analysis, compared to the FEM. This appealing feature is beneficial to adaptive mesh refinement, where a straightforward subdivision of individual polygonal element usually results in hanging nodes, as displayed in Fig. (**1**). Non-conforming mesh containing hanging nodes can be handled in the HFS-FEM in a straightforward way, because the HFS-FEM simply interprets a hanging node in a non-conforming mesh as a division of an edge, and as a result, the number of edges in element sharing this node increase by one. This operation requires no modification to the HFS-FEM formulation and the element is treated normally. More importantly, the special fundamental solutions satisfying the local constraints can be used to construct specially-purposed elements, *i.e.*, special hole element, special inclusion element, as illustrated in literature [41, 44, 45, 47], which can extremely reduce the meshing and computing effort around the local defects.

It is worth pointing out that the HFS-FEM formulation is different to the HT-FEM, which uses a series of T-complete solutions satisfying the governing equation of problem to represent the interior approximation [24, 26-28]. However, it is observed that the required T-complete solutions are difficult to be derived for most of physical problems. Moreover, T-complete functions typically are power functions of the distance variable r, *i.e.*, for two-dimensional potential problem, the T-complete functions in polar coordinates (r, θ) can be written as [24, 51]:

$$\left\{1,\ r^m \cos\left(m\theta\right),\ r^m \sin\left(m\theta\right)\right\}, \qquad m = 1, 2, 3, \cdots \tag{1}$$

so the exponential growth of T-complete functions commonly leads to an ill-conditioned system of equations.

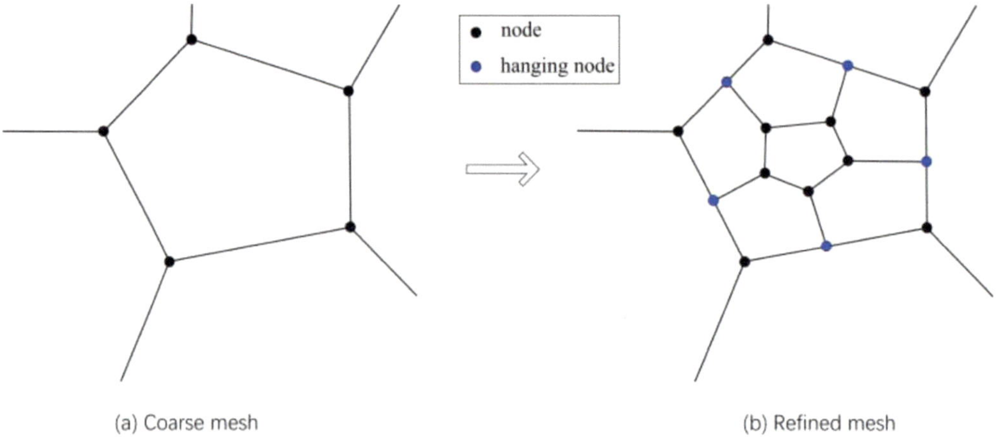

Fig. (1). Schematic of subdivision of a polygonal element with hanging nodes.

In this chapter, a two-dimensional steady-state heat conduction in fully anisotropic solids is thoroughly studied using the n-sided polygonal hybrid finite element method with flexible conforming or non-conforming, convex or non-convex polygon discretization. The main novelty of this work comes from the employment of the anisotropic fundamental solutions to achieve the exact satisfaction of the anisotropic governing equations for heat conduction, and then the one-dimensional line integrals over the element boundary for the computation of element stiffness matrix, which greatly serves the construction of arbitrarily shaped polygonal element and the improvement of computational efficiency. Also, the role of polygon-shaped hybrid finite elements would be strengthened in the analysis by the simple and straightforward refinement manipulation, as shown in Fig. (**1**). Besides, it is worth noting that the present method can be easily adapted for the three-dimensional problems in the same theoretical framework, given that the three-dimensional fundamental solutions and the two-dimensional shape functions are introduced for approximating the intra- and inter-element fields, respectively. As a result, the two-dimensional area integrals along the element boundary are involved in computing the element stiffness matrix.

GOVERNING EQUATIONS OF ANISOTROPIC HEAT CONDUCTION

Heat Balance Equation

In the absence of internal heat sources, the generalized partial differential equation governing steady-state energy balance for heat conduction behavior at an arbitrary point $\mathbf{x} = (x_1, x_2)$ in a bounded homogeneous anisotropic media $\Omega \subset \mathbb{R}^2$ is written as

$$\frac{\partial q_1(\mathbf{x})}{\partial x_1} + \frac{\partial q_2(\mathbf{x})}{\partial x_2} = 0, \qquad \forall \mathbf{x} \in \Omega \tag{2}$$

where q_1 and q_2 represent the components of heat flux vector, respectively.

Fourier's Law

Generally, the heat flux components are related to the gradient of temperature variable through the following Fourier's law of heat conduction

$$\mathbf{q}(\mathbf{x}) = \begin{Bmatrix} q_1(\mathbf{x}) \\ q_2(\mathbf{x}) \end{Bmatrix} = -\begin{bmatrix} k_{11} & k_{12} \\ k_{12} & k_{22} \end{bmatrix} \begin{Bmatrix} \dfrac{\partial T(\mathbf{x})}{\partial x_1} \\ \dfrac{\partial T(\mathbf{x})}{\partial x_2} \end{Bmatrix}. \tag{3}$$

where T is the unknown temperature variable. $k_{ij} > 0$ denote the general anisotropic coefficients in a thermal conductivity matrix

$$\mathbf{k} = \begin{bmatrix} k_{11} & k_{12} \\ k_{21} & k_{22} \end{bmatrix} \tag{4}$$

satisfying the Onsager symmetry relation

$$k_{ij} = k_{ji} \tag{5}$$

and the positive definite condition

$$\Lambda = \det(\mathbf{k}) = k_{11}k_{22} - k_{12}^2 > 0 \tag{6}$$

The smaller the value of $\det(\mathbf{k})$ is, the more asymmetric the temperature field and the behavior of heat flux vectors are, and the more difficult the derivation of solution will be.

Substituting Eq. (3) into the energy banlance equation (1) gives the following generalized partial differential governing equation with respect to the unknown temperature variable

$$k_{11}\frac{\partial^2 T(\mathbf{x})}{\partial x_1^2}+2k_{12}\frac{\partial^2 T(\mathbf{x})}{\partial x_1 \partial x_2}+k_{22}\frac{\partial^2 T(\mathbf{x})}{\partial x_2^2}=0 \quad \textbf{(7)}$$

Boundary Conditions

To keep the problem solvable, the following three types of boundary conditions are considered

Dirichlet boundary condition (i.e. prescribed temperature):

$$T=\overline{T} \qquad \text{on } \Gamma_T \quad \textbf{(8)}$$

Neumann boundary condition (i.e. prescribed normal heat flux):

$$q_n=\overline{q} \qquad \text{on } \Gamma_q \quad \textbf{(9)}$$

Robin boundary condition (i.e. prescribed convection):

$$q_n=h_c\left(T-T_c\right) \qquad \text{on } \Gamma_c \quad \textbf{(10)}$$

where $\boldsymbol{q_n}$ denotes the normal heat flux defined on the boundary. $\overline{\boldsymbol{T}}$ and $\overline{\boldsymbol{q}}$ are specified values given on the temperature boundary $\boldsymbol{\Gamma_T}$ and the heat flux boundary $\boldsymbol{\Gamma_q}$, respectively. $\boldsymbol{\Gamma_c}$ is the convection boundary part, and $\boldsymbol{h_c}$ denotes the convection coefficient, $\boldsymbol{T_c}$ is the surrounding temperature. $\boldsymbol{\Gamma_T \cup \Gamma_q \cup \Gamma_c = \Gamma}$ and $\boldsymbol{\Gamma_T \cap \Gamma_q = \emptyset}$, $\boldsymbol{\Gamma_q \cap \Gamma_c = \emptyset}$, $\boldsymbol{\Gamma_T \cap \Gamma_c = \emptyset}$.

Mathmatically, the normal heat flux can be expressed by the heat flux components $\boldsymbol{q_i}$ $(\boldsymbol{i=1,2})$, that is

$$q_n=q_1 n_1+q_2 n_2 \quad \textbf{(11)}$$

in which n_i are components of the unit outward normal vector $\mathbf{n}$ to the boundary, as indicated in Fig. (**2**).

So, the normal heat flux can be further written as

$$q_n(\mathbf{x}) = -\{n_1 \quad n_2\}\begin{bmatrix} k_{11} & k_{12} \\ k_{12} & k_{22} \end{bmatrix}\begin{Bmatrix} \dfrac{\partial T(\mathbf{x})}{\partial x_1} \\ \dfrac{\partial T(\mathbf{x})}{\partial x_2} \end{Bmatrix}. \tag{12}$$

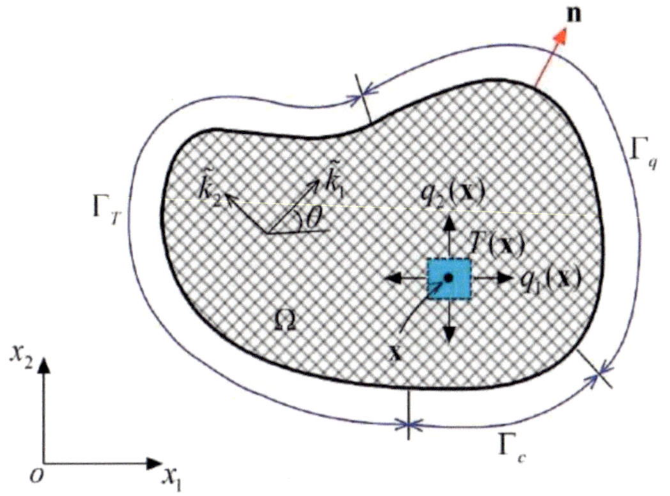

Fig. (2). Schematic of anisotropic heat conduction in a bounded domain.

Transformation of Thermal Conductivity Matrix

Particularly, for orthotropic materials with two principal directions, which have a ply angle $\boldsymbol{\theta}$ to the global $\boldsymbol{x_1}$ axis, as depicted in Fig. (**2**), the relation between the orthotropic parameters $\widetilde{\boldsymbol{k}}_{\boldsymbol{i}}$ $(\boldsymbol{i} = \mathbf{1}, \mathbf{2})$ and the generalized anisotropic parameters $\boldsymbol{k_{ij}}$ $(\boldsymbol{i} = \mathbf{1}, \mathbf{2})$ can be given by [20, 52]

$$\begin{bmatrix} k_{11} & k_{12} \\ k_{12} & k_{22} \end{bmatrix} = \begin{bmatrix} \cos\theta & \sin\theta \\ -\sin\theta & \cos\theta \end{bmatrix}^{\mathrm{T}} \begin{bmatrix} \tilde{k}_1 & 0 \\ 0 & \tilde{k}_2 \end{bmatrix} \begin{bmatrix} \cos\theta & \sin\theta \\ -\sin\theta & \cos\theta \end{bmatrix} \tag{13}$$

or alternatively

$$\begin{bmatrix} \tilde{k}_1 & 0 \\ 0 & \tilde{k}_2 \end{bmatrix} = \begin{bmatrix} \cos\theta & \sin\theta \\ -\sin\theta & \cos\theta \end{bmatrix} \begin{bmatrix} k_{11} & k_{12} \\ k_{12} & k_{22} \end{bmatrix} \begin{bmatrix} \cos\theta & \sin\theta \\ -\sin\theta & \cos\theta \end{bmatrix}^{\mathrm{T}} \tag{14}$$

Specially, the isotropic heat conduction equation can be obtained by setting $\boldsymbol{k_{11}} = \boldsymbol{k_{22}} = \boldsymbol{k},\ \boldsymbol{k_{12}} = \mathbf{0}$

$$k\frac{\partial^2 T(\mathbf{x})}{\partial x_1^2} + k\frac{\partial^2 T(\mathbf{x})}{\partial x_2^2} = 0 \tag{15}$$

Correspondingly, the heat flux component can be given by

$$\begin{Bmatrix} q_1(\mathbf{x}) \\ q_2(\mathbf{x}) \end{Bmatrix} = -\begin{bmatrix} k & 0 \\ 0 & k \end{bmatrix} \begin{Bmatrix} \dfrac{\partial T(\mathbf{x})}{\partial x_1} \\ \dfrac{\partial T(\mathbf{x})}{\partial x_2} \end{Bmatrix} = \begin{Bmatrix} -k\dfrac{\partial T(\mathbf{x})}{\partial x_1} \\ -k\dfrac{\partial T(\mathbf{x})}{\partial x_2} \end{Bmatrix}. \tag{16}$$

FUNDAMENTAL SOLUTIONS

For two-dimensional heat conduction in a fully anisotropic material with thermal conductivities $\boldsymbol{k_{ij}}\ (\boldsymbol{i} = \mathbf{1}, \mathbf{2})$, the temperature fundamental solution of the problem is defined as the induced temperature response at arbitrary field point $\mathbf{x} = (\boldsymbol{x_1}, \boldsymbol{x_2}) \in \mathbb{R}^2$ in an infinite two-dimensional space $\mathbb{R}^2$ when a unit point heat source is applied at a source point $\mathbf{y} = (\boldsymbol{y_1}, \boldsymbol{y_2}) \in \mathbb{R}^2$. This means that the caused temperature distribution represented as $\boldsymbol{T}^*(\mathbf{x}, \mathbf{y})$ is the solution of the following partial differential equation

$$k_{11}\frac{\partial^2 T^*(\mathbf{x},\mathbf{y})}{\partial x_1^2} + 2k_{12}\frac{\partial^2 T^*(\mathbf{x},\mathbf{y})}{\partial x_1 \partial x_2} + k_{22}\frac{\partial^2 T^*(\mathbf{x},\mathbf{y})}{\partial x_2^2} + \delta(\mathbf{x},\mathbf{y}) = 0, \quad \mathbf{x},\mathbf{y} \in \mathbb{R}^2 \tag{17}$$

where $\boldsymbol{\delta}(\mathbf{x}, \mathbf{y})$ is Dirac's delta function satisfying the following property

$$\int_\Omega h(\mathbf{x})\delta(\mathbf{x},\mathbf{y})\mathrm{d}\Omega = \begin{cases} h(\mathbf{y}), & \text{for } \mathbf{y} \in \Omega \\ 0, & \text{for } \mathbf{y} \notin \Omega \end{cases} \tag{18}$$

Obviously, the temperature solution of Eq. (17) is defined everywhere except when the source point **y** coincides with the field point **x**, where it is singular. Thus the source point **y** is sometimes called the singularity. In order to obtain this solution, Fourier transform or Radon transform can be applied on both sides of Eq. (17), and the concentrated excitation is reproduced by contour integral in the neighborhood of the source point **y** [53, 54]. As a result, the temperature fundamental solution can be written as

$$T^*(\mathbf{x},\mathbf{y}) = -\frac{1}{2\pi\sqrt{\Lambda}} \ln R(\mathbf{x},\mathbf{y}), \qquad \mathbf{x} \neq \mathbf{y} \tag{19}$$

in which $R(\mathbf{x},\mathbf{y})$ is the geodesic distance between the field point **x** and the source point **y**

$$R(\mathbf{x},\mathbf{y}) = \sqrt{(\mathbf{x}-\mathbf{y})\cdot\mathbf{k}^{-1}(\mathbf{x}-\mathbf{y})} = \sqrt{k_{22}r_1^2 - 2k_{12}r_1r_2 + k_{11}r_2^2} \tag{20}$$

and

$$\mathbf{x}-\mathbf{y} = \begin{Bmatrix} x_1 - y_1 \\ x_2 - y_2 \end{Bmatrix} = \begin{Bmatrix} r_1 \\ r_2 \end{Bmatrix} \tag{21}$$

In Eq. (20), the inverse of the material matrix **k** is

$$\mathbf{k}^{-1} = \frac{1}{\Lambda}\begin{bmatrix} k_{22} & -k_{12} \\ -k_{12} & k_{11} \end{bmatrix} \tag{22}$$

Subsequently, the derivatives of the induced temperature field can be given by

$$\begin{aligned} \frac{\partial T^*(\mathbf{x},\mathbf{y})}{\partial x_1} &= -\frac{1}{2\pi\sqrt{\Lambda}} \frac{k_{22}r_1 - k_{12}r_2}{R^2} \\ \frac{\partial T^*(\mathbf{x},\mathbf{y})}{\partial x_2} &= -\frac{1}{2\pi\sqrt{\Lambda}} \frac{k_{11}r_2 - k_{12}r_1}{R^2} \end{aligned} \tag{23}$$

from which the induced heat flux components can be written as

$$\begin{Bmatrix} q_1^*(\mathbf{x},\mathbf{y}) \\ q_2^*(\mathbf{x},\mathbf{y}) \end{Bmatrix} = -\begin{bmatrix} k_{11} & k_{12} \\ k_{12} & k_{22} \end{bmatrix} \begin{Bmatrix} \dfrac{\partial T^*(\mathbf{x},\mathbf{y})}{\partial x_1} \\ \dfrac{\partial T^*(\mathbf{x},\mathbf{y})}{\partial x_2} \end{Bmatrix} = \frac{\sqrt{\Lambda}}{2\pi R^2} \begin{Bmatrix} r_1 \\ r_2 \end{Bmatrix} \tag{24}$$

Specially, the fundamental solutions for the isotropic heat conduction can be obtained by simplifying the anisotropic solutions (19) and (24) with the relationship $\boldsymbol{k_{11}} = \boldsymbol{k_{22}} = \boldsymbol{k},\ \boldsymbol{k_{12}} = \mathbf{0}$, that is

$$T^*(\mathbf{x},\mathbf{y}) = -\frac{1}{2\pi k}\ln r \tag{25}$$

and

$$\begin{Bmatrix} q_1^*(\mathbf{x},\mathbf{y}) \\ q_2^*(\mathbf{x},\mathbf{y}) \end{Bmatrix} = -\begin{bmatrix} k & 0 \\ 0 & k \end{bmatrix} \begin{Bmatrix} \dfrac{\partial T^*(\mathbf{x},\mathbf{y})}{\partial x_1} \\ \dfrac{\partial T^*(\mathbf{x},\mathbf{y})}{\partial x_2} \end{Bmatrix} = \frac{1}{2\pi r^2}\begin{Bmatrix} r_1 \\ r_2 \end{Bmatrix} \tag{26}$$

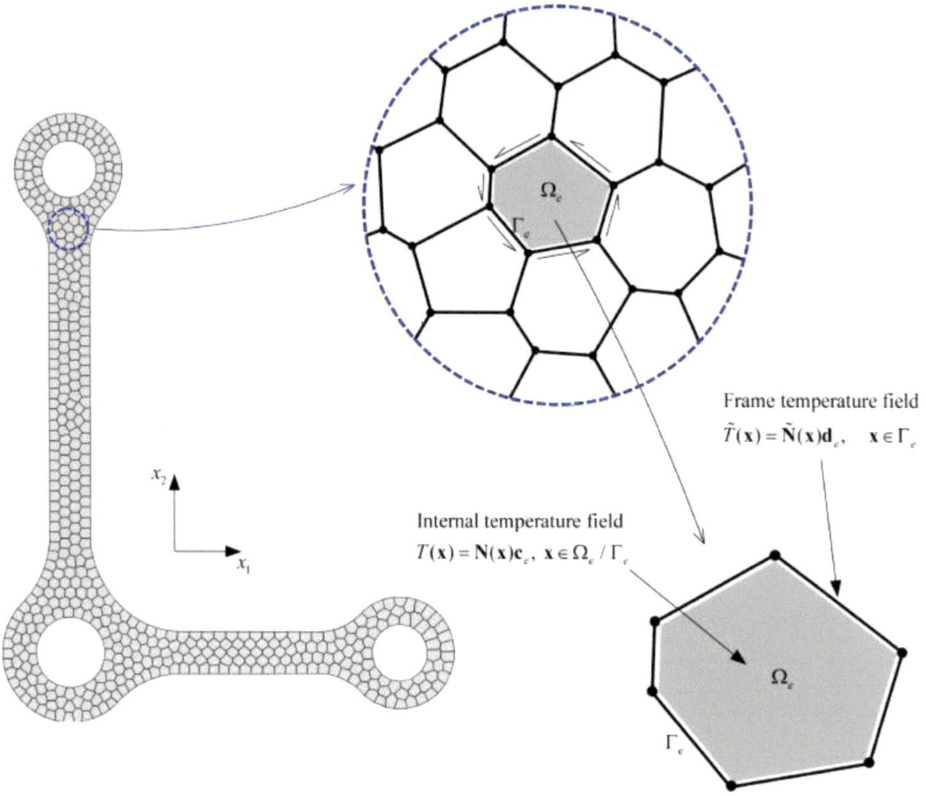

Fig. (3). Temperature approximations for a hybrid finite element.

HYBRID FINITE ELEMENT FORMULATION

In this section, the n-sided polygonal element is first formulated to solve the general anisotropic heat conduction problem. To achieve polygonal mesh division of a given domain, the Voronoi diagram can be employed. The Voronoi diagram is a fundamental polygonal geometrical subdivision of a given region, where the point set with more nearest neighbors can make up the vertices of the diagram [55, 56]. The quality of the generated polygonal mesh strongly depends on the distribution of scattered seeding points. In order to improve the quality of polygonal mesh to increase the accuracy of the solution, the centroidal Vornoi technique [57, 58] can be used, in which the seed point generating each Voronoi cell can be used as its center of mass. Fig. (**3**) displays a polygonal mesh generated by the centroidal Vornoi technique in a complex L-shaped domain. For a particular polygonal element e displayed in Fig. (**3**), we need separately define an internal temperature field, a frame temperature field and a hybrid variation functional at element level.

Internal Temperature Field

The internal temperature field, also named as intra-element temperature field defined inside the element e, can be approximated by the linear combination of anisotropic fundamental solutions at different source points $\mathbf{y}_i\ (i = 1, 2, \cdots, m)$, that is

$$T(\mathbf{x}) = \sum_{i=1}^{m} c_i T^*(\mathbf{x}, \mathbf{y}_i) = \mathbf{N}(\mathbf{x})\mathbf{c}_e, \qquad \mathbf{x} \in \Omega_e / \Gamma_e,\ \mathbf{y}_i \notin \Omega_e \tag{27}$$

where m is the number of source points locating outside the element domain, $c_i\ (i = 1, 2, \cdots, m)$ present unknown coefficients, $T^*(\mathbf{x}, \mathbf{y}_i)$ is the induced temperature fundamental solution at the point $\mathbf{x}$ caused by the unit point source at the point $\mathbf{y}_i$, and

$$\mathbf{N}(\mathbf{x}) = \begin{Bmatrix} T^*(\mathbf{x}, \mathbf{y}_1) & T^*(\mathbf{x}, \mathbf{y}_2) & \cdots & T^*(\mathbf{x}, \mathbf{y}_m) \end{Bmatrix} \tag{28}$$

$$\mathbf{c}_e = \begin{Bmatrix} c_1 & c_2 & \cdots & c_m \end{Bmatrix}^{\mathrm{T}} \tag{29}$$

Next, differentiating the temperate field in Eq. (27) yields the following expression of normal heat flux

$$q_n(\mathbf{x}) = \mathbf{Q}(\mathbf{x})\mathbf{c}_e, \qquad \mathbf{x} \in \Omega_e / \Gamma_e,\ \mathbf{y}_i \notin \Omega_e \tag{30}$$

where

$$\mathbf{Q}(\mathbf{x}) = -\{n_1 \quad n_2\} \begin{bmatrix} k_{11} & k_{12} \\ k_{21} & k_{22} \end{bmatrix} \begin{bmatrix} \dfrac{\partial \mathbf{N}(\mathbf{x})}{\partial x_1} \\ \dfrac{\partial \mathbf{N}(\mathbf{x})}{\partial x_2} \end{bmatrix} \tag{31}$$

with

$$\frac{\partial \mathbf{N}(\mathbf{x})}{\partial x_i} = \left\{ \frac{\partial T^*(\mathbf{x}, \mathbf{y}_1)}{\partial x_i} \quad \frac{\partial T^*(\mathbf{x}, \mathbf{y}_2)}{\partial x_i} \quad \cdots \quad \frac{\partial T^*(\mathbf{x}, \mathbf{y}_m)}{\partial x_i} \right\}, \quad (i = 1, 2) \tag{32}$$

Frame Temperature Field

It is observed that the internal temperature approximation (27) is chosen so as to priori satisfy the governing partial differential equation (7) inside the element domain. However, it is non-conforming across the inter-element boundary. To overcome this drawback, in this study, the hybrid technique is employed. In this technique, the element e is linked to adjacent elements through an auxiliary conforming frame field independently defined on the element boundary which has the same form as in the conventional FEM and BEM. Here, the frame temperature field $\tilde{\boldsymbol{T}}$ is given by

$$\tilde{T}(\mathbf{x}) = \tilde{\mathbf{N}}(\xi)\mathbf{d}_e, \quad \mathbf{x} \in \Gamma_e \tag{33}$$

where

$$\tilde{\mathbf{N}}(\mathbf{x}) = \left[\tilde{\mathbf{N}}_1(\xi) \quad \tilde{\mathbf{N}}_2(\xi) \quad \cdots \quad \tilde{\mathbf{N}}_p(\xi) \right], \quad \xi \in [-1, 1] \tag{34}$$

is the shape function vector,

$$\mathbf{d}_e = \left[T_1 \quad T_2 \quad \cdots \quad T_p \right]^{\mathrm{T}} \tag{35}$$

is the nodal temperature vector, p represents the number of nodes for the polygonal element.

The local natural coordinate $\boldsymbol{\xi} \in [-\mathbf{1}, \mathbf{1}]$ is related to the global coordinate **x** on the element boundary $\boldsymbol{\Gamma_e}$ by the following expression in matrix form

$$\mathbf{x}(\xi) = \tilde{\mathbf{N}}(\xi)\mathbf{x}_e, \quad \mathbf{x} \in \Gamma_e \tag{36}$$

where $\mathbf{x}_e$ is the nodal coordinate matrix

$$\mathbf{x}_e = \begin{bmatrix} \mathbf{x}_1^{\mathrm{T}} \\ \mathbf{x}_2^{\mathrm{T}} \\ \vdots \\ \mathbf{x}_p^{\mathrm{T}} \end{bmatrix} = \begin{bmatrix} x_{11} & x_{12} \\ x_{21} & x_{22} \\ \vdots & \vdots \\ x_{p1} & x_{p2} \end{bmatrix} \tag{37}$$

For the n-sided polygonal element with n edges and p nodes, if there are two nodes on each edge, we have $\boldsymbol{n} = \boldsymbol{p}$. For such case, the linear interpolation scheme is employed. For example, when the boundary point **x** locates on the ith edge ($\boldsymbol{i} = \mathbf{1}, \mathbf{2}, \cdots, \boldsymbol{n} - \mathbf{1}$),

$$\begin{aligned} &\tilde{\mathbf{N}}_i(\xi) = \tilde{N}_1(\xi), \;\; \tilde{\mathbf{N}}_{i+1}(\xi) = \tilde{N}_2(\xi), \\ &\tilde{\mathbf{N}}_j(\xi) = 0 \quad (j \neq i,\, i+1) \end{aligned} \tag{38}$$

while when the boundary point x locates on the nth edge,

$$\begin{aligned} &\tilde{\mathbf{N}}_n(\xi) = \tilde{N}_1(\xi), \;\; \tilde{\mathbf{N}}_1(\xi) = \tilde{N}_2(\xi), \\ &\tilde{\mathbf{N}}_j(\xi) = 0 \quad (j \neq n,\, 1) \end{aligned} \tag{39}$$

where $\tilde{\boldsymbol{N}}_i$ $(\boldsymbol{i} = \mathbf{1}, \mathbf{2})$ stand for the conventional one-dimensional shape functions expressed in terms of the local natural coordinate $\boldsymbol{\xi}$, that is

$$\tilde{N}_1(\xi) = \frac{1-\xi}{2}, \quad \tilde{N}_2(\xi) = \frac{1+\xi}{2}, \qquad (-1 \leq \xi \leq 1) \tag{40}$$

Certainly, a more accurate approximation of the frame temperature solution can be taken in Eqs. (33) and (36) by employing the following quadratic shape functions on each element edge [19]

$$\tilde{N}_1(\xi) = -\frac{\xi(1\ \ \xi)}{2}, \quad \tilde{N}_2(\xi) = 1 - \xi^2, \quad \tilde{N}_3(\xi) = \frac{\xi(1\ \ \xi)}{2}, \quad (-1 \le \xi \le 1) \tag{41}$$

For this case, there are three nodes on each element edge, so $\boldsymbol{p} = \mathbf{2}\boldsymbol{n}$. When the boundary point **x** locates on the *i*th edge ($\boldsymbol{i} = \mathbf{1}, \mathbf{2}, \cdots, \boldsymbol{n} - \mathbf{1}$),

$$\begin{gathered} \tilde{\mathbf{N}}_{2i-1}(\xi) = \tilde{N}_1(\xi), \ \tilde{\mathbf{N}}_{2i}(\xi) = \tilde{N}_2(\xi), \ \tilde{\mathbf{N}}_{2i+1}(\xi) = \tilde{N}_3(\xi), \\ \tilde{\mathbf{N}}_j(\xi) = 0 \quad (j \ne 2i-1, 2i, 2i+1) \end{gathered} \tag{42}$$

while when the boundary point **x** locates on the *n*th edge,

$$\begin{gathered} \tilde{\mathbf{N}}_{2n-1}(\xi) = \tilde{N}_1(\xi), \ \tilde{\mathbf{N}}_{2n}(\xi) = \tilde{N}_2(\xi), \ \tilde{\mathbf{N}}_1(\xi) = \tilde{N}_3(\xi), \\ \tilde{\mathbf{N}}_j(\xi) = 0 \quad (j \ne 2n-1, 2n, 1) \end{gathered} \tag{43}$$

Hybrid Variational Functional

To enforce the inter-element continuity and apply the specific boundary conditions, the following double-variable hybrid functional for the element *e* is defined by an integral form [59]

$$\begin{aligned} \Pi_e = &-\frac{1}{2}\int_{\Omega_e}\left[k_{11}\left(\frac{\partial T}{\partial x_1}\right)^2 + 2k_{12}\frac{\partial T}{\partial x_1}\frac{\partial T}{\partial x_2} + k_{22}\left(\frac{\partial T}{\partial x_2}\right)^2\right]\mathrm{d}\Omega \\ &-\int_{\Gamma_{qe}}\bar{q}\tilde{T}\mathrm{d}\Gamma + \int_{\Gamma_e} q_n\left(\tilde{T} - T\right)\mathrm{d}\Gamma - \frac{1}{2}\int_{\Gamma_{ce}} h_\infty\left(\tilde{T} - T_\infty\right)^2\mathrm{d}\Gamma \end{aligned} \tag{44}$$

which can also be written in tensor form for the sake of convenience

$$\Pi_e = -\frac{1}{2}\int_{\Omega_e} k_{ij}T_{,i}T_{,j}\mathrm{d}\Omega - \int_{\Gamma_{qe}}\bar{q}\tilde{T}\mathrm{d}\Gamma + \int_{\Gamma_e} q_n\left(\tilde{T} - T\right)\mathrm{d}\Gamma - \frac{1}{2}\int_{\Gamma_{ce}} h_\infty\left(\tilde{T} - T_\infty\right)^2\mathrm{d}\Gamma \tag{45}$$

In Eqs. (44) and (45), the boundary Γ_e of the element e is assumed to consist of the following parts

$$\Gamma_e = \Gamma_{Te} \cup \Gamma_{qe} \cup \Gamma_{Ie} \tag{46}$$

in which

$$\Gamma_{ue} = \Gamma_e \cap \Gamma_u, \ \Gamma_{qe} = \Gamma_e \cap \Gamma_q, \ \Gamma_{ce} = \Gamma_e \cap \Gamma_c \tag{47}$$

and Γ_{Ie} represents the inter-element boundary of the element *e* shown in Fig. (**4**).

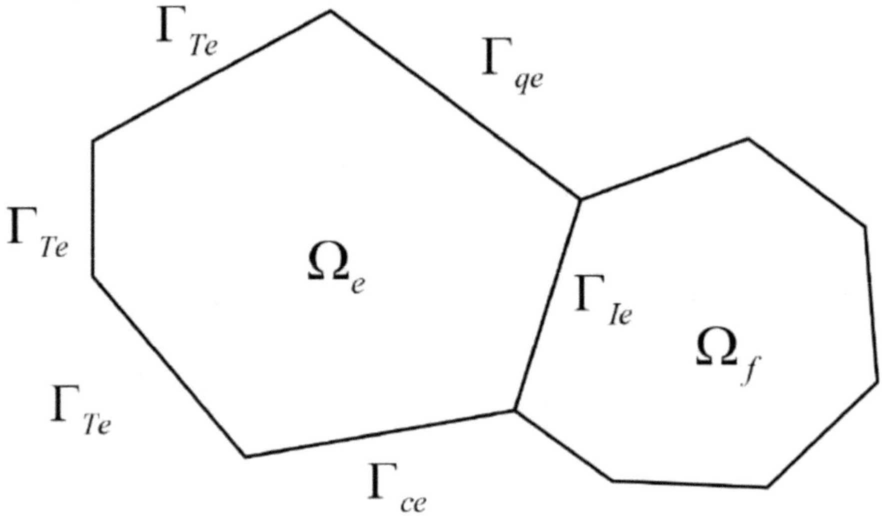

Fig. (4). Element boundary for a hybrid finite element.

Additionally, the element *e* is connected to the adjacent element, *i.e. f*, thus the continuity condition is required on the common boundary $\boldsymbol{\Gamma_{Ief}}$ between the two adjacent elements *e* and *f*

$$\left.\begin{aligned} T_e &= T_f \quad \text{(conformity)} \\ q_{ne} + q_{nf} &= 0 \quad \text{(reciprocity)} \end{aligned}\right\} \text{ on } \Gamma_{Ief} = \Gamma_e \cap \Gamma_f \tag{48}$$

In the functional (45), the governing equation (7) is assumed to be satisfied, *a priori*, for deriving the present hybrid finite element model. Next we will show that the stationary condition of the functional (45) leads to the governing equation (Euler equation), the boundary conditions and the continuity conditions.

Proof

The first-order variational of the functional (45) yields

$$\begin{aligned}\delta\Pi_e = &-\int_{\Omega_e} k_{ij} T_{,i} \delta T_{,j} \mathrm{d}\Omega - \int_{\Gamma_{qe}} \bar{q} \delta \tilde{T} \mathrm{d}\Gamma + \int_{\Gamma_e} \left(\delta \tilde{T} - \delta T\right) q_n \mathrm{d}\Gamma \\ &+ \int_{\Gamma_e} \left(\tilde{T} - T\right) \delta q_n \mathrm{d}\Gamma - \int_{\Gamma_{ce}} h_\infty \left(\tilde{T} - T_\infty\right) \delta \tilde{T} \mathrm{d}\Gamma\end{aligned} \quad (49)$$

Using the relations

$$q_n = -k_{ij} T_{,j} n_i = -k_{ij} T_{,i} n_j \quad (50)$$

$$k_{ij} \delta T_{,i} T_{,j} = k_{ij} T_{,i} \delta T_{,j} \quad (51)$$

and the Gauss divergence theorem for any smooth function $\boldsymbol{f}$ in the domain

$$\int_\Omega f_{,i} \mathrm{d}\Omega = \int_\Gamma f n_i \mathrm{d}\Gamma \quad (52)$$

we have

$$\begin{aligned}-\int_{\Omega_e} k_{ij} T_{,i} \delta T_{,j} \mathrm{d}\Omega &= -\int_{\Omega_e} \left\{ (k_{ij} T_{,i} \delta T)_{,j} - (k_{ij} T_{,i})_{,j} \delta T \right\} \mathrm{d}\Omega \\ &= \int_{\Gamma_e} -k_{ij} T_{,i} n_j \delta T \mathrm{d}\Gamma + \int_{\Omega_e} (k_{ij} T_{,i})_{,j} \delta T \mathrm{d}\Omega \\ &= \int_{\Gamma_e} q_n \delta T \mathrm{d}\Gamma + \int_{\Omega_e} (k_{ij} T_{,i})_{,j} \delta T \mathrm{d}\Omega\end{aligned} \quad (53)$$

Further, the variational functional can be written as

$$\begin{aligned}\delta\Pi_e = &\int_{\Omega_e} (k_{ij} T_{,i})_{,j} \delta T \mathrm{d}\Omega + \int_{\Gamma_{qe}} \left(q_n - \bar{q}\right) \delta \tilde{T} \mathrm{d}\Gamma + \int_{\Gamma_{Te}} q_n \delta \tilde{T} \mathrm{d}\Gamma + \int_{\Gamma_{Ie}} q_n \delta \tilde{T} \mathrm{d}\Gamma \\ &+ \int_{\Gamma_e} \left(\tilde{T} - T\right) \delta q_n \mathrm{d}\Gamma + \int_{\Gamma_{ce}} \left\{ q_n - h_\infty \left(\tilde{T} - T_\infty\right) \right\} \delta \tilde{T} \mathrm{d}\Gamma\end{aligned} \quad (54)$$

For the displacement-based method, the potential conformity should be satisfied in advance, that is,

$$\begin{aligned}&\delta \tilde{T} = 0 && \text{on } \Gamma_{Te} && (\because \tilde{T} = \bar{T}) \\ &\delta \tilde{T}^e = \delta \tilde{T}^f && \text{on } \Gamma_{Ief} && (\because \tilde{T}^e = \tilde{T}^f)\end{aligned} \quad (55)$$

Susbstituting Eq. (55) into Eq. (54) yields

$$\delta\Pi_e = \int_{\Omega_e} (k_{ij}T_{,i})_{,j}\,\delta T\mathrm{d}\Omega + \int_{\Gamma_{qe}} \left(q_n - \bar{q}\right)\delta\tilde{T}\mathrm{d}\Gamma + \int_{\Gamma_{Ie}} q_n\delta\tilde{T}\mathrm{d}\Gamma$$
$$+\int_{\Gamma_e}\left(\tilde{T}-T\right)\delta q_n\mathrm{d}\Gamma + \int_{\Gamma_{ce}}\left\{q_n - h_\infty\left(\tilde{T}-T_\infty\right)\right\}\delta\tilde{T}\mathrm{d}\Gamma \quad \textbf{(56)}$$

Using the stationary condition of the functional

$$\delta\Pi_e = 0 \quad \textbf{(57)}$$

the Euler equation in the domain **Ω** and the boundary conditions can be obtained

$$\begin{aligned} &\left(k_{ij}T_{,i}\right)_{,j} = 0 && \text{in } \Omega_e \\ &q_n = \bar{q} && \text{on } \Gamma_{qe} \\ &q_n = h_\infty\left(\tilde{T}-T_\infty\right) && \text{on } \Gamma_{ce} \end{aligned} \quad \textbf{(58)}$$

Element Stiffness Equation

With the Gauss divergence theorem (52) again to the functional (44), we have

$$\Pi_e = \frac{1}{2}\int_{\Omega_e}\left(k_{11}\frac{\partial^2 T}{\partial x_1^2} + 2k_{12}\frac{\partial^2 T}{\partial x_1\partial x_2} + k_{22}\frac{\partial^2 T}{\partial x_2^2}\right)T\mathrm{d}\Omega$$
$$-\int_{\Gamma_{qe}}\bar{q}\tilde{T}\mathrm{d}\Gamma - \frac{1}{2}\int_{\Gamma_e} q_n T\mathrm{d}\Gamma + \int_{\Gamma_e} q_n\tilde{T}\mathrm{d}\Gamma - \frac{1}{2}\int_{\Gamma_{ce}} h_\infty\left(\tilde{T}-T_\infty\right)^2\mathrm{d}\Gamma \quad \textbf{(59)}$$

Considering the fact that the approximated internal temperature field *priori* satisfies the governing equation (7), the functional (59) can be simplified as

$$\Pi_e = -\frac{1}{2}\int_{\Gamma_e} q_n T\mathrm{d}\Gamma - \int_{\Gamma_{qe}}\bar{q}\tilde{T}\mathrm{d}\Gamma + \int_{\Gamma_e} q_n\tilde{T}\mathrm{d}\Gamma - \frac{1}{2}\int_{\Gamma_{ce}} h_\infty\left(\tilde{T}-T_\infty\right)^2\mathrm{d}\Gamma \quad \textbf{(60)}$$

Then, the substitution of Eqs. (27) and (33) into Eq. (60) yields

$$\Pi_e = -\frac{1}{2}\mathbf{c}_e^{\mathrm{T}}\mathbf{H}_e\mathbf{c}_e - \mathbf{d}_e^{\mathrm{T}}\mathbf{g}_e + \mathbf{c}_e^{\mathrm{T}}\mathbf{G}_e\mathbf{d}_e - \frac{1}{2}\mathbf{d}_e^{\mathrm{T}}\mathbf{F}_e\mathbf{d}_e + \mathbf{d}_e^{\mathrm{T}}\mathbf{f}_e - \mathbf{a}_e \quad \textbf{(61)}$$

with

$$\mathbf{H}_e = \int_{\Gamma_e} \mathbf{Q}^{\mathrm{T}}\mathbf{N}\mathrm{d}\Gamma, \quad \mathbf{G}_e = \int_{\Gamma_e} \mathbf{Q}^{\mathrm{T}}\tilde{\mathbf{N}}\mathrm{d}\Gamma, \quad \mathbf{g}_e = \int_{\Gamma_{qe}} \tilde{\mathbf{N}}^{\mathrm{T}}\overline{q}\mathrm{d}\Gamma$$
$$\mathbf{F}_e = \int_{\Gamma_{ce}} h_\infty \tilde{\mathbf{N}}^{\mathrm{T}}\tilde{\mathbf{N}}\mathrm{d}\Gamma, \quad \mathbf{f}_e = \int_{\Gamma_{ce}} h_\infty T_\infty \tilde{\mathbf{N}}^{\mathrm{T}}\mathrm{d}\Gamma, \quad \mathbf{a}_e = \int_{\Gamma_{ce}} \frac{h_\infty T_\infty^2}{2}\mathrm{d}\Gamma \tag{62}$$

which can be evaluated by the standard one-dimensional Gaussian quadrature rules along the element boundary.

The minimization of the hybrid functional $\mathbf{\Pi}_e$ with respect to the unknown coefficients $\mathbf{c}_e$ and $\mathbf{d}_e$ respectively leads to

$$\frac{\partial \Pi_e}{\partial \mathbf{c}_e^{\mathrm{T}}} = -\mathbf{H}_e\mathbf{c}_e + \mathbf{G}_e\mathbf{d}_e = \mathbf{0}$$
$$\frac{\partial \Pi_e}{\partial \mathbf{d}_e^{\mathrm{T}}} = \mathbf{G}_e^{\mathrm{T}}\mathbf{c}_e - \mathbf{g}_e - \mathbf{F}_e\mathbf{d}_e + \mathbf{f}_e = \mathbf{0} \tag{63}$$

from which one obtains the optional relationship between $\mathbf{c}_e$ and $\mathbf{d}_e$ for the enforcement of inter-element continuity on the common element boundary

$$\mathbf{c}_e = \mathbf{H}_e^{-1}\mathbf{G}_e\mathbf{d}_e \tag{64}$$

and the element stiffness equations

$$\mathbf{k}_e\mathbf{d}_e = \mathbf{g}_e - \mathbf{f}_e \tag{65}$$

with the symmetric coefficient matrix $\mathbf{k}_e$

$$\mathbf{k}_e = \mathbf{G}_e^{\mathrm{T}}\mathbf{H}_e^{-1}\mathbf{G}_e - \mathbf{F}_e \tag{66}$$

Assembling the element stiffness equation (65) can give the final global stiffness equation (67) with the symmetric and sparse coefficient matrix $\mathbf{K}$

$$\mathbf{KD} = \mathbf{F} \tag{67}$$

which can be solved for determining the nodal temperature vector $\mathbf{D}$ after the specific nodal temperature constraint is introduced. Furthermore, the unknown coefficients $\mathbf{c}_e$ can be obtained using Eq. (64) to further evaluate the temperature and heat flux components at arbitrary point in the element domain.

Recovery of Constant Temperature Mode

However, it is noticed that the constant temperature mode associated with a vanishing heat flux is fully discarded from the fundamental solution so that the matrix $\mathbf{H}_e$ is of full rank and keeps invertible [24, 51]. Therefore, the temperature at any point inside the element e cannot be evaluated directly by Eq. (27). Here, to recover the discarded term in each element, Eq. (27) is modified by adding a constant term c_0

$$T(\mathbf{x}) = \mathbf{N}(\mathbf{x})\mathbf{c}_e + c_0 \tag{68}$$

Following the general treatment in [24, 51], the least square matching of T and $\tilde{T}$ at the p element nodes gives

$$\sum_{i=1}^{p}\left(\mathbf{N}\mathbf{c}_e + c_0 - \tilde{T}\right)^2\Big|_{\text{node } i} = \min \tag{69}$$

from which the stability term c_0 can be determined as

$$c_0 = \frac{1}{p}\sum_{i=1}^{p}\left(\tilde{T} - \mathbf{N}\mathbf{c}_e\right)\Big|_{\text{node } i} \tag{70}$$

Generation of Source Points

From Eq. (66), it is observed that the inverse of the matrix $\mathbf{H}_e$ have to be evaluated in the computation of the elementary stiffness matrix $\mathbf{k}_e$. The necessary condition for the matrix $\mathbf{H}_e$ to be of full rank is that the number of approximating kernels in Eq. (27) should satisfy the following condition

$$m \geq n_{dof} - 1 \tag{71}$$

where n_{dof} is the total number of degrees of freedom of the polygonal element, which is specially equal to the number of nodes p of the element for the heat conduction problem.

Practically, an optimal number of source points are chosen based on the number of nodes p of the polygonal element, that is $m = p$, so that the rank condition (71) can be automatically achieved for the heat conduction problem. For such typical case, as indicated in Fig. (**5**), the simplest way to generate these source points is

$$\mathbf{y}_i = \mathbf{x}_i^b + \gamma(\mathbf{x}_i^b - \mathbf{x}_c), \quad i = 1, 2, \cdots, p \tag{72}$$

where $\mathbf{x}_i^b$ is the coordinates of the boundary node i and $\mathbf{x}_c$ is the coordinates of the central point of the element. Here, a dimensionless positive parameter γ is introduced to control the distance between the source point and the element boundary. For the polygonal element of interest, the location of centroid $\mathbf{x}_c$ can be evaluated using

$$\mathbf{x}_c = \frac{1}{p}\sum_{i=1}^{p}\mathbf{x}_i^b \tag{73}$$

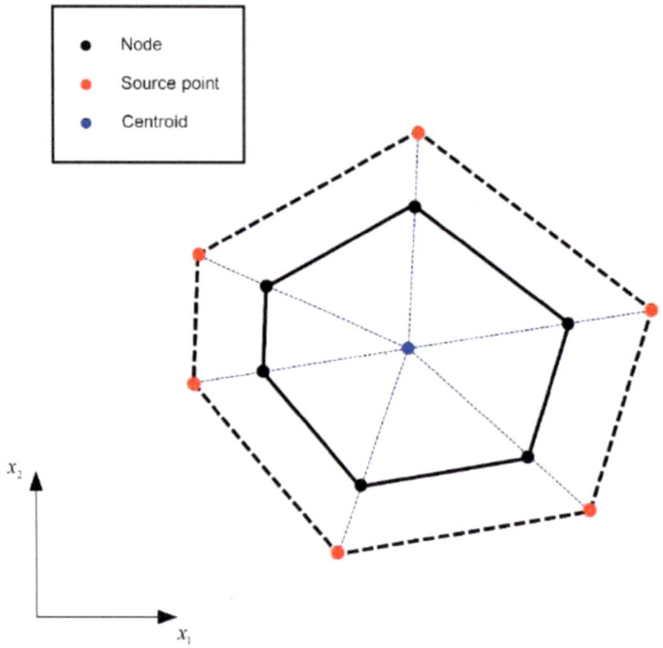

Fig. (5). Schematic of the generation of source points by element nodes.

NUMERICAL EXAMPLES

In this section, four examples are solved to demonstrate the efficiency, accuracy and convergence properties of the present algorithm using n-sided hybrid finite elements. All physical quantities used have consistent units, *i.e.* the units of geometrical length, thermal conductivity, convection coefficient, temperature, heat flux, are m, W/(m·K), W/(m^2·K), K or °C, W/m^2, respectively. In order to properly assess the numerical results, the error norm or average relative error (Arerr) on an arbitrary smooth variable f is defined as

$$\text{Arerr}(f) = \sqrt{\frac{\sum_{i=1}^{m}\left(f_h - f_e\right)_i^2}{\sum_{i=1}^{m}\left(f_e\right)_i^2}} \tag{74}$$

where m is the number of test points in the computing domain, f_h the hybrid finite element solution, and f_e the exact solution.

Anisotropic Heat Conduction in Circular Plate

In the first example, the anisotropic heat transfer in a unit circle is considered to test the accuracy, convergence and computing efficiency of the present method. The thermal conductivities of the anisotropic material are

$$\mathbf{k} = \begin{bmatrix} 1 & 2 \\ 2 & 5 \end{bmatrix} \tag{75}$$

Correspondingly, the exact solution of the problem is given by

, 3

$$T\left(x_1\ x_2\right) = x_1^2 - x_2^2 + x_1 x_2 \tag{76}$$

which is also used to apply the Dirichlet boundary condition on the circular boundary, as shown in Fig. (**6**).

The circular region is modeled with 4-sided hybrid finite element firstly for the purpose of comparison to the conventional quadrilateral finite element. The number of 4-sided hybrid elements are 21, 46 and 150, which correspond to 28, 57 and 169 nodes, respectively, as indicated in Fig. (**7**). In the present hybrid element, one of open issues is the position of source points located outside the element. Fig. (**8**) gives the variations of temperature norm at all nodes to the dimensional parameter γ controlling the distance of source points to the element boundary, and it is observed that there are plateau stages to achieve steady results for all meshing schemes. In practical computation, the value of the dimensionless parameter γ can be taken to be 20 for the quadrilateral hybrid element. Additionally, a good convergence is demonstrated in Fig. (**8**) as the error norm in temperature decreases with the increase of number of nodes. Subsequently, the results of temperature and heat flux components are tabulated in Table **1**, from which it is indicated that the present hybrid finite element can produce better results of heat flux than the conventional finite element.

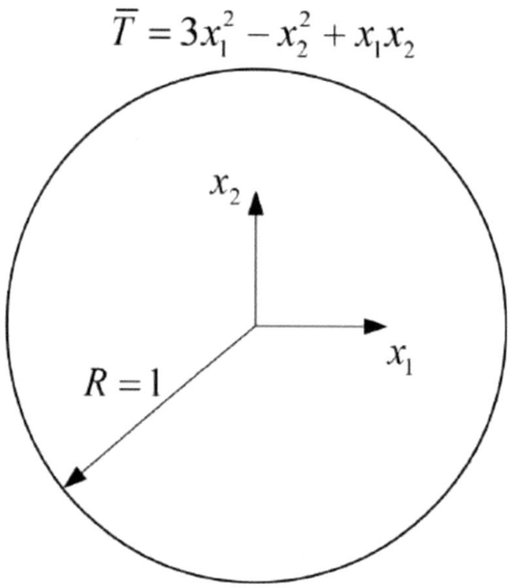

Fig. (6). The unit circular computing domain and the applied boundary conditions.

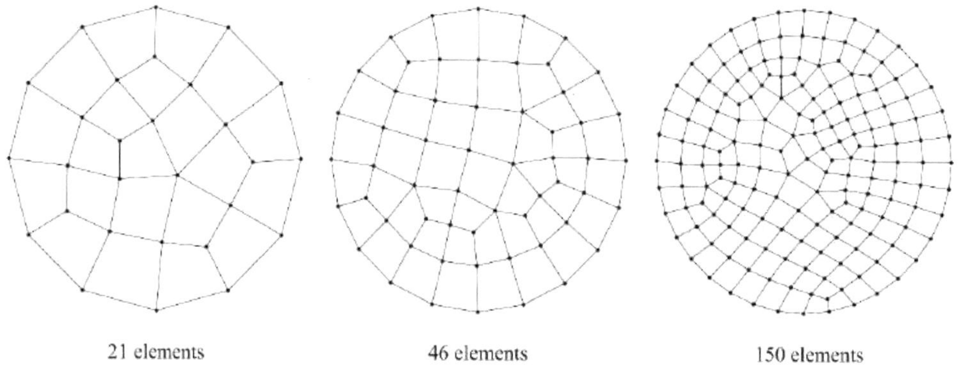

Fig. (7). Different divisions using quadrilateral hybrid elements for the circular region.

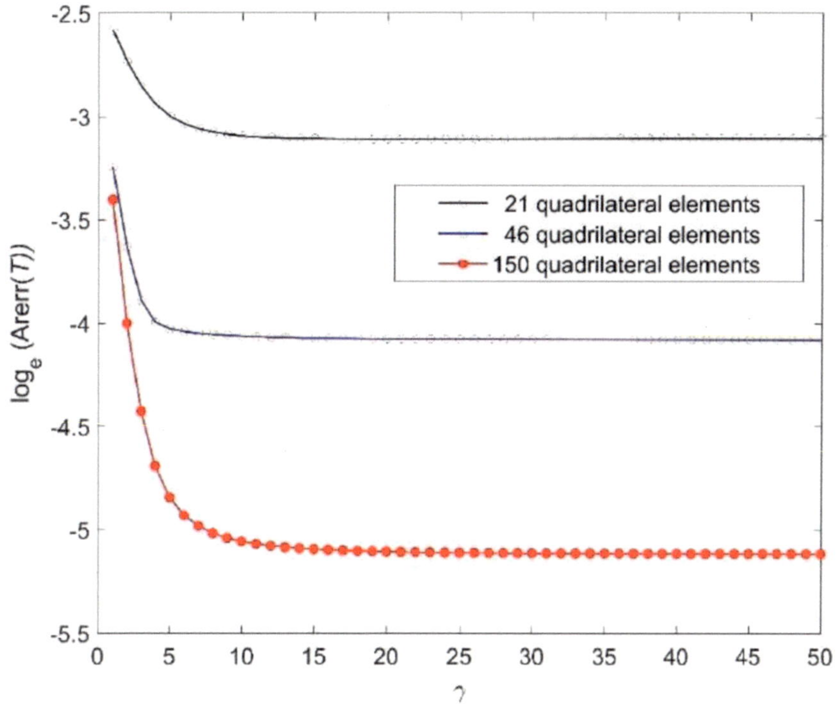

Fig. (8). Effect of the dimensionless parameter γ for the present quadrilateral hybrid elements.

Table 1. Comparison of results from the quadrilateral finite element and hybrid element.

Elements	**Finite Element**			**Hybrid Element**		
	Arerr(T)	Arerr(q_1)	Arerr(q_2)	Arerr(T)	Arerr(q_1)	Arerr(q_2)
21	3.89E-2	2.13E-1	2.14E-1	4.18E-2	1.25E-1	1.28E-1
46	1.41E-2	1.18E-1	1.20E-1	1.69E-2	7.14E-2	7.40E-2
150	4.82E-3	5.92E-2	5.85E-2	5.00E-3	3.95E-2	4.11E-2

Apart from the quadrilateral hybrid element, another interesting issue of the present hybrid method is that the construction of polygonal element with arbitrary numbers of sides can be flexibly implemented without any difficulty. To illustrate this feature, the circular region is modeled using polygonal elements, as shown in Fig. (**9**). The number of nodes involved are 38, 90, 164 for the cases of 20, 46, 83 polygonal elements, respectively. Correspondingly, the variations of temperature norm are displayed in Fig. (**10**), from which it is seen that relatively steady results can be achieved when the value of the parameter γ exceeds 5. Moreover, a good convergence is observed in Fig. (**10**) with the increase of the number of elements. For performing more detailed investigation of convergence and computing efficiency, Figs. **11** and **12** displays the variations of error norm in temperature and heat flux component against the total degrees of freedom (tdofs) of meshing. In Figs. (**11** and **12**), both the quadrilateral hybrid element and the polygonal hybrid element show good convergence in temperature and heat flux. Typically, in Fig.(**11**), the polygonal hybrid element shows slightly lower accuracy than the quadrilateral hybrid element. This can be attributed to the fact that the polygonal mesh possessing multiple-node connection feature includes less boundary nodes than the quadrilateral mesh when the same number of nodes are involved, so the variation of boundary conditions may not be fully captured for the polygonal mesh. For instance, the ratio of boundary nodes to all nodes respectively is 0.3158, 0.2111, 0.1829 only for the three polygonal meshes in Fig. (**9**), while it is 0.4286, 0.3509, 0.2130 for the quadrilateral meshes in Fig. (**7**). On the other hand, the quadrilateral and polygonal meshes in Fig. (**11**) have similar convergence rate (the average slope), which corresponds to 2.23, and 2.51, respectively. Comparatively, the convergence rate for the heat flux component in Fig. **12** is 1.28 and 1.79 for the quadrilateral and polygonal hybrid elements, respectively.

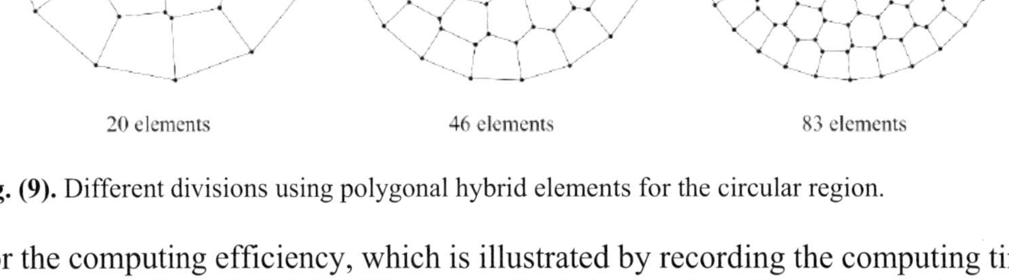

Fig. (9). Different divisions using polygonal hybrid elements for the circular region.

For the computing efficiency, which is illustrated by recording the computing time of the generation of global stiffness matrix, it is found in Fig. (**13**) that both the polygonal and quadrilateral hybrid elements are remarkably faster than the conventional finite element. This can be attributed to the element boundary integration scheme in the present method. Finally, the distributions of the temperature T and the heat flux component q_1 are plotted in Fig. (**14**) when 83 polygonal hybrid finite elements are used. From Fig. (**14**), very good agreements for temperature and heat flux distributions are observed between the numerical results and the exact solutions.

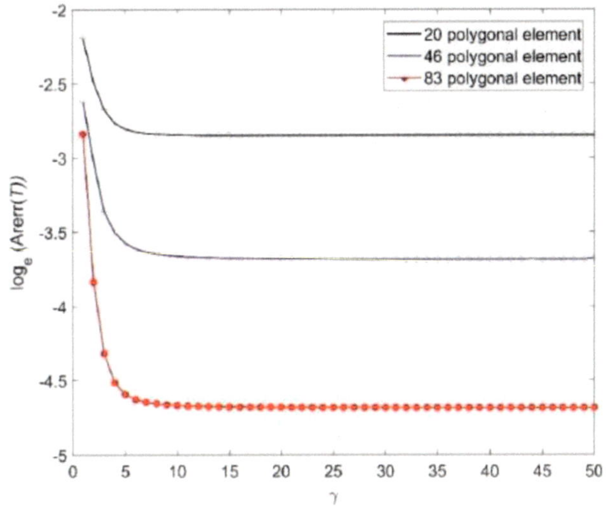

Fig. (10). Effect of the dimensionless parameter γ for polygonal hybrid elements.

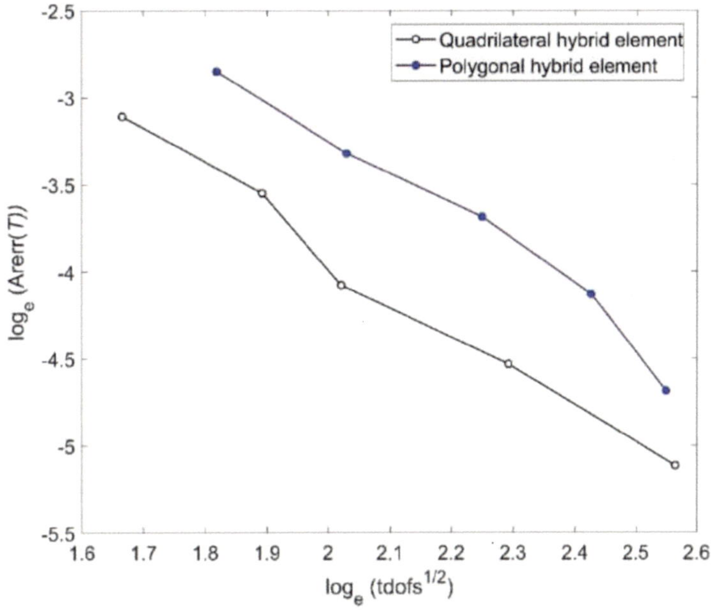

Fig. (11). Convergence results of the error norm in temperature for the present hybrid elements.

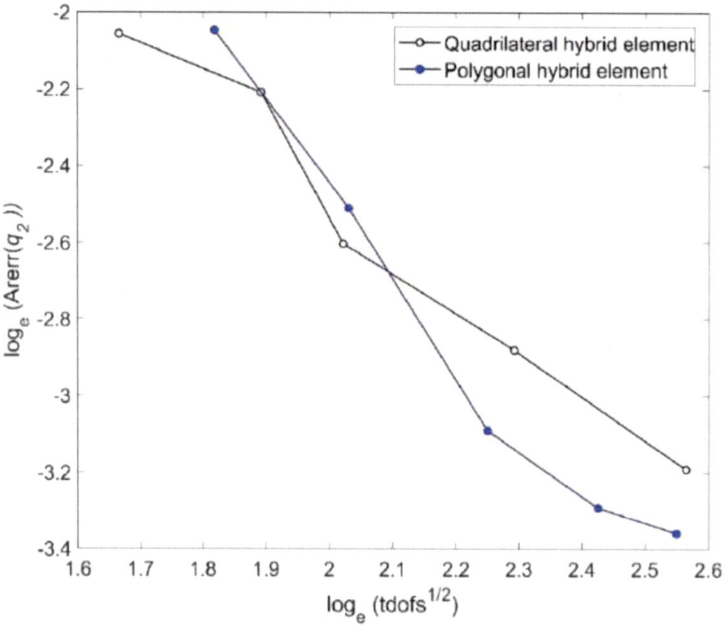

Fig. (12). Convergence results of the error norm in the heat flux component q_2 for the present hybrid elements.

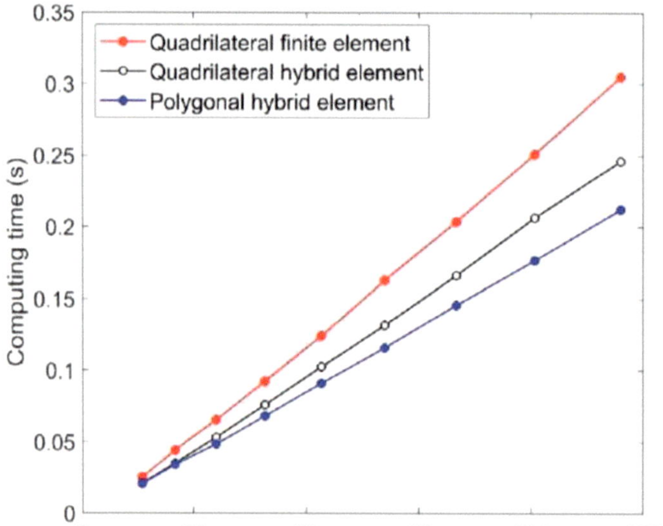

Fig. (13). Comparison of computing efficiency for finite element and hybrid element.

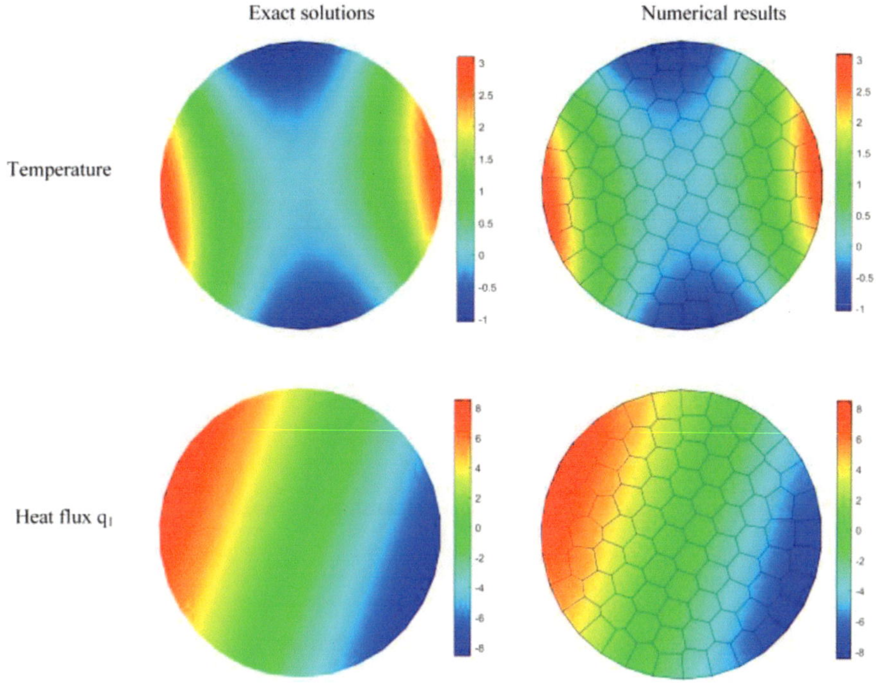

Fig. (14). Distributions of temperature and heat flux component in the circular region.

Anisotropic Heat Conduction in Square Plate

The second example is designed to investigate the capability of the present method over the non-convex elements and the non-conforming discretization in a unit square domain. Under the given anisotropic thermal conductivities

$$\mathbf{k} = \begin{bmatrix} 1 & 0 \\ 0 & 2 \end{bmatrix} \tag{77}$$

and the specified temperature and heat flux boundary conditions (see Fig. **15**), the exact solution of the problem is

$$T(x_1, x_2) = 2x_1^2 - x_2^2 \tag{78}$$

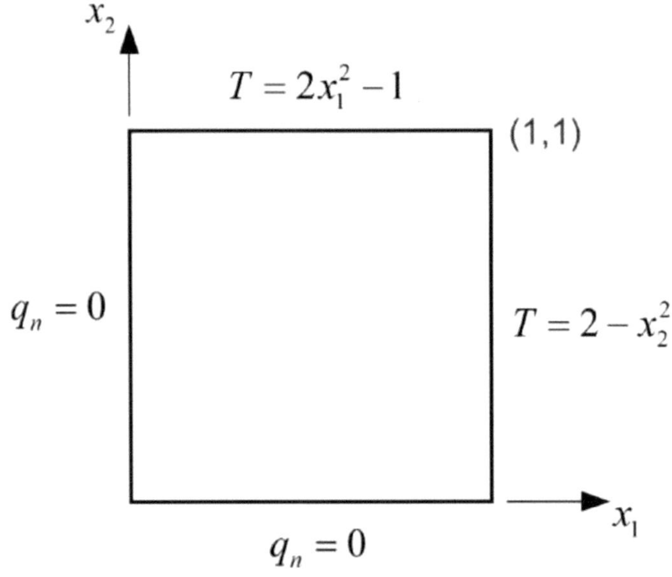

Fig. (15). The anisotropic square plate under specified boundary conditions.

We now study a quantitative test where a non-convex polygon discretization in Fig. **(16)** is made in the square domain. The discretization includes four 8-sided polygons with 21 nodes. The parameter *d* is used to control the shape of non-convex polygons and the parameter *a* denotes the side length of regular element. Typically, the polygons become regular when d/a=0. Here, the result from the regular elements is taken as reference value. Correspondingly, the average relative error of temperature variable for various d is listed in Table **2**, from which very small deviations to the reference result are observed for all cases of non-convex polygons. Therefore, the present method can effectively handle the non-convex behavior of polygonal elements.

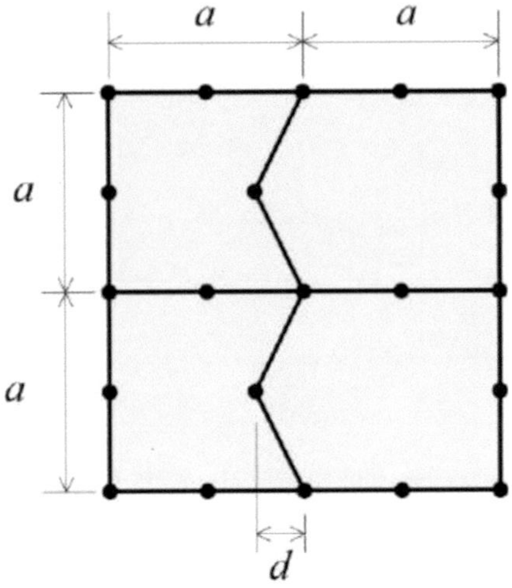

Fig. (16). The non-convex polygon discretization in the anisotropic square plate.

Table 2. Comparison of non-convex and normal polygonal elements.

d/a	0.0	0.2	0.4	0.6
Arerr(T)	0.01976	0.01996	0.01991	0.01974

Next the numerical convergence of the relative error in temperature norm is investigated by comparing the results from three mesh divisions, as shown in Fig. (**17**). Total 22, 42 and 62 nodes are involved for the three meshing configurations, including 12, 17 and 21 boundary nodes, respectively. The results of temperature in Fig. (**18**) show that relatively steady results can be achieved when the value of the parameter γ exceeds 5. Moreover, it is seen from Fig. (**18**) that the numerical results from the present method converge to the exact solutions as the mesh is refined. To clearly demonstrate the rate of convergence in temperature, the logarithm relation of the total degrees of freedom (tdofs) of hybrid mesh and the relative error norm Arerr(T) is plotted in Fig. (**19**), from which the average slope of the convergence line is evaluated as 1.71.

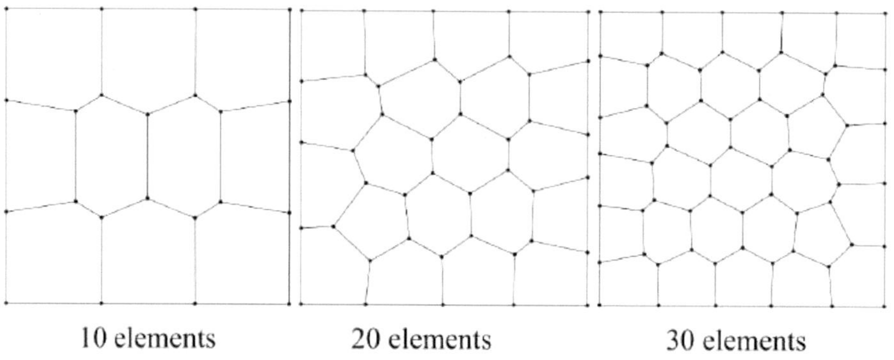

Fig. (17). Three divisions using polygonal hybrid elements for the square region.

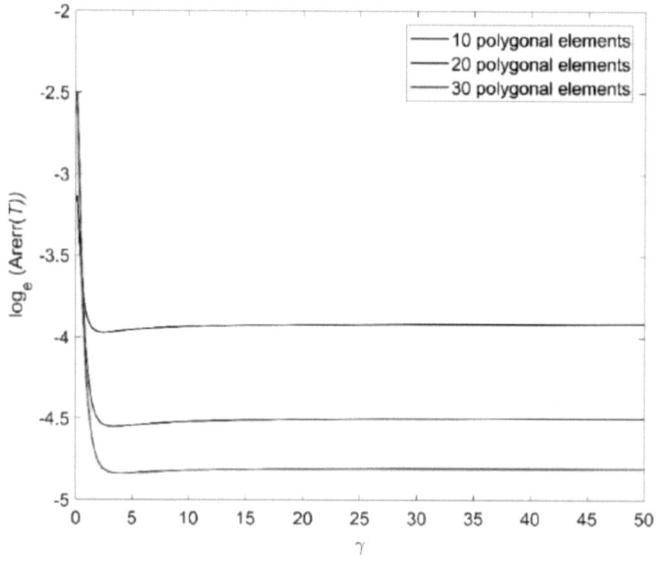

Fig. (18). Effect of the dimensionless parameter γ for polygonal hybrid elements.

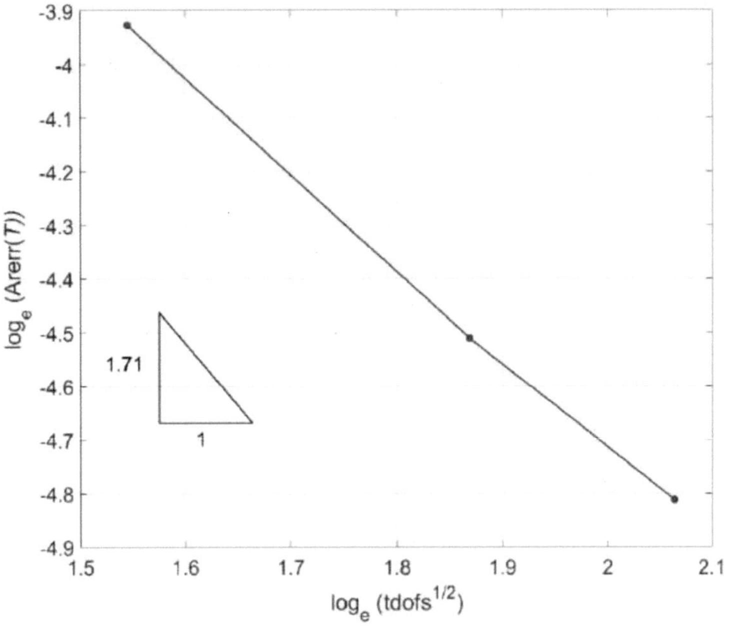

Fig. (19). Rate of convergence in temperature for the present polygonal hybrid element in Example 2.

Subsequently, an assessment of non-conforming discretization is carried out in the implementation of the present method, which has a clear edge over the conventional FEM in this aspect. Fig. **(20)** shows a non-conforming discretization generated by applying the polytree rule [60] to the initial polygonal mesh including 20 elements and 42 nodes, in which a *father* element with n nodes is subdivided into $n+1$ *children* elements, and then each new element can been divided again to generate other children elements. In Fig. **(20)**, the blue points denote the newly generated interior and hanging nodes and the hanging nodes are defined at the intersection between element edges. For such non-conforming mesh, the HFS-FEM employs a more straightforward and normal way to handle the elements containing hanging nodes. In the HFS-FEM, a hanging node is simply understood as a subdivision of an edge, and as a result, the element including m hanging nodes and n nodes is regarded as a normal element with ($n+m$) nodes, and the element boundary integrals in the element stiffness matrix are normally evaluated edge by edge without any special treatment. With the locally refined mesh in Fig. **(20)**, the model is solved with $\gamma = 15$ and the relative error norms of temperature T and heat flux component q_1 are 0.0101 and 0.0648, respectively, which are smaller than the reference results 0.0110 and 0.0919 from the initial mesh, as expected. Finally, the temperature distribution in the square domain is plotted in Fig. **(21)** with 30 polygonal elements in Fig. (**17**), and it is found that the numerical results show great agreement to the exact solutions. Also, in order to demonstrate the applicability of the present method for treating the convection condition, the left side of the square is changed into a convective side with a convection coefficient 50 and an environmental temperature 25. With the locally refined mesh division developed based on the 30 polygonal elements given in Fig. (**17**), the temperature distribution is plotted in Fig. (**22**), from which it is observed that the simulated results from the HFS-FEM show good consistence to the reference results from ABAQUS with 400 linear quadrilateral elements.

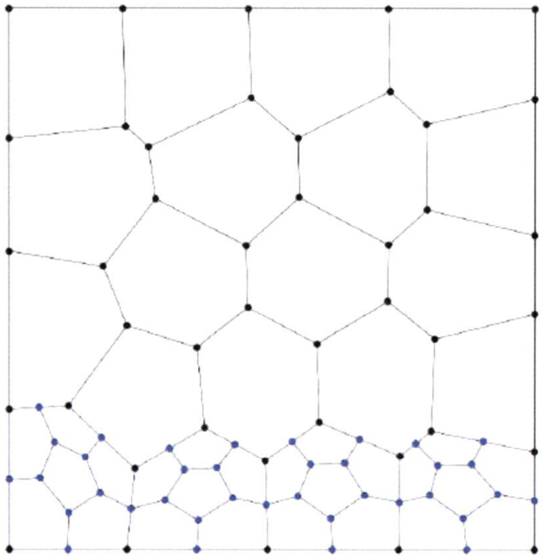

Fig. (20). Non-conforming polygon discretization in the square.

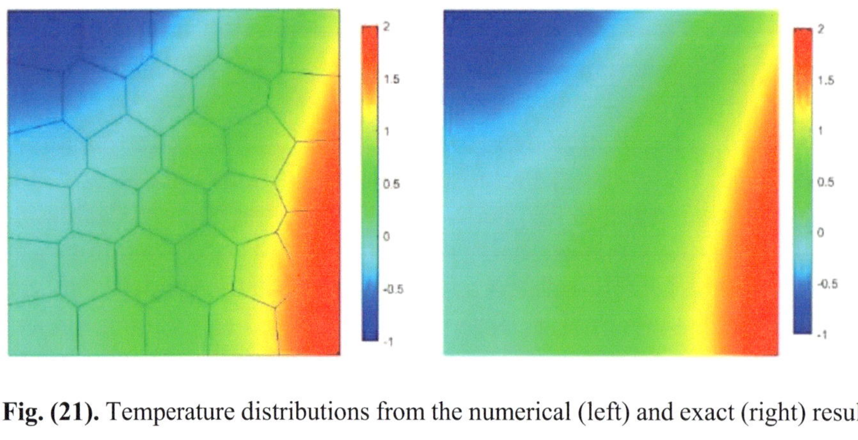

Fig. (21). Temperature distributions from the numerical (left) and exact (right) results in the square.

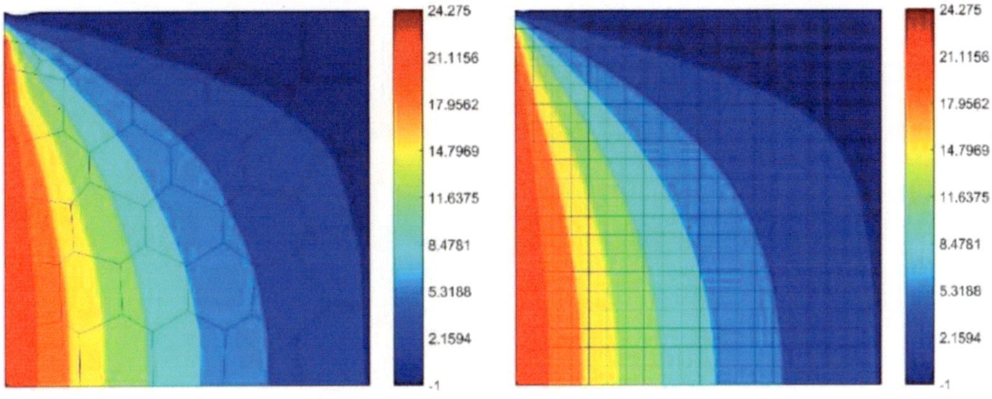

Fig. (22). Temperature distributions from the simulated (left) and reference (right) results in the square with applied convection condition.

Anisotropic Heat Conduction In A Retangular Plate

The third example is used to investigate the effect of ply angle θ of orthotropic principal direction. For simplicity, a rectangular plate with dimensions [0, 3]×[0, 1] is taken into consideration. The temperature values 100°C and 0°C are applied on the left and right side of the plate, respectively, and others keep insulated, as shown in Fig. (**23**).

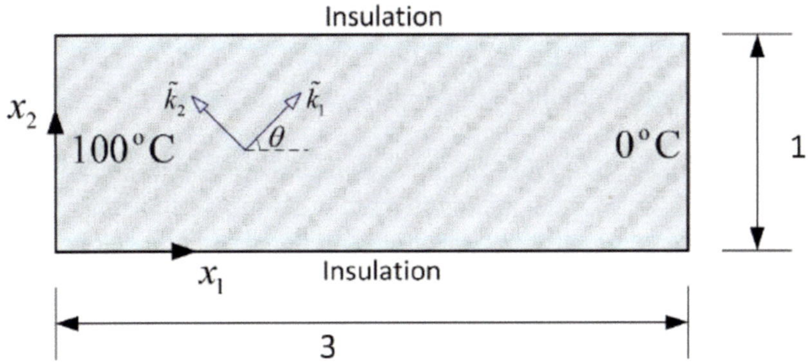

Fig. (23). Anisotropic rectangular domain with specific boundary conditions.

The used orthotropic thermal conductivities are

$$\tilde{\mathbf{k}} = \begin{bmatrix} 46.2 & 0 \\ 0 & 0.72 \end{bmatrix} \text{W/m/K} \tag{79}$$

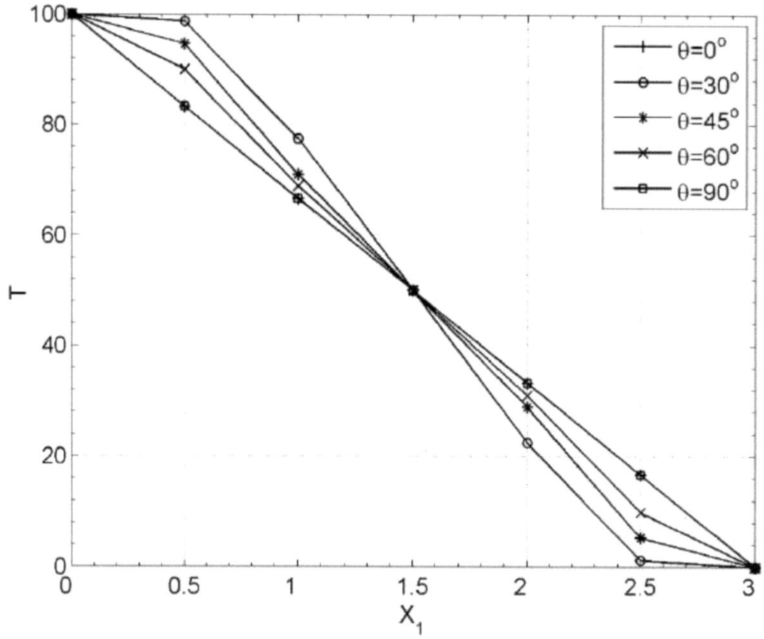

Fig. (24). Ply angle effect on temperature along the line $x_1 = 0.5$.

Total 3 anisotropic quadratic hybrid finite elements are used and the variation of temperature along the line $x_1 = 0.5$ is plotted in Fig. (**24**) for different ply angle θ. It is observed that the ply angle θ is important for the temperature variation in the rectangular plate. For the cases of $\theta = 0$°C and 90°C, which correspond to pure orthotropic materials, the temperature field shows same linear distribution along the x_1 direction, whereas for other cases, which correspond to full anisotropic materials, the temperature field displays nonuniform variation along the two coordinate directions due to the existence of k_{12}. Moreover, compared to the cases of pure orthotropy, the temperature symmetry on the axis $x_1 = 0.5$ is destroyed for the cases of full anisotropy, although the geometrical symmetry and the boundary condition symmetry keep true. Besides, the temperature distributions in the rectangular plate are plotted in Fig. (**25**) for various ply angles. It is observed that the ply angle significantly affects the distribution of temperature.

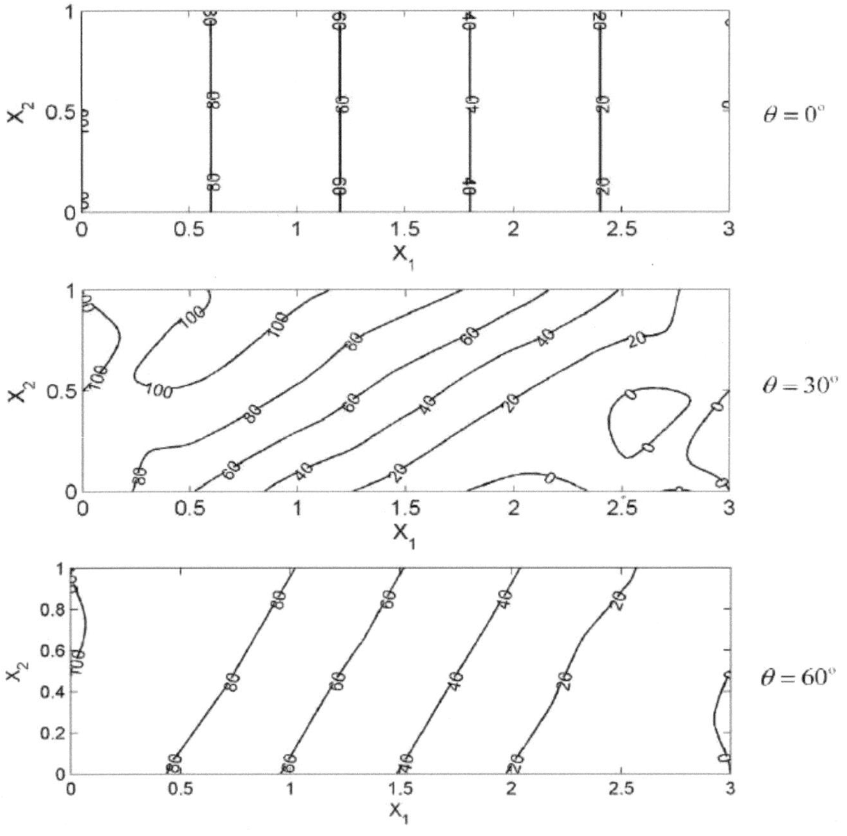

Fig. (25). Ply angle effect on temperature in the rectangular plate.

Anisotropic Heat Conduction in Plane Hook Domain

As the last example, an anisotropic hook plate with relatively complex geometrical configuration is modelled to validate the capability of the HFS-FEM to handle complex structure with the present *n*-sided polygonal elements. As shown in Fig. **(26)**, the temperature boundary condition is assumed as $T = 0$ on the outer face and $T = 3$ on the inner face, respectively. The given orthotropic material parameters are

$$\tilde{\mathbf{k}} = \begin{bmatrix} 1 & 0 \\ 0 & 5 \end{bmatrix} \text{W/m/K} \tag{80}$$

which means that the anisotropy of model can be represented by adjusting the ply angle θ in Eq. (11).

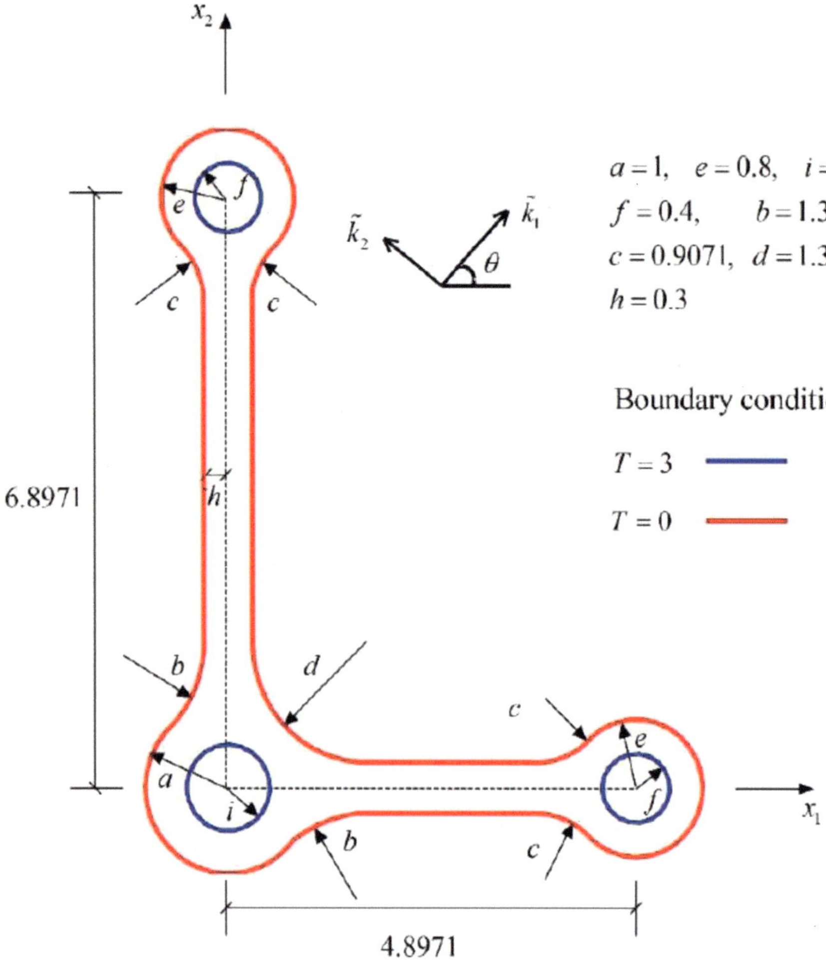

Fig. (26). Anisotropic plane hook domain.

In this calculation, the hook domain is modeled with 500 polygonal elements and 1026 nodes. Moreover, the finite element solutions from ABAQUS are provided for comparison, in which the hook region is discretized with 600 quadrilateral elements and 1028 nodes. The color maps of temperature distribution are shown in Fig. (**27**) for different ply angles and it is observed that the numerical results from the HFS-FEM have good agreement to the FEM results.

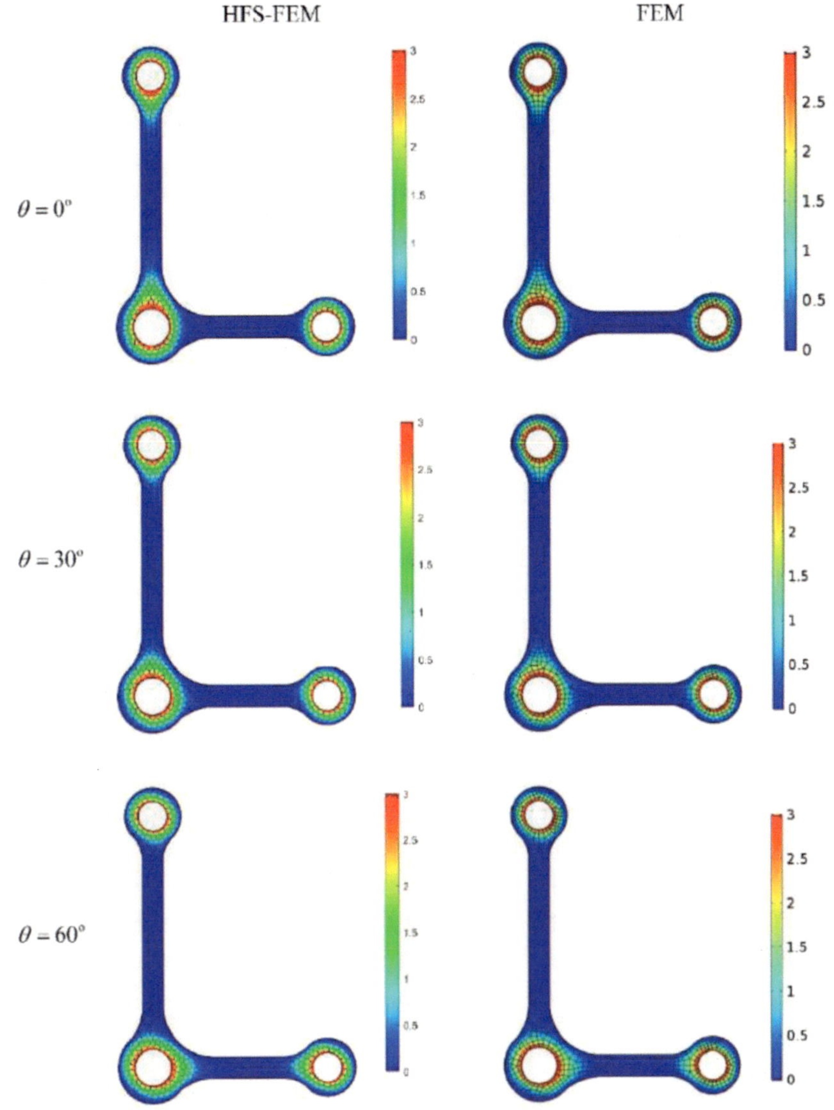

Fig. (27). The caused temperature distributions in the hook region for different ply angles.

CONCLUDING REMARKS

In this chapter, the performance of accuracy, convergence, computational efficiency and easiness for treating non-convext and non-conforming mesh are studied for the present polygonal hybrid finite element method in the context of two-dimensional anisotropic heat conduction. In this method, the anisotropic fundamental solutions of the problem are employed for approximating element interior temperature field so as to exactly satisfy the governing equation of problem, while one-dimensional linear shape functions are used for approximating element frame temperature field to enforce the continuity across the element boundary and the specified boundary conditions. As a result, this method involving element boundary integrals only permits to construct arbitrary polygonal elements with more sides to discretize complex computing domain, regardless of the types of polygonal elements, *i.e.* those with non-convex shapes and/or non-conforming connections. Numerical experiments demonstrate such features of this polygonal-mesh based technique.

It can be concluded that (1) The HFS-FEM with polygon discretization is practically applicable for two-dimensional anisotropic heat conduction; (2) Compared to the conventional FEM, the HFS-FEM has higher accuracy for heat flux component; (3) Both the quadrilateral hybrid elements and the polygonal hybrid elements exhibit better computing efficiency than the conventional finite element; (4) The method shows certain insensitivity to non-convex mesh; (5) The method exhibits great capability to exploit non-conforming mesh to reduce remeshing effort, which is typically important for crack propagation.

However, compared to the well-known conventional FEM, we have to point out that the present method is strongly dependent of fundamental solutions of problem, so it is not applicable to those without explicit expressions of fundamental solutions. Also, special treatment is typically required for inhomogeneous heat transfer problems, *i.e.* transient heat transfer and/or coupled thermo-mechanical processes, by introducing radial basis functions in element level for inhomogeneous term. This will be extendedly studied in the future. Actually, we think that the most appealing feature of the present method is the applicability of special polygonal element for homogeneous problems, as demonstrated in [41, 44, 45, 47], rather than for inhomogeneous problems.

CONSENT FOR PUBLICATION

Not applicable.

CONFLICT OF INTEREST

The author declares no conflict of interest, financial or otherwise.

ACKNOWLEDGEMENTS

The research was funded by Henan Province University Innovation Group Support Program (No. 19IRTSTHN020).

REFERENCES

[1] H. Wang, Y. P. Lei, J. S. Wang, Q. H. Qin, and Y. Xiao, "Theoretical and computational modeling of clustering effect on effective thermal conductivity of cement composites filled with natural hemp fibers," *Journal of Composite Materials,* vol. 50, no. 11, pp. 1509-1521, 2016.04 2016.

[2] H. Wang, Y. Xiao, and Q. H. Qin, "2D hierarchical heat transfer computational model of natural fiber bundle reinforced composite," *Scientia Iranica,* vol. 23, no. 1, pp. 268-276, 2016.01 2016.

[3] W. Q. Lin, Y. X. Zhang, and H. Wang, "Thermal conductivity of unidirectional composites consisting of randomly dispersed glass fibers and temperature-dependent polyethylene matrix," *Science and Engineering of Composite Materials,* vol. 26, no. 1, pp. 412-422, 27 Aug 2019 2019.

[4] H. Wang, F. Hou, and C. Chang, "Experimental and computational modeling of thermal conductivity of cementitious syntactic foams filled with hollow glass microspheres," *Construction and Building Materials,* vol. 265, p. 120739, 30 December 2020.

[5] L. Chen, N. Song, L. Shi, and P. Ding, "Anisotropic thermally conductive composite with wood-derived carbon scaffolds," *Composites Part A: Applied Science and Manufacturing,* vol. 112, pp. 18-24, 2018/09/01/ 2018.

[6] J. Li *et al.*, "A novel CNTs array-PDMS composite with anisotropic thermal conductivity for optoacoustic transducer applications," *Composites Part B: Engineering,* vol. 196, p. 108073, 2020/09/01/ 2020.

[7] K. D. Cole and D. H. Y. Yen, "Green's functions, temperature and heat flux in the rectangle," *International Journal of Heat and Mass Transfer,* vol. 44, no. 20, pp. 3883-3894, 2001/10/01/ 2001.

[8] R. L. McMasters and J. V. Beck, "A Two-Dimensional Cylindrical Transient Conduction Solution Using Green's Functions," *Journal of Heat Transfer,* vol. 136, no. 10, p. 101301 (8 pages), 2014.

[9] K. A. Woodbury, H. Najafi, and J. V. Beck, "Exact analytical solution for 2-D transient heat conduction in a rectangle with partial heating on one edge," *International Journal of Thermal Sciences,* vol. 112, pp. 252-262, 2017.

[10] H. Gu and J. F. Hunt, "Two-dimensional finite element heat transfer model of softwood. Part II. Macrostructural effects," *Wood Science and Technology,* vol. 38, no. 4, pp. 599-608, 2006.

[11] M. Trcala, "A 3D transient nonlinear modelling of coupled heat, mass and deformation fields in anisotropic material," *International Journal of Heat and Mass Transfer,* vol. 55, no. 17-18, pp. 4588-4596, 2012.

[12] M. I. Azis and D. L. Clements, "Nonlinear transient heat conduction problems for a class of inhomogeneous anisotropic materials by BEM," *Engineering Analysis with Boundary Elements,* vol. 32, no. 12, pp. 1054-1060, 2008.

[13] S. Ishiguro, H. Nakajima, and M. Tanaka, "Analysis of Two-Dimensional Steady-State Heat Conduction in Anisotropic Solids by Boundary Element Method Using Analog Equation Method and Green's Theorem," *Transactions of the Japan Society of Mechanical Engineers Series A,* vol. 74, no. 740, pp. 477-483 2008.

[14] N. Salam, A. Haddade, D. L. Clements, and M. I. Azis, "A boundary element method for a class of elliptic boundary value problems of functionally graded media," *Engineering Analysis with Boundary Elements,* vol. 84, pp. 186-190, 2017/11/01/ 2017.

[15] F. J. Wang, Q. S. Hua, and C. S. Liu, "Boundary function method for inverse geometry problem in two-dimensional anisotropic heat conduction equation," *Applied Mathematics Letters,* vol. 84, pp. 130-136, 2018.

[16] H. Wang, Q. H. Qin, and Y. L. Kang, "A meshless model for transient heat conduction in functionally graded materials," *Computational Mechanics,* vol. 38, no. 1, pp. 51-60, 2005.

[17] J. Sladek, V. Sladek, P. H. Wen, and B. Hon, "Inverse heat conduction problems in three-dimensional anisotropic functionally graded solids," *Journal of Engineering Mathematics,* vol. 75, no. 1, pp. 157-171, 2012.

[18] J. P. Zhang, G. Q. Zhou, S. G. Gong, and S. S. Wan, "Transient heat transfer analysis of anisotropic material by using Element-Free Galerkin method," *International Communications in Heat and Mass Transfer,* vol. 84, pp. 134-143, 2017.

[19] H. Wang, Q. H. Qin, and Y. L. Kang, "A new meshless method for steady-state heat conduction problems in anisotropic and inhomogeneous media," *Archive of Applied Mechanics,* vol. 74, no. 8, pp. 563-579, Jun 2005.

[20] H. Wang and Q. H. Qin, *Methods of Fundamental Solutions in Solid Mechanics*. New York: Elsevier, 2019.

[21] L. Cao, Q. H. Qin, and N. Zhao, "An RBF–MFS model for analysing thermal behaviour of skin tissues," *International Journal of Heat and Mass Transfer,* vol. 53, no. 7, pp. 1298-1307, 2010.

[22] H. Wang and Q. H. Qin, "Meshless approach for thermo-mechanical analysis of functionally graded materials," *Engineering Analysis with Boundary Elements,* vol. 32, no. 9, pp. 704-712, 2008.

[23] J. Jirousek and Q. H. Qin, "Application of hybrid-Trefftz element approach to transient heat conduction analysis," *Computers & Structures,* vol. 58, pp. 195-201, 1996.

[24] Q. H. Qin, *Trefftz Finite and Boundary Element Method.* Southampton: WIT Press, 2000.

[25] H. Wang, Q. H. Qin, and D. Arounsavat, "Application of hybrid Trefftz finite element method to non-linear problems of minimal surface," *International Journal for Numerical Methods in Engineering,* vol. 69, no. 6, pp. 1262-1277, Feb 2007.

[26] K. Y. Wang, P. C. Li, and D. Z. Wang, "Trefftz-type FEM for solving orthotropic potential problems," *Latin American Journal of Solids and Structures,* vol. 11, pp. 2537-2554, 2014.

[27] Z. She, K. Y. Wang, and P. C. Li, "Hybrid Trefftz polygonal elements for heat conduction problems with inclusions/voids," *Computers & Mathematics with Applications,* 2019/03/28/ 2019.

[28] Z. She, K. Y. Wang, and P. C. Li, "Thermal analysis of multilayer coated fiber-reinforced composites by the hybrid Trefftz finite element method," *Composite Structures,* vol. 224, p. 110992, 2019/05/16/ 2019.

[29] Y.-h. Cui, Q. H. Qin, and J.-S. Wang, "Application of HT finite element method to multiple crack problems of Mode I, II and III," *Chinese Journal of Engineering Mechanics,* vol. 23, no. 3, pp. 104-110, 2006.

[30] M. Dhanasekar, K. Y. Wang, Q. H. Qin, and Y. L. Kang, "Contact analysis using Trefftz and interface finite elements," *Computer Assisted Mechanics and Engineering Sciences,* vol. 13, no. 3, pp. 457-471, 2006.

[31] S. Lu, J. Liu, G. Lin, and P. Zhang, "Modified scaled boundary finite element analysis of 3D steady-state heat conduction in anisotropic layered media," *International Journal of Heat and Mass Transfer,* vol. 108, pp. 2462-2471, 2017.

[32] N. Sukumar and A. Tabarraei, "Conforming polygonal finite elements," *International Journal for Numerical Methods in Engineering,* vol. 61, no. 12, pp. 2045-2066, 2004.

[33] A. Tabarraei and N. Sukumar, "Application of polygonal finite elements in linear elasticity," *International Journal of Computational Methods,* vol. 3, no. 4, pp. 503-520, 2006.

[34] G. Manzini, A. Russo, and N. Sukumar, "New perspectives on polygonal and polyhedral finite element methods," *Mathematical Models and Methods in Applied Sciences,* vol. 24, no. 08, pp. 1665-1699, 2014.

[35] H. Wang and Q. H. Qin, "Hybrid FEM with fundamental solutions as trial functions for heat conduction simulation," *Acta Mechanica Solida Sinica,* vol. 22, no. 5, pp. 487-498, 2009.

[36] Q. H. Qin and H. Wang, "Fundamental solution based FEM for nonlinear thermal radiation problem," in *12th International Conference on Boundary Element and Meshless Techniques (BeTeq 2011), ed. EL Albuquerque, MH Aliabadi, EC Ltd, Eastleigh, UK*, 2011, pp. 113-118.

[37] H. Wang, Y. T. Gao, and Q. H. Qin, "Green's function based finite element formulations for isotropic seepage analysis with free surface," *Latin American Journal of Solids and Structures,* vol. 12, pp. 1991-2005, 2015.

[38] H. Wang and Q. H. Qin, "Fundamental solution-based hybrid finite element analysis for non-linear minimal surface problems," *Recent Developments in Boundary Element Methods: A Volume to Honour Professor John T. Katsikadelis,* p. 309, 2010.

[39] Z. W. Zhang, H. Wang, and Q. H. Qin, "Analysis of transient bioheat transfer in the human eye using hybrid finite element model," in *Applied Mechanics and Materials*, 2014, vol. 553, pp. 356-361.

[40] H. Wang and Q. H. Qin, "Fundamental-solution-based finite element model for plane orthotropic elastic bodies," *European Journal of Mechanics-A/Solids,* vol. 29, no. 5, pp. 801-809, 2010.

[41] H. Wang and Q. H. Qin, "Fundamental-solution-based hybrid FEM for plane elasticity with special elements," *Computational Mechanics,* vol. 48, no. 5, pp. 515-528, 2011.

[42] H. Wang and Q. H. Qin, "A Fundamental Solution-Based Finite Element Model for Analyzing Multi-Layer Skin Burn Injury," *Journal of Mechanics in Medicine and Biology,* vol. 12, no. 5, p. 1250027, 2012.

[43] H. Wang and Q. H. Qin, "Boundary integral based graded element for elastic analysis of 2D functionally graded plates," *European Journal of Mechanics - A/Solids,* vol. 33, pp. 12-23, 2012.

[44] H. Wang and Q. H. Qin, "A new special element for stress concentration analysis of a plate with elliptical holes," *Acta Mechanica,* vol. 223, no. 6, pp. 1323-1340, 2012.

[45] H. Wang and Q. H. Qin, "A new special coating/fiber element for analyzing effect of interface on thermal conductivity of composites," *Applied Mathematics and Computation,* vol. 268, pp. 311-321, 2015.

[46] H. Wang, Q. H. Qin, and C. Y. Lee, "n-sided polygonal hybrid finite elements with unified fundamental solution kernels for topology optimization," *Applied Mathematical Modelling,* vol. 66, pp. 97-117, 2019.

[47] H. Wang, Q. H. Qin, and Y. Xiao, "Special n -sided Voronoi fiber/matrix elements for clustering thermal effect in natural-hemp-fiber-filled cement composites," *International Journal of Heat and Mass Transfer,* vol. 92, pp. 228-235, 2016.

[48] J. C. Zhou, K. Y. Wang, and P. C. Li, "A hybrid fundamental-solution-based 8-node element for axisymmetric elasticity problems," *Engineering Analysis with Boundary Elements,* vol. 101, pp. 297-309, 2019/04/01/ 2019.

[49] J. C. Zhou, K. Y. Wang, and P. C. Li, "Hybrid fundamental solution based finite element method for axisymmetric potential problems with arbitrary boundary conditions," *Computers & Structures,* vol. 212, pp. 72-85, 2019/02/01/ 2019.

[50] J. C. Zhou, K. Y. Wang, P. C. Li, and X. D. Miao, "Hybrid fundamental solution based finite element method for axisymmetric potential problems," *Engineering Analysis with Boundary Elements,* vol. 91, pp. 82-91, 2018/06/01/ 2018.

[51] Q. H. Qin and H. Wang, *Matlab and C Programming for Trefftz Finite Element Methods*. Boca Raton: CRC Press, 2009.

[52] O. C. Zienkiewicz, R. L. Taylor, and J. Z. Zhu, *The Finite Element Method: Its Basic and Fundamentals*. New York: Elsevier, 2005.

[53] R. J. Marczak and M. Denda, "New derivations of the fundamental solution for heat conduction problems in three-dimensional general anisotropic media," *International Journal of Heat and Mass Transfer,* vol. 54, no. 15, pp. 3605-3612, 2011/07/01/ 2011.

[54] F. C. Buroni, R. J. Marczak, M. Denda, and A. Sáez, "A formalism for anisotropic heat transfer phenomena: Foundations and Green's functions," *International Journal of Heat and Mass Transfer,* vol. 75, pp. 399-409, 2014/08/01/ 2014.

[55] A. Okabe, B. Boots, K. Sugihara, S. N. Chiu, and D. G. Kendall, *Spatial Tessellations: Concepts and Applications of Voronoi Diagrams*. Wiley, 2000.

[56] F. Aurenhammer, R. Klein, and D. T. Lee, *Voronoi Diagrams and Delaunay Triangulations*. Singapore: World Scientific, 2013, pp. 377-388.

[57] Q. Du, V. Faber, and M. Gunzburger, "Centroidal Voronoi Tessellations: Applications and Algorithms," *SIAM Review,* vol. 41, no. 4, pp. 637-676, 1999.

[58] Q. Du and D. Wang, "Anisotropic Centroidal Voronoi Tessellations and Their Applications," *SIAM Journal on Scientific Computing,* vol. 26, no. 3, pp. 737-761, 2005.

[59] R. F. Cao, X. J. Zhao, W. Q. Lin, and H. Wang, "n-sided polygonal hybrid finite elements involving element boundary integrals only for anisotropic thermal analysis," *Archives of Mechanics,* vol. 72, no. 2, pp. 109-137, 2020/04 2020.

[60] K. N. Chau, K. N. Chau, T. Ngo, K. Hackl, and H. Nguyen-Xuan, "A polytree-based adaptive polygonal finite element method for multi-material topology optimization," *Computer Methods in Applied Mechanics and Engineering,* vol. 332, pp. 712-739, Apr 2018.

CHAPTER 9

Hybrid Fundamental-Solution-Based FEM for Piezoelectric Materials

Changyong Cao[1*] and **Qing-Hua Qin**[2]

[1]*Laboratory for Soft Machines & Electronics, Departments of Mechanical Engineering, Electrical and Computer Engineering, Michigan State University, East Lansing, MI 48824, USA*

[2] *Department of Engineering, Shenzhen MSU-BIT University, Shenzhen, Guangdong 518172, China*

Abstract: This chapter deals with the hybrid fundamental-solution-based finite element method (HFS-FEM) for modeling piezoelectric materials. The linear piezoelectric theory is reviewed and applied in deriving the corresponding HFS-FEM. A modified variational functional is proposed and discussed for the new model. The assumed intra-element displacement fields are presented to satisfy a priori the governing equations of the problem using a linear combination of the fundamental solutions at a number of source points located outside the element domain. To ensure the continuity of fields over inter-element boundaries, conventional shape functions are employed to construct the independent element frame displacement fields defined over the element boundary.

Keywords: HFS-FEM, Piezoelectric material, Fundamental solution.

INTRODUCTION

Piezoelectric materials are one of the most widely used smart materials that have the capability of converting electrical energy into mechanical energy and vice versa. The reciprocal energy conversion of piezoelectric materials makes them very attractive for a broad range of applications in electromechanical devices, such as sensors, actuators, transducers, and frequency generators. During the past decades, a variety of piezoelectric composites have also been developed by dispersing piezoceramics particles into non-piezoelectric polymers or elastomers. Superior properties can be obtained due to the flexibility and robustness of the polymer matrix and the excellent energy conversion of embedded responsive materials.

Both analytical and numerical studies have been conducted to improve the understanding of the electromechanical coupling mechanism of piezoelectric materials and to explore their potential applications in practical applications. For

*[*]Corresponding author Changyong Cao:** Laboratory for Soft Machines & Electronics, Michigan State University, East Lansing, MI 48824, USA; Tel: +1-517-353-9504, Fax:+1-517-353-8999; E-mail: ccao@msu.edu

example, Barnett and Lothe [1] extended the Stroh formalism [2] to an eight-dimensional framework to solve the problem of a line dislocation and a line charge in anisotropic piezoelectric materials. Chung and Ting [3] studied the two-dimensional problem of an elliptic hole in a solid of anisotropic piezoelectric material through the Stroh formalism. Dunn and Taya [4], Qin *et al* [5] and Zhao *et al* [6] performed micromechanics analysis for piezoelectric composites to predict their effective properties. Sosa [7] derived the analytical solution for transversely isotropic piezoelectric plates that contains an elliptic hole with impermeable electric boundary conditions using the theory of complex variables. Later, Xu and Rajapakse [8] presented an analytical solution for plates with arbitrarily oriented elliptical voids based on Sosa's work. Park and Sun [9] derived the closed-form solutions for an infinite piezoelectric medium with a horizontal center crack using the Stroh formalism. Based on the Stroh's formalism, Qin and Yu [10] also derived the explicit solutions for an arbitrarily oriented crack terminating at the interface between dissimilar piezoelectric materials.

These analytical methods, however, are only limited to relatively simple geometries due to the involved challenges of material anisotropy and electromechanical coupling effect. As a result, numerical methods are commonly utilized in solving practical problems. For instance, finite element method (FEM) has been used for the design and analysis of smart devices or structures made of piezoelectric materials [11-13]. Additionally, boundary element method (BEM) has also been explored for dealing with piezoelectric problems [14-18]. Compared with FEM, BEM is more computationally efficient and highly accurate in simulating linear fracture problems. In this regard, Lee and Jiang [16] developed a boundary element formulation for piezoelectric solids using the method of weighted residuals and Green's functions for a transversely isotropic piezoelectric plate. Pan [19] presented a single-domain BEM for fracture analysis of two-dimensional anisotropic piezoelectric solids based on the Green's functions derived from Stroh's formalism. More recently, Xu and Rajapakse [18] proposed a BEM of cracking problem based on the Green's functions derived from Lekhnitskii's formalism.

In addition, Trefftz method such as HT-FEM, Trefftz BEM, and Trefftz collocation method has also been successfully applied to the problems of piezoelasticity. One representative study by Qin [20] derived the corresponding Trefftz finite element formulation for anti-plane problems of piezoelectric materials by constructing a pair of dual variational functional. Wang *et al.* [21] analyzed singular electromechanical stress fields in piezoelectrics by combining the eigensolution approach and Trefftz finite element models. Later, Qin [22] presented a family of modified variational principles for piezoelectricity and successfully applied them to establish Trefftz finite element formulations. Jin *et al.* [23] formulated the Trefftz collocation method and the Trefftz Galerkin method for plane piezoelectricity problems based on the solution sets derived by a complex variable method. Sheng *et al.* [24] developed a multi-domain Trefftz boundary collocation method for plane piezoelectricity with defects according to the plane piezoelectricity solution derived from Lekhnitskii's formalism. Recently, HFS-FEM was developed as an alternative to the HT-FEM for solving two-dimensional or three-dimensional elastic and thermal problems for isotropic [25, 26] and orthotropic materials [27]. This new method inherits the advantages of the HT-FEM over the FEM and the BEM, such as the higher accuracy with coarse meshes of high-degree elements, the enhanced insensitivity to mesh distortion, the flexibility of using different element shapes, and the possibility of representing various local effects accurately [28].

It should be noted that all the formulations presented in this chapter are taken from the previous publication of Cao *et al.* [29] in which a HFS-FEM for modeling 2D piezoelectric materials was presented [29, 30]. In this approach, the foundational solutions for transversely isotropic piezoelectric materials are employed to approximate the intra-element displacement and electrical potential (DEP) field. A modified variational functional that satisfies the governing equation, boundary conditions, and continuity conditions is proposed to derive the element stiffness matrix. The stationary conditions of the variational functional and its theorem on the existence of extremum are also derived to show its suitability for implementing the proposed HFS-FEM. Finally, a numerical example is shown to demonstrate the accuracy and efficiency of the proposed method. More numerical examples associated with this method for the coupling behaviors of piezoelectric materials under variant different conditions can be found in the literature [29].

This chapter is organized as follows: a brief review of the basic equations for plane piezoelectric elasticity problem is presented in Section "Basic equations for piezoelectric materials" , the approximation of the intra-element field and auxiliary frame field are described in Section "Assumed fields". Then, a detailed derivation of the proposed HFS-FEM for plane piezoelasticity and the corresponding algorithm is given in Section "HFS-FEM FORMULATIONS BASED ON VARIATIONAL PRINCIPLE". A numerical example is presented in Section "NUMERICAL RESULTS AND DISCUSSIONS" to demonstrate the performance of the proposed element model and some concluding remarks are provided in Section "CONCLUDING REMARKS".

BASIC EQUATIONS AND ASSUMED FIELDS

Basic Equations for Piezoelectric Materials

For a linear piezoelectric material without considering the body forces and electric charge density, the differential governing equations of the problem in the Cartesian coordinate system x_i (i=1, 2, 3) can be expressed as

$$\sigma_{ij,j} = 0, \qquad D_{i,i} = 0 \quad \text{in } \Omega \tag{1}$$

where σ_{ij} is the stress tensor, D_i is the electric displacement vector, Ω is the solution domain. If taking strain and electric field as the independent variables, the constitutive equations can be presented as

$$\sigma_{ij} = c_{ijkl}\varepsilon_{kl} - e_{kij}E_k\,, \quad D_i = e_{ikl}\varepsilon_{kl} + \kappa_{ik}E_k \tag{2}$$

where c_{ijkl} is the elasticity tensor measured under zero electric field, e_{ikl} and κ_{ij} are the piezoelectric tensor and dielectric tensor measured under zero strain, respectively. ε_{ij} and E_i are the elastic strain tensor and the electric field vector, respectively.

The relation between the strain tensor ε_{ij} and the displacement u_i is given by

$$\varepsilon_{ij} = \frac{1}{2}(u_{i,j} + u_{j,i}) \tag{3}$$

and the electric field components E_i are related to the electric potential ϕ by

$$E_i = -\phi_{,i} \tag{4}$$

The boundary conditions of the boundary value problem (1)-(4) can be defined by:

$$u_i = \bar{u}_i \qquad \text{on } \Gamma_u \tag{5}$$

$$t_i = \sigma_{ij}n_j = \bar{t}_i \qquad \text{on } \Gamma_t \tag{6}$$

$$D_n = D_i n_i = -\bar{q}_n = \bar{D}_n \quad \text{on } \Gamma_D \tag{7}$$

$$\phi = \bar{\phi} \qquad \text{on } \Gamma_\phi \tag{8}$$

where $\bar{u}_i$, $\bar{t}_i$, $\bar{q}_n$ and $\bar{\varphi}$ are the prescribed boundary displacement, the prescribed traction vector, the prescribed surface charge, and the prescribed electric potential, respectively. In addition, n_i is the unit outward normal to the boundary and $\Gamma=\Gamma_u+\Gamma_t=\Gamma_D+\Gamma_\phi$ is the boundary of the solution domain Ω.

For the transversely isotropic material, if x_1-x_2 is taken as the isotropic plane, the plane strain conditions are $\varepsilon_{22}=\varepsilon_{32}=\varepsilon_{12}=0$, and $E_2=0$. Then, Eq.(2) can be further reduced to

$$\begin{Bmatrix} \sigma_{11} \\ \sigma_{33} \\ \sigma_{13} \end{Bmatrix} = \begin{bmatrix} c_{11} & c_{13} & 0 \\ c_{13} & c_{33} & 0 \\ 0 & 0 & c_{44} \end{bmatrix} \begin{Bmatrix} \varepsilon_{11} \\ \varepsilon_{33} \\ 2\varepsilon_{13} \end{Bmatrix} - \begin{bmatrix} 0 & e_{31} \\ 0 & e_{33} \\ e_{15} & 0 \end{bmatrix} \begin{Bmatrix} E_1 \\ E_3 \end{Bmatrix}$$
$$\begin{Bmatrix} D_1 \\ D_3 \end{Bmatrix} = \begin{bmatrix} 0 & 0 & e_{15} \\ e_{31} & e_{33} & 0 \end{bmatrix} \begin{Bmatrix} \varepsilon_{11} \\ \varepsilon_{33} \\ 2\varepsilon_{13} \end{Bmatrix} + \begin{bmatrix} \kappa_{11} & 0 \\ 0 & \kappa_{33} \end{bmatrix} \begin{Bmatrix} E_1 \\ E_3 \end{Bmatrix} \tag{9}$$

For a plain stress piezoelectric problem (*i.e.*, $\sigma_{22}=\sigma_{32}=\sigma_{12}=0$ and $D_2=0$), the constitutive equations can be obtained by replacing the coefficients $c_{11}, c_{13}, c_{33}, c_{44}, e_{15}, e_{31}, e_{33}, \kappa_{11}, \kappa_{33}$ in Eq. (9) as

$$\begin{aligned} & c_{11}^* = c_{11} - c_{12}^2/c_{11},\ c_{13}^* = c_{13} - c_{12}c_{13}/c_{11}, \\ & c_{33}^* = c_{33} - c_{13}^2/c_{11},\ c_{44}^* = c_{44},\ e_{15}^* = e_{15}, \\ & e_{31}^* = e_{31} - c_{12}e_{31}/c_{11},\ e_{33}^* = e_{33} - c_{13}e_{31}/c_{11}, \\ & \kappa_{11}^* = \kappa_{11},\ \kappa_{33}^* = \kappa_{33} + e_{31}^2/c_{11} \end{aligned} \tag{10}$$

Assumed Fields

The main idea of the proposed HFS-FEM originates from the hybrid Trefftz finite element method (HT-FEM), which utilizes two independent approximate fields: intra-element field and an independent frame field along the element boundary. However, unlike HT-FEM, the intra-element fields of HFS-FEM are constructed based on the fundamental solutions, rather than a truncated T-complete function set. For piezoelectric problems, HFS-FEM is based on the assumption of two distinct displacement and electric potential (DEP) fields: intra-element DEP field u and an independent DEP frame field $\tilde{\mathbf{u}}$ along element boundaries [22, 28].

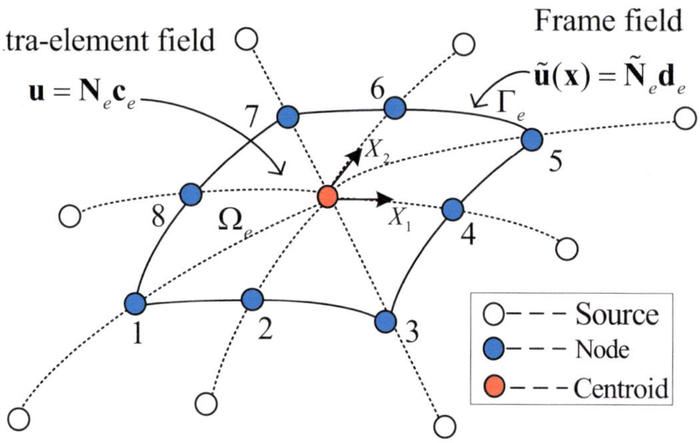

Fig. (1). Intra-element fields and frame fields of a particular element in HFS-FEM.

Intra-element Field

The intra-element DEP field u fulfills the governing differential Eq. (1) and is approximated by a linear combination of the foundational solutions at different source points located outside of the element domain

$$\mathbf{u_e} = \begin{Bmatrix} u_1 \\ u_2 \\ \phi \end{Bmatrix} = \sum_{j=1}^{n_s} \begin{bmatrix} u_{11}^*(\mathbf{x},\mathbf{y}_{sj}) & u_{21}^*(\mathbf{x},\mathbf{y}_{sj}) & u_{31}^*(\mathbf{x},\mathbf{y}_{sj}) \\ u_{12}^*(\mathbf{x},\mathbf{y}_{sj}) & u_{22}^*(\mathbf{x},\mathbf{y}_{sj}) & u_{32}^*(\mathbf{x},\mathbf{y}_{sj}) \\ u_{13}^*(\mathbf{x},\mathbf{y}_{sj}) & u_{23}^*(\mathbf{x},\mathbf{y}_{sj}) & u_{33}^*(\mathbf{x},\mathbf{y}_{sj}) \end{bmatrix} \begin{Bmatrix} c_{1j} \\ c_{2j} \\ c_{3j} \end{Bmatrix} = \mathbf{N_e c_e} \ (\mathbf{x} \in \Omega_e, \mathbf{y}_{sj} \notin \Omega_e) \quad \textbf{(11)}$$

where n_s is the number of source points outside the element domain, which is equal to the number of nodes of an element in the present research based on the generation approach of the source points (see Fig. **1**), $\mathbf{c_e}$ an unknown coefficient vector (not nodal displacements), $\mathbf{N_e}$ the fundamental solution matrix, which can be written as

$$\mathbf{N_e} = \begin{bmatrix} u_{11}^*(\mathbf{x},\mathbf{y}_{s1}) & u_{21}^*(\mathbf{x},\mathbf{y}_{s1}) & u_{31}^*(\mathbf{x},\mathbf{y}_{s1}) & \ldots & u_{33}^*(\mathbf{x},\mathbf{y}_{sn_s}) & u_{33}^*(\mathbf{x},\mathbf{y}_{sn_s}) & u_{33}^*(\mathbf{x},\mathbf{y}_{sn_s}) \\ u_{12}^*(\mathbf{x},\mathbf{y}_{s1}) & u_{22}^*(\mathbf{x},\mathbf{y}_{s1}) & u_{32}^*(\mathbf{x},\mathbf{y}_{s1}) & \ldots & u_{33}^*(\mathbf{x},\mathbf{y}_{sn_s}) & u_{33}^*(\mathbf{x},\mathbf{y}_{sn_s}) & u_{33}^*(\mathbf{x},\mathbf{y}_{sn_s}) \\ u_{13}^*(\mathbf{x},\mathbf{y}_{s1}) & u_{23}^*(\mathbf{x},\mathbf{y}_{s1}) & u_{33}^*(\mathbf{x},\mathbf{y}_{s1}) & \ldots & u_{33}^*(\mathbf{x},\mathbf{y}_{sn_s}) & u_{33}^*(\mathbf{x},\mathbf{y}_{sn_s}) & u_{33}^*(\mathbf{x},\mathbf{y}_{sn_s}) \end{bmatrix} \quad \textbf{(12)}$$

$$\mathbf{c_e} = [c_{11} \quad c_{21} \quad c_{31} \quad \ldots \quad c_{1n_s} \quad c_{2n_s} \quad c_{3n_s}]^T \quad \textbf{(13)}$$

in which $\mathbf{x}$ and $\mathbf{y}_{sj}$ are respectively the field point (*i.e.*, the nodal points of the element in this work) and source point in the local coordinate system (X_1, X_2). The component $u_{ij}^*(\mathbf{x},\mathbf{y}_{sj})$ is the induced displacement component (i=1, 2) or electric potential (i=3) in i-direction at the field point x due to a unit point load (j=1, 2) or point charge (j=3) applied in j-direction at the source point $\mathbf{y}_{sj}$. The fundamental solution $u_{ij}^*(\mathbf{x},\mathbf{y}_{sj})$ is given as [31]

$$\begin{cases} u_{11}^* = \dfrac{1}{\pi M_{11}} \sum_{j=1}^{3} s_{j1} t_{(2j)1}^1 \ln r_j \\ u_{12}^* = \dfrac{1}{\pi M_{12}} \sum_{j=1}^{3} s_{j2} t_{(2j)2}^1 arctg \dfrac{x_1 - x_{s1}}{s_j(x_3 - x_{s3})} \\ u_{13}^* = \dfrac{1}{\pi M_{13}} \sum_{j=1}^{3} s_{j3} t_{(2j)3}^1 arctg \dfrac{x_1 - x_{s1}}{s_j(x_3 - x_{s3})} \end{cases} \tag{14}$$

$$\begin{cases} u_{21}^* = \dfrac{1}{\pi M_{11}} \sum_{j=1}^{3} d_{j1} t_{(2j)1}^1 arctg \dfrac{x_1 - x_{s1}}{s_j(x_3 - x_{s3})} \\ u_{22}^* = \dfrac{1}{\pi M_{12}} \sum_{j=1}^{3} d_{j2} t_{(2j)2}^1 \ln r_j \\ u_{23}^* = \dfrac{1}{\pi M_{13}} \sum_{j=1}^{3} d_{j3} t_{(2j)3}^1 \ln r_j \end{cases} \tag{15}$$

$$\begin{cases} u_{31}^* = \dfrac{1}{\pi M_{11}} \sum_{j=1}^{3} g_{j1} t_{(2j)1}^1 arctg \dfrac{x_1 - x_{s1}}{s_j(x_3 - x_{s3})} \\ u_{32}^* = \dfrac{1}{\pi M_{12}} \sum_{j=1}^{3} g_{j2} t_{(2j)2}^1 \ln r_j \\ u_{33}^* = \dfrac{1}{\pi M_{13}} \sum_{j=1}^{3} g_{j3} t_{(2j)3}^1 \ln r_j \end{cases} \tag{16}$$

where $r_j = \sqrt{(x_1 - x_{s1})^2 + s_j^2(x_3 - x_{s3})^2}$, and s_j is the three different roots of the characteristic equation $as_i^6 - bs_i^4 + cs_i^3 - d = 0$. The coefficients $s_{ij}, d_{ij}, g_{ij}, t_{ij}$ are given in the Appendix. The source point $\mathbf{y}_{sj}(j=1,2,\ldots,n_s)$ can be generated using the following method [27]

$$\mathbf{y}_s = \mathbf{x}_0 + \gamma(\mathbf{x}_0 - \mathbf{x}_c) \tag{17}$$

where γ is a dimensionless coefficient to determine the distance between the source point $\mathbf{y}_s$ and the geometrical centroid of the element $\mathbf{x}_c$, and $\mathbf{x}_0$ is the point on the element boundary (*i.e.*, the nodal points in this work) as shown in Fig. (**1**).

Utilizing Eq.(3) and the intra-element DEP field u in Eq.(11), the corresponding stress and electric displacement in Eq.(9) can be further written as

$$\boldsymbol{\sigma} = \mathbf{T_e c_e} \tag{18}$$

where $\boldsymbol{\sigma} = \begin{bmatrix}\sigma_{11} & \sigma_{22} & \sigma_{12} & D_1 & D_2\end{bmatrix}^T$ and

$$\mathbf{T_e} = \begin{bmatrix} \sigma_{11}^*(\mathbf{x},\mathbf{y}_{s1}) & \sigma_{12}^*(\mathbf{x},\mathbf{y}_{s1}) & \sigma_{13}^*(\mathbf{x},\mathbf{y}_{s1}) & \ldots & \sigma_{11}^*(\mathbf{x},\mathbf{y}_{sn_s}) & \sigma_{12}^*(\mathbf{x},\mathbf{y}_{sn_s}) & \sigma_{13}^*(\mathbf{x},\mathbf{y}_{sn_s}) \\ \sigma_{21}^*(\mathbf{x},\mathbf{y}_{s1}) & \sigma_{22}^*(\mathbf{x},\mathbf{y}_{s1}) & \sigma_{23}^*(\mathbf{x},\mathbf{y}_{s1}) & \ldots & \sigma_{21}^*(\mathbf{x},\mathbf{y}_{sn_s}) & \sigma_{22}^*(\mathbf{x},\mathbf{y}_{sn_s}) & \sigma_{23}^*(\mathbf{x},\mathbf{y}_{sn_s}) \\ \sigma_{31}^*(\mathbf{x},\mathbf{y}_{s1}) & \sigma_{32}^*(\mathbf{x},\mathbf{y}_{s1}) & \sigma_{33}^*(\mathbf{x},\mathbf{y}_{s1}) & \ldots & \sigma_{31}^*(\mathbf{x},\mathbf{y}_{sn_s}) & \sigma_{32}^*(\mathbf{x},\mathbf{y}_{sn_s}) & \sigma_{33}^*(\mathbf{x},\mathbf{y}_{sn_s}) \\ \sigma_{41}^*(\mathbf{x},\mathbf{y}_{s1}) & \sigma_{42}^*(\mathbf{x},\mathbf{y}_{s1}) & \sigma_{43}^*(\mathbf{x},\mathbf{y}_{s1}) & \ldots & \sigma_{41}^*(\mathbf{x},\mathbf{y}_{sn_s}) & \sigma_{42}^*(\mathbf{x},\mathbf{y}_{sn_s}) & \sigma_{43}^*(\mathbf{x},\mathbf{y}_{sn_s}) \\ \sigma_{51}^*(\mathbf{x},\mathbf{y}_{s1}) & \sigma_{52}^*(\mathbf{x},\mathbf{y}_{s1}) & \sigma_{53}^*(\mathbf{x},\mathbf{y}_{s1}) & \ldots & \sigma_{51}^*(\mathbf{x},\mathbf{y}_{sn_s}) & \sigma_{52}^*(\mathbf{x},\mathbf{y}_{sn_s}) & \sigma_{53}^*(\mathbf{x},\mathbf{y}_{sn_s}) \end{bmatrix} \tag{19}$$

in which $\sigma_{ij}^*(\mathbf{x},\mathbf{y}_{sj})$ denotes the corresponding stress components (*i*=1,2,3) or electric displacement (*i*=4,5) along i-direction at the field point x due to a unit point load (*j*=1,2) or a unit point charge (*j*=3) applied in j-direction at the source point ys and can be derived from Eqs.(14)-(16), and are listed below, which are derived by substituting the fundamental solutions into the constitutive equations [32].

$$\begin{cases} \sigma_{11}^* = \dfrac{1}{\pi M_{11}} \displaystyle\sum_{j=1}^{3} \left[\left(c_{11}s_{j1} - c_{13}d_{j1}s_j - e_{31}g_{j1}s_j\right) t_{(2j)1}^1 \dfrac{x_1 - x_{s1}}{r_j^2} \right] \\ \sigma_{12}^* = \dfrac{1}{\pi M_{12}} \displaystyle\sum_{j=1}^{3} \left[\left(c_{11}s_{j2} + c_{13}d_{j2}s_j + e_{31}g_{j2}s_j\right) t_{(2j)2}^1 \dfrac{s_j\left(x_3 - x_{s3}\right)}{r_j^2} \right] \\ \sigma_{13}^* = \dfrac{1}{\pi M_{13}} \displaystyle\sum_{j=1}^{3} \left[\left(c_{11}s_{j3} + c_{13}d_{j3}s_j + e_{31}g_{j3}s_j\right) t_{(2j)3}^1 \dfrac{s_j\left(x_3 - x_{s3}\right)}{r_j^2} \right] \end{cases} \tag{20}$$

$$\begin{cases} \sigma_{21}^{*} = \dfrac{1}{\pi M_{11}} \displaystyle\sum_{j=1}^{3} \left[\left(c_{13} s_{j1} - c_{33} d_{j1} s_j - e_{33} g_{j1} s_j \right) t_{(2j)1}^{1} \dfrac{x_1 - x_{s1}}{r_j^2} \right] \\ \sigma_{22}^{*} = \dfrac{1}{\pi M_{12}} \displaystyle\sum_{j=1}^{3} \left[\left(c_{13} s_{j2} + c_{33} d_{j2} s_j + e_{33} g_{j2} s_j \right) t_{(2j)2}^{1} \dfrac{s_j \left(x_3 - x_{s3} \right)}{r_j^2} \right] \\ \sigma_{23}^{*} = -\dfrac{1}{\pi M_{13}} \displaystyle\sum_{j=1}^{3} \left[\left(c_{13} s_{j3} + c_{33} d_{j3} s_j + e_{33} g_{j3} s_j \right) t_{(2j)3}^{1} \dfrac{s_j \left(x_3 - x_{s3} \right)}{r_j^2} \right] \end{cases} \tag{21}$$

$$\begin{cases} \sigma_{31}^{*} = \dfrac{1}{\pi M_{11}} \displaystyle\sum_{j=1}^{3} \left[\left(c_{44} s_{j1} s_j + c_{44} d_{j1} + e_{15} g_{j1} \right) t_{(2j)1}^{1} \dfrac{s_j \left(x_3 - x_{s3} \right)}{r_j^2} \right] \\ \sigma_{32}^{*} = \dfrac{1}{\pi M_{12}} \displaystyle\sum_{j=1}^{3} \left[\left(-c_{44} s_{j2} s_j + c_{44} d_{j2} + e_{15} g_{j2} \right) t_{(2j)2}^{1} \dfrac{x_1 - x_{s1}}{r_j^2} \right] \\ \sigma_{33}^{*} = -\dfrac{1}{\pi M_{13}} \displaystyle\sum_{j=1}^{3} \left[\left(-c_{44} s_{j3} s_j + c_{44} d_{j3} + e_{15} g_{j3} \right) t_{(2j)3}^{1} \dfrac{x_1 - x_{s1}}{r_j^2} \right] \end{cases} \tag{22}$$

$$\begin{cases} \sigma_{41}^{*} = \dfrac{1}{\pi M_{11}} \displaystyle\sum_{j=1}^{3} \left[\left(e_{15} s_{j1} s_j + e_{15} d_{j1} - \lambda_{11} g_{j1} \right) t_{(2j)1}^{1} \dfrac{s_j \left(x_3 - x_{s3} \right)}{r_j^2} \right] \\ \sigma_{42}^{*} = \dfrac{1}{\pi M_{12}} \displaystyle\sum_{j=1}^{3} \left[\left(-e_{15} s_{j2} s_j + e_{15} d_{j2} - \lambda_{11} g_{j2} \right) t_{(2j)2}^{1} \dfrac{x_1 - x_{s1}}{r_j^2} \right] \\ \sigma_{43}^{*} = -\dfrac{1}{\pi M_{13}} \displaystyle\sum_{j=1}^{3} \left[\left(-e_{15} s_{j3} s_j + e_{15} d_{j3} - \lambda_{11} g_{j3} \right) t_{(2j)3}^{1} \dfrac{x_1 - x_{s1}}{r_j^2} \right] \end{cases} \tag{23}$$

$$\begin{cases} \sigma_{51}^{*} = \dfrac{1}{\pi M_{11}} \displaystyle\sum_{j=1}^{3} \left[\left(e_{31} s_{j1} - e_{33} d_{j1} s_j + \lambda_{33} g_{j1} s_j \right) t_{(2j)1}^{1} \dfrac{x_1 - x_{s1}}{r_j^2} \right] \\ \sigma_{52}^{*} = \dfrac{1}{\pi M_{12}} \displaystyle\sum_{j=1}^{3} \left[\left(e_{31} s_{j2} + e_{33} d_{j2} s_j - \lambda_{33} g_{j2} s_j \right) t_{(2j)2}^{1} \dfrac{s_j \left(x_3 - x_{s3} \right)}{r_j^2} \right] \\ \sigma_{53}^{*} = -\dfrac{1}{\pi M_{13}} \displaystyle\sum_{j=1}^{3} \left[\left(e_{31} s_{j3} + e_{33} d_{j3} s_j - \lambda_{33} g_{j3} s_j \right) t_{(2j)3}^{1} \dfrac{s_j \left(x_3 - x_{s3} \right)}{r_j^2} \right] \end{cases} \tag{24}$$

in which the coefficients $s_{ij}, d_{ij}, g_{ij}, t_{ij}, M_{11}, M_{12}, M_{13}$ are defined as below [31].

$$\begin{aligned} a_{i1} &= -c_{13}m_1 + m_6 s_i^2 + m_7 s_i^4 \\ b_{i1} &= m_8 s_i + m_9 s_i^3 + c_{44} m_3 s_i^5 \\ p_{i1} &= -e_{31} m_1 + m_{10} s_i^2 + m_{11} s_i^4 \\ d_{i1} &= -h_1 s_i + h_2 s_i^3 \\ s_{i1} &= m_1 - m_2 s_i^2 + m_3 s_i^4 \\ g_{i1} &= m_4 s_i - m_5 s_i^3 \end{aligned} \tag{25}$$

$$\begin{aligned} a_{i2} &= h_7 s_i^5 + h_6 s_i^3 + h_5 s_i \\ b_{i2} &= h_{10} s_i^4 - h_9 s_i^2 - h_8 \\ p_{i2} &= h_{12} s_i^3 + h_{11} s_i \\ d_{i2} &= c_{44} f_{33} s_i^4 - h_3 s_i^2 + c_{11} f_{11} \\ s_{i2} &= h_2 s_i^3 - h_1 s_i \\ g_{i2} &= c_{44} e_{33} s_i^4 - h_4 s_i^2 + c_{11} e_{15} \end{aligned} \tag{26}$$

$$\begin{aligned} a_{i3} &= n_2 s_i + n_3 s_i^3 \\ b_{i3} &= -n_4 s_i^2 + n_5 s_i^4 \\ p_{i3} &= n_6 s_i + n_7 s_i^3 + n_8 s_i^5 \\ d_{i3} &= -c_{11} e_{15} + h_4 s_i^2 - c_{44} e_{33} s_i^4 \\ s_{i3} &= -m_4 s_i + m_5 s_i^3 \\ g_{i3} &= c_{11} c_{44} - n_1 s_i^2 + c_{44} c_{33} s_i^4 \end{aligned} \tag{27}$$

$$\begin{aligned} M_{11} &= -2[b_{11} t_{21}^1 + b_{21} t_{41}^1 + b_{31} t_{61}^1] \\ M_{12} &= -2[a_{12} t_{22}^1 + a_{22} t_{42}^1 + a_{32} t_{62}^1] \\ M_{13} &= -2[p_{13} t_{23}^1 + p_{23} t_{43}^1 + p_{33} t_{63}^1] \end{aligned} \tag{28}$$

$$\begin{aligned} &t_{21}^1 = d_{21} g_{31} - d_{31} g_{21}, \quad t_{41}^1 = d_{31} g_{11} - d_{11} g_{31}, \quad t_{61}^1 = d_{11} g_{21} - d_{21} g_{11}, \\ &t_{22}^1 = p_{32} s_{22} - p_{22} s_{32}, \quad t_{42}^1 = p_{12} s_{32} - p_{32} s_{12}, \quad t_{62}^1 = p_{22} s_{12} - p_{12} s_{22}, \\ &t_{23}^1 = a_{23} s_{33} - a_{33} s_{23}, \quad t_{43}^1 = a_{33} s_{13} - a_{13} s_{33}, \quad t_{63}^1 = a_{13} s_{23} - a_{23} s_{13}, \end{aligned} \tag{29}$$

$$\begin{aligned}
n_1 &= c_{11}c_{33} - c_{13}c_{13} - 2c_{13}c_{44}; & n_5 &= c_{44}(m_5 + c_{44}e_{33} - c_{33}e_{15}); \\
n_2 &= -c_{13}m_4 + c_{11}(c_{44}e_{33} - c_{33}e_{15}); & n_6 &= -e_{33}m_4 + c_{11}(e_{15}e_{33} + c_{44}\kappa_{33}); \\
n_3 &= c_{13}m_5 + c_{33}h_4 - e_{33}n_1; & n_7 &= e_{31}m_5 + e_{33}h_4 + n_1\kappa_{33}; \\
n_4 &= c_{44}(m_4 + h_4) - e_{15}n_1; & n_8 &= -c_{44}e_{33}e_{33} - c_{33}c_{44}\kappa_{33};
\end{aligned} \tag{30}$$

$$\begin{aligned}
h_1 &= (c_{13} + c_{44})\kappa_{11} + (e_{15} + e_{31})e_{15}; \\
h_2 &= (c_{13} + c_{44})\kappa_{33} + (e_{15} + e_{31})e_{33}; \\
h_3 &= c_{11}\kappa_{33} + c_{44}\kappa_{11} + (e_{15} + e_{31})(e_{15} + e_{31}); \\
h_4 &= c_{11}e_{33} - c_{13}(e_{15} + e_{31}) - c_{44}e_{31}; \\
h_5 &= c_{11}(c_{33}\kappa_{11} + e_{15}e_{33}) - c_{13}h_1; \\
h_6 &= c_{13}h_2 - c_{33}h_3 - e_{33}h_4; \\
h_7 &= c_{44}(c_{33}\kappa_{33} + e_{33}e_{33}); \\
h_8 &= c_{11}(c_{44}\kappa_{11} + e_{15}e_{15}); \\
h_9 &= c_{44}(h_1 - h_3) - e_{15}h_4; \\
h_{10} &= c_{44}(h_2 - c_{44}\kappa_{33} - e_{15}e_{33}); \\
h_{11} &= c_{11}(\kappa_{11}e_{33} - \kappa_{33}e_{15}) - e_{31}h_1; \\
h_{12} &= e_{31}h_2 - e_{33}h_3 + \kappa_{33}h_4;
\end{aligned} \tag{31}$$

$$\begin{aligned}
m_1 &= c_{44}\kappa_{11} + e_{15}e_{15}; \\
m_2 &= c_{44}\kappa_{33} + c_{33}\kappa_{11} + 2e_{15}e_{33}; \\
m_3 &= c_{33}\kappa_{33} + e_{33}e_{33}; \\
m_4 &= c_{44}e_{31} - c_{13}e_{15}; \\
m_5 &= -(c_{13} + c_{44})e_{33} + (e_{15} + e_{31})c_{33}; \\
m_6 &= c_{13}m_2 - c_{33}h_1 + e_{33}m_4; \\
m_7 &= -c_{13}m_3 + c_{33}h_2 - e_{33}m_5; \\
m_8 &= c_{44}m_1 - c_{44}h_1 + e_{15}m_4; \\
m_9 &= -c_{44}m_2 + c_{44}h_2 - e_{15}m_5; \\
m_{10} &= e_{31}m_2 - e_{33}h_1 - \kappa_{33}m_4; \\
m_{11} &= -e_{31}m_3 + e_{33}h_2 + \kappa_{33}m_5;
\end{aligned} \tag{32}$$

From Eqs.(2), (6) and (7), the generalized traction forces and electric displacement can be written as

$$\begin{Bmatrix} t_1 \\ t_2 \\ D_n \end{Bmatrix} = \begin{Bmatrix} \mathbf{Q}_1 \\ \mathbf{Q}_2 \\ \mathbf{Q}_3 \end{Bmatrix} \mathbf{c}_e = \mathbf{Q}_e \mathbf{c}_e \tag{33}$$

where

$$\mathbf{Q}_e = \mathbf{n}\mathbf{T}_e \tag{34}$$

$$\mathbf{n} = \begin{bmatrix} n_1 & 0 & n_2 & 0 & 0 \\ 0 & n_2 & n_1 & 0 & 0 \\ 0 & 0 & 0 & n_1 & n_2 \end{bmatrix} \tag{35}$$

Auxiliary Frame Field

The unknown coefficient $\mathbf{c}_e$ may be calculated from the conditions on the external boundary and/or the continuity conditions on the inter-element boundary. In a hybrid technique, the meshed elements are linked through an auxiliary conforming displacement frame. This means that a conforming DEP field is independently defined on the element boundary to enforce the field continuity between elements and to link the coefficient $\mathbf{c}_e$, appearing in Eq.(11), with nodal DEP $\mathbf{d}_e$. For the 2D piezoelectric problems, the frame field is assumed as

$$\tilde{\mathbf{u}}(\mathbf{x}) = \begin{Bmatrix} \tilde{u}_1 \\ \tilde{u}_2 \\ \tilde{\phi} \end{Bmatrix} = \begin{Bmatrix} \tilde{\mathbf{N}}_1 \\ \tilde{\mathbf{N}}_2 \\ \tilde{\mathbf{N}}_3 \end{Bmatrix} \mathbf{d}_e = \tilde{\mathbf{N}}_e \mathbf{d}_e, \qquad (\mathbf{x} \in \Gamma_e) \tag{36}$$

where the symbol "~" is used to specify that the field is only defined on the element boundary, $\mathbf{d}_e = \mathbf{d}_e(\mathbf{c}_e)$ stands for the vector of the nodal displacements which are the final unknowns of the problem, Γ_e represents the boundary of element e, and $\tilde{\mathbf{N}}_e$ is a matrix of the corresponding shape functions which are the same as those used in the conventional FEM. For example, for the side 3-4-5 of a particular quadratic element as shown in Fig. (**1**), the shape function matrix $\tilde{\mathbf{N}}_e$ and nodal vector $\mathbf{d_e}$ can be given as

$$\tilde{\mathbf{N}}_e = \begin{bmatrix} \underbrace{0\cdots0}_{6} & \tilde{N}_1 & 0 & 0 & \tilde{N}_2 & 0 & 0 & \tilde{N}_3 & 0 & 0 & \underbrace{0\cdots0}_{9} \\ 0\cdots0 & 0 & \tilde{N}_1 & 0 & 0 & \tilde{N}_2 & 0 & 0 & \tilde{N}_3 & 0 & 0\cdots0 \\ 0\cdots0 & 0 & 0 & \tilde{N}_1 & 0 & 0 & \tilde{N}_2 & 0 & 0 & \tilde{N}_3 & 0\cdots0 \end{bmatrix} \quad \textbf{(37)}$$

$$\mathbf{d_e} = \begin{bmatrix} u_{11} & u_{21} & \phi_1 & \cdots & u_{14} & u_{24} & \phi_4 & \cdots & u_{18} & u_{28} & \phi_8 \end{bmatrix}^T \quad \textbf{(38)}$$

where the shape functions are expressed by natural coordinate ξ

$$\tilde{N}_1 = -\frac{\xi(1-\xi)}{2}, \quad \tilde{N}_2 = 1-\xi^2, \quad \tilde{N}_3 = \frac{\xi(1+\xi)}{2} \qquad (\xi \in [-1,1]) \quad \textbf{(39)}$$

HFS-FEM FORMULATIONS BASED ON VARIATIONAL PRINCIPLE

Variational Principles

Based on the assumption of the two distinct DEP fields, the Euler equations of the proposed variational functional should meet the following inter-element continuity conditions besides Eqs.(1)-(8):

$$u_{ie} = u_{if} \quad \phi_e = \phi_f \qquad (\text{on } \Gamma_e \cap \Gamma_f, \text{ conformity}) \quad \textbf{(40)}$$

$$t_{ie} + t_{if} = 0 \quad D_{ne} + D_{nf} = 0 \quad (\text{on } \Gamma_e \cap \Gamma_f, \text{ reciprocity}) \quad \textbf{(41)}$$

where '*e*' and '*f*' stand for any two neighboring elements. Eqs.(1)-(8) together with Eq.(40) and (41) can now be taken as the basis to establish the modified variational principle for the HFS-FEM of piezoelectric materials [22, 28].

A new modified variational functional should be developed for the HFS-FEM because the stationary conditions of the traditional potential or complementary variational functional cannot satisfy the inter-element continuity conditions required. Without considering the body forces and electric charge density, the hybrid functional Π_{me} for a particular element e can be constructed as

$$\Pi_{me} = \Pi_e + \int_{\Gamma_e} t_i(\tilde{u}_i - u_i)d\Gamma + \int_{\Gamma_e} D_n(\tilde{\phi} - \phi)d\Gamma \tag{42}$$

where

$$\Pi_e = \frac{1}{2}\iint_{\Omega_e} (\sigma_{ij}\varepsilon_{ij} + D_i E_i)d\Omega - \int_{\Gamma_t} \bar{t}_i \tilde{u}_i d\Gamma - \int_{\Gamma_D} \bar{D}_n \tilde{\phi} d\Gamma \tag{43}$$

and the boundary Γ_e of the element e is

$$\Gamma_e = \Gamma_{eu} \cup \Gamma_{et} \cup \Gamma_{el} = \Gamma_{e\phi} \cup \Gamma_{eD} \cup \Gamma_{el} \tag{44}$$

and

$$\Gamma_{eu} = \Gamma_e \cap \Gamma_u, \Gamma_{et} = \Gamma_e \cap \Gamma_t, \Gamma_{e\phi} = \Gamma_e \cap \Gamma_\phi, \Gamma_{eD} = \Gamma_e \cap \Gamma_D \tag{45}$$

and Γ_{el} is the inter-element boundary of element *e*. Compared to the functional employed in the conventional FEM, the present hybrid functional is constructed by adding two integral terms related to the intra-element and element frame DEP fields to guarantee the continuity of displacement and electrical potential on the common boundary of two adjacent elements.

It can be proved that the stationary conditions of the above functional (42) leads to Eqs.(1)-(8). To this end, the variation of Π_m can generate

$$\delta\Pi_{me} = \delta\Pi_e + \int_{\Gamma_e} [(\tilde{u}_i - u_i)\delta t_i + t_i(\delta\tilde{u}_i - \delta u_i)]d\Gamma + \int_{\Gamma_e} [(\tilde{\phi} - \phi)\delta D_n + D_n(\delta\tilde{\phi} - \delta\phi)]d\Gamma \tag{46}$$

in which the first term is

$$\begin{aligned} \delta\Pi_e &= \iint_{\Omega_e} \sigma_{ij}\delta\varepsilon_{ij}d\Omega) + \iint_{\Omega_e} D_i\delta E_i d\Omega - \int_{\Gamma_{et}} \bar{t}_i\delta\tilde{u}_i d\Gamma - \int_{\Gamma_{eD}} \bar{D}_n\delta\tilde{\phi} d\Gamma \\ &= \iint_{\Omega_e} \sigma_{ij}\delta u_{i,j} d\Omega + \iint_{\Omega_e} D_i\delta\phi_{,i} d\Omega - \int_{\Gamma_{et}} \bar{t}_i\delta\tilde{u}_i d\Gamma - \int_{\Gamma_{eD}} \bar{D}_n\delta\tilde{\phi} d\Gamma \end{aligned} \tag{47}$$

Applying Gaussian theorem

$$\iint_{\Omega_e} f_{,i} d\Omega = \int_{\Gamma_e} f \cdot n_i d\Gamma \tag{48}$$

and the definitions of traction force and electrical displacement

$$t_i = \sigma_{ij} n_j, \quad D_n = D_i n_i \tag{49}$$

we can obtain the following expression

$$\begin{aligned}\delta\Pi_e = &-\iint_{\Omega_e} \sigma_{ij,j}\delta u_i d\Omega - \iint_{\Omega_e} D_{i,i}\delta\phi d\Omega + \int_{\Gamma_e} t_i \delta u_i d\Gamma + \int_{\Gamma_e} D_n \delta\phi d\Gamma \\ &-\int_{\Gamma_{et}} \bar{t}_i \delta\tilde{u}_i d\Gamma - \int_{\Gamma_{eD}} \bar{D}_n \delta\tilde{\phi} d\Gamma\end{aligned} \tag{50}$$

Then, substituting Eq. (50) into Eq. (46) gives

$$\begin{aligned}\delta\Pi_{me} = &-\iint_{\Omega_e} \sigma_{ij,j}\delta u_i d\Omega - \iint_{\Omega_e} D_{i,i}\delta\phi d\Omega - \int_{\Gamma_{et}} \bar{t}_i \delta\tilde{u}_i d\Gamma - \int_{\Gamma_{eD}} \bar{D}_n \delta\tilde{\phi} d\Gamma \\ &+\int_{\Gamma_e} [(\tilde{u}_i - u_i)\delta t_i + t_i \delta\tilde{u}_i] d\Gamma + \int_{\Gamma_e} [(\tilde{\phi} - \phi)\delta D_n + D_n \delta\tilde{\phi}] d\Gamma\end{aligned} \tag{51}$$

Considering the fact that

$$\int_{\Gamma_{eu}} t_i \delta\tilde{u}_i d\Gamma = 0, \ \int_{\Gamma_{e\phi}} D_n \delta\tilde{\phi} d\Gamma = 0 \tag{52}$$

we can finally get the following form

$$\begin{aligned}\delta\Pi_{me} = &-\iint_{\Omega_e} \sigma_{ij,j}\delta u_i d\Omega - \iint_{\Omega_e} D_{i,i}\delta\phi d\Omega + \int_{\Gamma_{et}} (t_i - \bar{t}_i)\delta\tilde{u}_i d\Gamma + \int_{\Gamma_{eD}} (D_n - \bar{D}_n)\delta\tilde{\phi} d\Gamma \\ &+\int_{\Gamma_e} (\tilde{u}_i - u_i)\delta t_i d\Gamma + \int_{\Gamma_e} (\tilde{\phi} - \phi)\delta D_n d\Gamma + \int_{\Gamma_I} t_i \delta\tilde{u}_i d\Gamma + \int_{\Gamma_I} D_n \delta\tilde{\phi} d\Gamma\end{aligned} \tag{53}$$

Therefore, the Euler equations for Eq.(53) result in Eqs.(1)-(8) and Eq.(40) because the quantities δu_i, δt_i, $\delta\phi$, δD_n, $\delta\tilde{u}_i$, and $\delta\tilde{\phi}$ may be arbitrary. As for the continuity condition of Eq.(41), it can be seen from the following variational of two adjacent elements such as **e** and **f**

$$\begin{aligned}&\delta\Pi_{m(e\cup f)} \\ &= -\iint_{\Omega_e\cup\Omega_f} \sigma_{ij,j}\delta u_i d\Omega - \iint_{\Omega_e\cup\Omega_f} D_{i,i}\delta\phi d\Omega + \int_{\Gamma_{et}+\Gamma_{et}} (t_i - \bar{t}_i)\delta\tilde{u}_i d\Gamma + \int_{\Gamma_{eD}+\Gamma_{fD}} (D_n - \bar{D}_n)\delta\tilde{\phi} d\Gamma \\ &+ \int_{\Gamma_e+\Gamma_f} (\tilde{u}_i - u_i)\delta t_i d\Gamma + \int_{\Gamma_e+\Gamma_f} (\tilde{\phi} - \phi)\delta D_n d\Gamma + \int_{\Gamma_{efI}} \left(t_{ie} + t_{if}\right)\delta\tilde{u}_i + \int_{\Gamma_{efI}} \left(D_{ne} + D_{nf}\right)\delta\tilde{\phi} d\Gamma\end{aligned} \tag{54}$$

This indicates that the stationary condition of the functional satisfies both the required boundary and inter-element continuity equations. In addition, the existence of extremum of functional (42) can be easily proved by the so-called "second variational approach" as well, which indicates functional (42) have a local extreme. Therefore, it is concluded that modified functional (42) can be used to devive the hybrid finite element formulations.

Element Stiffness Equation

The element stiffness equation can be generated by setting $\delta\Pi_{me} = 0$. To simplify the derivation, all domain integrals in Eq. (42) are first transformed into boundary ones. With the Gaussian theorem, the functional in Eq. (42) can be simplified as

$$\begin{aligned}\Pi_{me} &= \frac{1}{2}\iint_{\Omega_e}(\sigma_{ij}\varepsilon_{ij} + D_i E_i)d\Omega - \int_{\Gamma_t}\bar{t}_i\tilde{u}_i d\Gamma - \int_{\Gamma_D}\bar{D}_n\tilde{\phi}d\Gamma + \int_{\Gamma_e}t_i(\tilde{u}_i - u_i)d\Gamma \\ &+ \int_{\Gamma_e}D_n(\tilde{\phi}-\phi)d\Gamma = \frac{1}{2}\left(\int_{\Gamma_e}t_i u_i d\Gamma - \iint_{\Omega_e}\sigma_{ij,j}u_i d\Omega\right) + \frac{1}{2}\left(\int_{\Gamma_e}D_n\phi d\Gamma\right. \\ &\left.- \iint_{\Omega_e}D_{i,i}\phi d\Omega\right) - \int_{\Gamma_t}\bar{t}_i\tilde{u}_i d\Gamma - \int_{\Gamma_D}\bar{D}_n\tilde{\phi}d\Gamma + \int_{\Gamma_e}t_i(\tilde{u}_i - u_i)d\Gamma + \int_{\Gamma_e}D_n(\tilde{\phi}-\phi)d\Gamma\end{aligned} \quad \textbf{(55)}$$

Due to the satisfaction of the equilibrium equation with the constructed intra-element fields, the following expression for the HT finite element model can be obtained

$$\begin{aligned}\Pi_{me} &= \frac{1}{2}\int_{\Gamma_e}t_i u_i d\Gamma + \frac{1}{2}\int_{\Gamma_e}D_n\phi d\Gamma - \int_{\Gamma_t}\bar{t}_i\tilde{u}_i d\Gamma - \int_{\Gamma_D}\bar{D}_n\tilde{\phi}d\Gamma + \int_{\Gamma_e}t_i(\tilde{u}_i - u_i)d\Gamma + \\ &\int_{\Gamma_e}D_n(\tilde{\phi}-\phi)d\Gamma = -\frac{1}{2}\int_{\Gamma_e}(t_i u_i + D_n\phi)d\Gamma + \int_{\Gamma_e}(t_i\tilde{u}_i + D_n\tilde{\phi})d\Gamma - \int_{\Gamma_t}\bar{t}_i\tilde{u}_i d\Gamma - \int_{\Gamma_D}\bar{D}_n\tilde{\phi}d\Gamma\end{aligned}$$

(56)

Substituting Eqs.(11), (36) and (33) into the above functional (56) yields the formulation as

$$\Pi_{me} = -\frac{1}{2}\mathbf{c}_e^{\ T}\mathbf{H}_e\mathbf{c}_e + \mathbf{c}_e^{\ T}\mathbf{G}_e\mathbf{d}_e - \mathbf{d}_e^{\ T}\mathbf{g}_e \quad \textbf{(57)}$$

where

$$\mathbf{H}_e = \int_{\Gamma_e}\mathbf{Q}_e^{\mathrm{T}}\mathbf{N}_e \mathrm{d}\Gamma, \quad \mathbf{G}_e = \int_{\Gamma_e}\mathbf{Q}_e^{\mathrm{T}}\tilde{\mathbf{N}}_e \mathrm{d}\Gamma, \quad \mathbf{g}_e = \int_{\Gamma_t}\tilde{\mathbf{N}}_e^{\mathrm{T}}\bar{\mathbf{t}}d\Gamma + \int_{\Gamma_D}\tilde{\mathbf{N}}_e^{\mathrm{T}}\bar{\mathbf{D}}d\Gamma \quad \textbf{(58)}$$

To enforce inter-element continuity on the common element boundary, the unknown vector $\mathbf{c}_e$ should be expressed in terms of nodal DOF $\mathbf{d}_e$. The stationary condition of the functional Π_{me} with respect to $\mathbf{c}_e$ and $\mathbf{d}_e$, respectively, yields

$$\frac{\partial \Pi_{me}}{\partial \mathbf{c}_e^{\mathrm{T}}} = -\mathbf{H}_e \mathbf{c}_e + \mathbf{G}_e \mathbf{d}_e = \mathbf{0} \tag{59}$$

$$\frac{\partial \Pi_{me}}{\partial \mathbf{d}_e^{\mathrm{T}}} = \mathbf{G}_e^T \mathbf{c}_e - \mathbf{g}_e = \mathbf{0} \tag{60}$$

from which the relationship between $\mathbf{c}_e$ and $\mathbf{d}_e$, and the stiffness equation can be obtained

$$\mathbf{c}_e = \mathbf{H}_e^{-1} \mathbf{G}_e \mathbf{d}_e \tag{61}$$

$$\mathbf{K}_e \mathbf{d}_e = \mathbf{g}_e \tag{62}$$

where $\mathbf{K}_e = \mathbf{G}_e^{\mathrm{T}} \mathbf{H}_e^{-1} \mathbf{G}_e$ is the element stiffness matrix.

Following the procedures in [28], the missing rigid-body motion can be recovered by setting the augmented internal field of a particular element **e** as

$$\mathbf{u}_e = \mathbf{N}_e \mathbf{c}_e + \begin{bmatrix} 1 & 0 & x_2 & 0 \\ 0 & 1 & -x_1 & 0 \\ 0 & 0 & 0 & 1 \end{bmatrix} \mathbf{c}_0 \tag{63}$$

where the undetermined rigid-body motion parameter $\mathbf{c}_0$ can be calculated using the least square matching of $\mathbf{u}_\mathbf{e}$ and $\tilde{\mathbf{u}}_\mathbf{e}$ at element nodes

$$\min = \sum_{i=1}^{n} \left[\left(u_{1i} - \tilde{u}_{1i}\right)^2 + \left(u_{2i} - \tilde{u}_{2i}\right)^2 + \left(\phi_i - \tilde{\phi}_i\right)^2 \right] \tag{64}$$

which finally gives

$$\mathbf{c}_0 = \mathbf{R}_e^{-1} \mathbf{r}_\mathbf{e} \tag{65}$$

$$\mathbf{R_e} = \sum_{i=1}^{n} \begin{bmatrix} 1 & 0 & x_{2i} & 0 \\ 0 & 1 & -x_{1i} & 0 \\ x_{2i} & -x_{1i} & x_{1i}^2 + x_{2i}^2 & 0 \\ 0 & 0 & 0 & 1 \end{bmatrix} \quad \textbf{(66)}$$

$$\mathbf{r_e} = \sum_{i=1}^{n} \begin{bmatrix} \Delta u_{e1i} \\ \Delta u_{e2i} \\ \Delta u_{e1i} x_{2i} - \Delta u_{e2i} x_{1i} \\ \Delta \phi_{ei} \end{bmatrix} \quad \textbf{(67)}$$

in which $\mathbf{\Delta u}_{ei} = \left(\tilde{\mathbf{u}}_e - \hat{\mathbf{u}}_e\right)\Big|_{\textbf{node i}}$ and n is the number of element nodes. As a consequence, $\mathbf{c}_0$ can be calculated by Eq.(65) once the nodal DEP fields $\mathbf{d}_e$ and the interpolation coefficients $\mathbf{c}_e$ are determined by Eqs. (62) and (61), respectively. Then, the complete DEP fields $\mathbf{u}_e$ can be calculated by Eq. (63).

NUMERICAL RESULTS AND DISCUSSIONS

Normalization

The order of magnitudes of the material constants and the corresponding field variables in piezoelectricity have a wide spectrum as large as 10^{19} in SI unit. This leads to an ill-conditioned matrix of the system [14, 23]. To resolve this problem, normalization of each quantity by its reference value should be utilized. The reference values for the stiffness, piezoelectric stress constant, dielectric constants and strain are $c_0 = 10^{11}\left(N/m^2\right)$, $e_0 = 10^1 (N/mV)$, $k_0 = 10^{-9}\left(C/mV\right)$, $\varepsilon_0 = 10^{-3}\left(V/m\right)$, respectively. The reference values of other quantities (Table **1**) are determined in terms of these four fundamental reference variables and the characteristic length $x_0 = 10^0\left(m\right)$ of the problem so that the normalized governing equations remain exactly in the same form as the original equations.

Numerical Results

The effectiveness and accuracy of the proposed method is validated by a numerical example in this section. The piezoelectric panel (Fig. **2**) is subjected to a linearly varying stress at one edge and the plain strain condition is considered. The stability, convergence and accuracy of the method are discussed.

Table 1. Reference values for material constants and field variables in piezoelectricity derived from basic reference variables: c_0, e_0, k_0, ε_0 and x_0.

Displacement	$u_0 = x_0\varepsilon_0 = 10^{-3}(m)$	Electric Potential	$\phi_0 = x_0 E_0 = 10^7(V)$
Stress	$\sigma_0 = c_0\varepsilon_0 = 10^8(N/m^2)$	Electric induction	$D_0 = k_0 E_0 = 10^{-2}(C/m^2)$
Compliance	$s_0 = \dfrac{\varepsilon_0}{\sigma_0} = 10^{-11}(m^2/N)$	Impermeability	$\beta_0 = \dfrac{E_0}{D_0} = 10^9(mV/C)$
Electric field	$E_0 = \dfrac{\sigma_0}{e_0} = 10^7(V/m)$	Piezoelectric strain constant	$g_0 = \dfrac{E_0}{\sigma_0} = 10^{-1}(mV/N)$

Bending of a Piezoelectric Panel

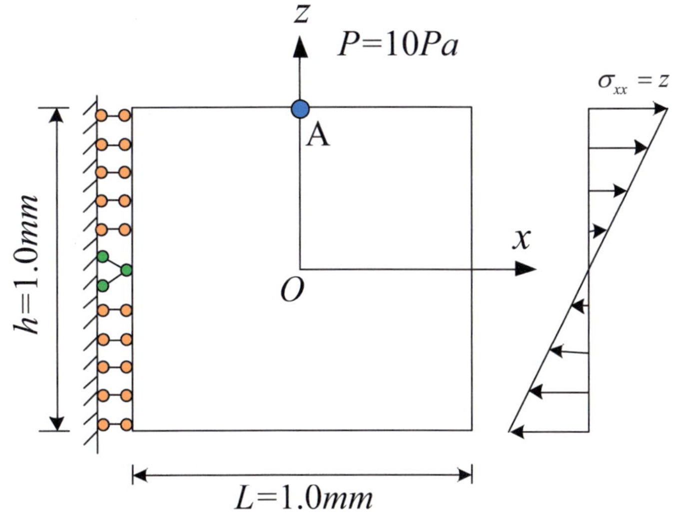

Fig. (2). Bending of a piezoelectric panel.

As shown in Fig. (**2**), a 1×1mm piezoelectric panel made of PZT-4 is subjected to a linearly varying stress at the right edge. The corresponding boundary conditions are

$$\sigma_{xx} = z, \quad \sigma_{xz} = D_x = 0, \qquad \text{at } x = L/2 \tag{68}$$

$$u_x = 0, u_z = 0, D_x = 0 \qquad \text{at } x = -L/2, z = 0 \tag{69}$$

$$u_x = 0, \sigma_{xz} = D_x = 0 \qquad \text{at } x = -L/2, z \neq 0 \tag{70}$$

$$\phi = 0, \sigma_{xz} = \sigma_z = 0 \qquad \text{at } x = h/2 \tag{71}$$

$$u_x = 0, \sigma_{xz} = \sigma_{zz} = D_z = 0 \qquad \text{at } x = h/2 \tag{72}$$

The properties of the material PZT-4 [33] are given in Table **2**. Sixteen quadratic elements are used to model the piezoelectric panel (Fig. **3**). For the proposed hybrid FEM, the position of source points have some effect on the accuracy and stability of the numerical results. Thus, we first determine the suitable range of the parameter γ to generate the source points according to the related element nodes. In this analysis, the variation of the stress σ_{xx} at point A (0, 0.5 mm), the condition number of matrix H and the condition number of global stiffness matrix K with parameter γ are shown in Fig. (**4**).

Table 2. Properties of the material PZT-4 used in the Example.

Parameters	Values	Parameters	Values
c_{11}	$13.9\times10^{10}\,\text{Nm}^{-2}$	e_{15}	13.44Cm^{-2}
c_{12}	$7.78\times10^{10}\,\text{Nm}^{-2}$	e_{31}	-6.98Cm^{-2}
c_{13}	$7.43\times10^{10}\,\text{Nm}^{-2}$	e_{33}	13.84Cm^{-2}
c_{33}	$11.3\times10^{10}\,\text{Nm}^{-2}$	κ_{11}	$6.0\times10^{-9}\,\text{C}/\text{Nm}$
c_{44}	$2.56\times10^{10}\,\text{Nm}^{-2}$	κ_{33}	$5.47\times10^{-9}\,\text{C}/\text{Nm}$

It can be seen from Fig. (**4**) that the results of both stress and condition number are stable when 3 <γ < 15 for the investigated range while they fluctuate dramatically when γ < 3. This may be explained by the fact that for small values of γ (γ < 3), the source points are close to the element boundary, which results in strong disturbance of singularity of the fundamental solutions on numerical accuracy. Conversely, for a larger γ, the source points are far from the element boundary so that the potential round-off error will lead to some components of the matrix H to have nearly the same value, which in turn leads to a larger condition number of matrix H, generating large round-off error in the inverse manipulation of matrix H [30]. Therefore, it is reasonable to have the parameter 3 < γ < 10. In this example, γ is chosen to be 5.

Fig. (3). Mesh used for the piezoelectric panel by HFS-FEM.

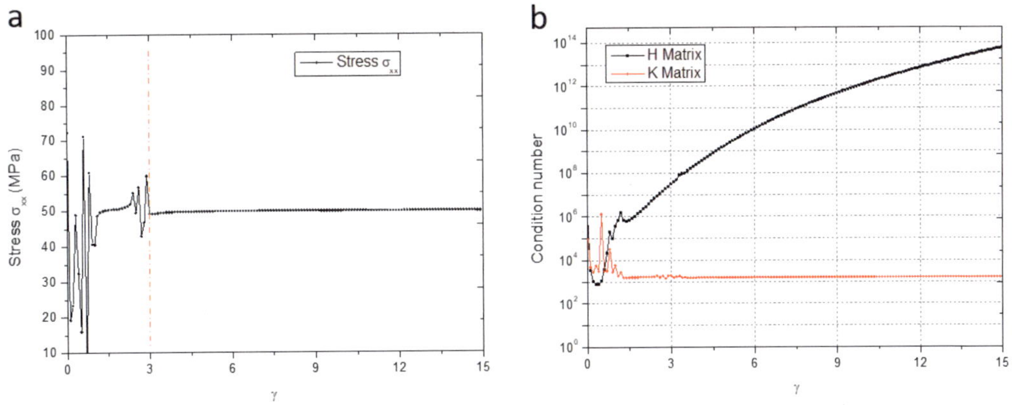

Fig. (4). (**a**) Variation of the stress σ_{xx} at point A (0, 0.5mm) with parameter γ . (**b**) Variation of the condition number of H matrix and K matrix with parameter γ .

Table **3** lists the displacement and electrical potential at five different reference points, in which the analytical results [33] are also presented for comparison. It can be seen that the proposed approach has very good accuracy, and the results agree well with the analytical solutions. Fig. (**5**) presents the contour plots for the displacements u_x, u_z, electrical potential ϕ, and stress σ_{xx}, which demonstrates the distribution of the concerned variables.

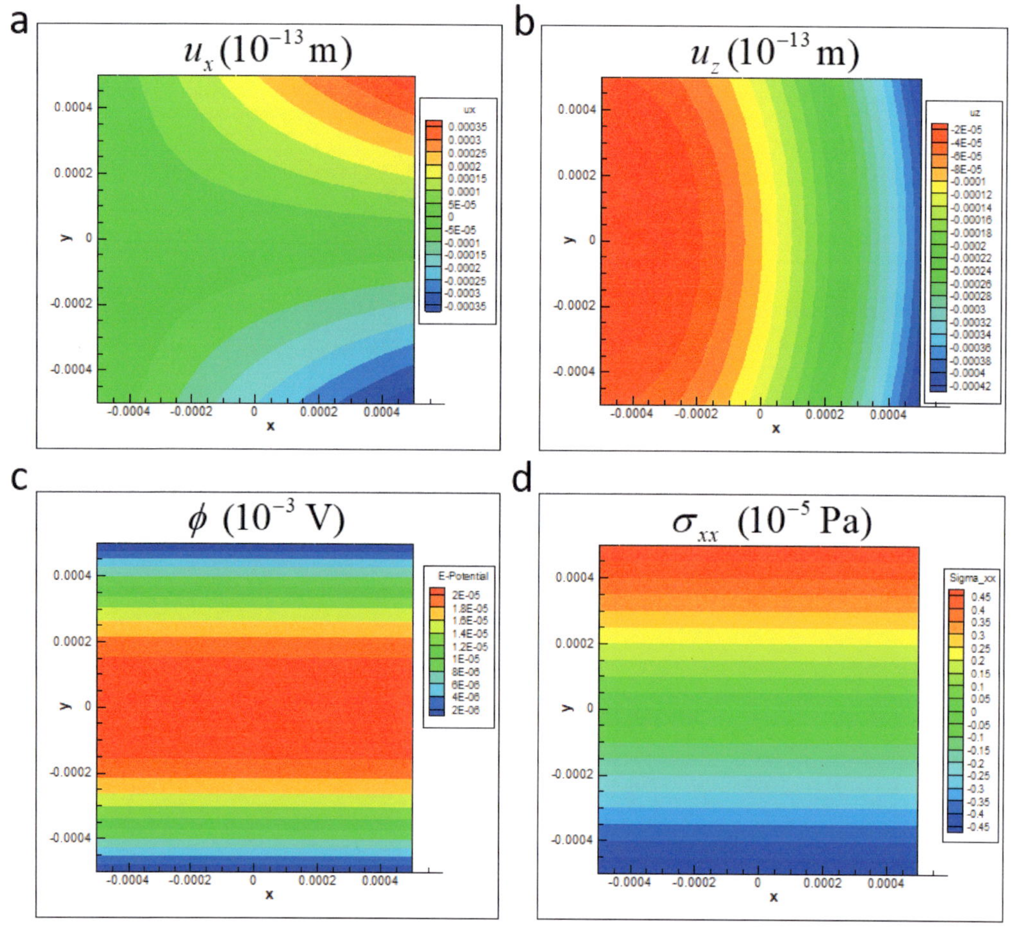

Fig. (5). Contour plots of (**a-b**) displacements, (**c**) electrical potential and (**d**) stress.

Table 3. Comparison of HFS-FEM and analytical results [33] for the piezoelectric panel.

Position (x, z) (mm, mm)	u_x (10^{-17}m)		u_z (10^{-17}m)		ϕ (10^{-8}V)	
	HFS-FEM	**Exact**	**HFS-FEM**	**Exact**	**HFS-FEM**	**Exact**
(0.0, 0.5)	1.9805	1.9805	-1.3691	-1.3691	0.0000	0.0000
(0.5, 0.0)	0.0000	0.0000	-3.9609	-3.9610	2.2225	2.2225
(0.25, 0.5)	2.9707	2.9708	-2.6069	-2.6069	0.0000	0.0000
(0.5, 0.25)	1.9804	1.9805	-4.0556	-4.0557	1.6667	1.6669
(0.5, 0.5)	3.9609	3.9610	-4.3398	-4.3399	0.0000	0.0000

CONCLUDING REMARKS

In this chapter the HFS-FEM for 2D piezoelectric materials is described for predicting the coupling behaviors of piezoelectric materials. The foundational solution of the transversely isotropic piezoelectric materials is utilized to approximate the DEP fields. The element stiffness matrix is further derived from a modified variational functional, whose minimization conditions leads to boundary conditions and continuity conditions between any two adjacent elements. A numerical example is finally demonstrated to validate the accuracy and efficiency of the proposed method in comparison with the analytical solutions. It has been shown that the new HFS-FEM has very good performance in analyzing the coupling behaviors of piezoelectric materials or structures under various conditions.

CONSENT FOR PUBLICATION

Not applicable.

CONFLICT OF INTEREST

The author declares no conflict of interest, financial or otherwise.

ACKNOWLEDGEMENT

Declared none.

REFERENCES

[1] D. M. Barnett and J. Lothe, "Dislocations and line charges in anisotropic piezoelectric insulators," *physica status solidi (b),* vol. 67, no. 1, pp. 105-111, 1975.

[2] A. N. Stroh, "Dislocations and cracks in anisotropic elasticity," (in English), *Philosophical Magazine,* Article vol. 3, no. 30, pp. 625-646, 1958.

[3] M. Y. Chung and T. C. T. Ting, "Piezoelectric solid with an elliptic inclusion or hole," *International Journal of Solids and Structures,* vol. 33, no. 23, pp. 3343-3361, 1996.

[4] M. L. Dunn and M. Taya, "Micromechanics predictions of the effective electroelastic moduli of piezoelectric composites," *International Journal of Solids and Structures,* vol. 30, no. 2, pp. 161-175, 1993.

[5] Q. H. Qin and S. W. Yu, "Effective moduli of piezoelectric material with microcavities," (in English), *International Journal of Solids and Structures,* vol. 35, no. 36, pp. 5085-5095, Dec 1998.

[6] P. Zhao, Y. Cao, and J. Li, "Nonlinear electromechanical coupling behavior of 1–3 piezoelectric composites," *Acta materialia,* vol. 59, no. 14, pp. 5534-5543, 2011.

[7] H. Sosa, "Plane problems in piezoelectric media with defects," *International Journal of Solids and Structures,* vol. 28, no. 4, pp. 491-505, 1991.

[8] X.-L. Xu and R. Rajapakse, "Analytical solution for an arbitrarily oriented void/crack and fracture of piezoceramics," *Acta materialia,* vol. 47, no. 6, pp. 1735-1747, 1999.

[9] S. Park and C. T. Sun, "Fracture criteria for piezoelectric ceramics," *Journal of the american ceramic society,* vol. 78, no. 6, pp. 1475-1480, 1995.

[10] Q.-H. Qin and S.-W. Yu, "An arbitrarily-oriented plane crack terminating at the interface between dissimilar piezoelectric materials," *International Journal of Solids and Structures,* vol. 34, no. 5, pp. 581-590, 1997.

[11] S. Kumar and R. Singh, "Energy release rate and crack propagation in piezoelectric materials. Part I: mechanical/electrical load," *Acta Materialia,* vol. 45, no. 2, pp. 849-857, 1997.

[12] Ł. Jański, M. Scherzer, P. Steinhorst, and M. Kuna, "Adaptive finite element computation of dielectric and mechanical intensity factors in piezoelectrics with impermeable cracks," *International journal for numerical methods in engineering,* vol. 81, no. 12, pp. 1492-1513, 2010.

[13] E. Béchet, M. Scherzer, and M. Kuna, "Application of the X-FEM to the fracture of piezoelectric materials," *International Journal for Numerical Methods in Engineering,* vol. 77, no. 11, pp. 1535-1565, 2009.

[14] M. Denda and C. Y. Wang, "3D BEM for the general piezoelectric solids," *Computer Methods in Applied Mechanics and Engineering,* vol. 198, no. 37-40, pp. 2950-2963, 2009.

[15] D. Haojiang, W. Guoqing, and C. Weiqiu, "A boundary integral formulation and 2D fundamental solutions for piezoelectric media," *Computer Methods in Applied Mechanics and Engineering,* vol. 158, no. 1-2, pp. 65-80, 1998.

[16] J. S. Lee and L. Z. Jiang, "A boundary integral formulation and 2D fundamental solution for piezoelastic media," *Mechanics Research Communications,* no. 21, pp. 47-54, 1994.

[17] F. García-Sánchez, C. Zhang, and A. Sáez, "2-D transient dynamic analysis of cracked piezoelectric solids by a time-domain BEM," *Computer Methods in Applied Mechanics and Engineering,* vol. 197, no. 33-40, pp. 3108-3121, 2008.

[18] X.-L. Xu and R. Rajapakse, "On a plane crack in piezoelectric solids," *International Journal of Solids and Structures,* vol. 38, no. 42-43, pp. 7643-7658, 2001.

[19] E. Pan, "A BEM analysis of fracture mechanics in 2D anisotropic piezoelectric solids," *Engineering Analysis with Boundary Elements,* vol. 23, no. 1, pp. 67-76, 1999.

[20] Q. H. Qin, "Solving anti-plane problems of piezoelectric materials by the Trefftz finite element approach," *Computational Mechanics,* vol. 31, no. 6, pp. 461-468, Aug 2003.

[21] H. Wang, K. Sze, and X. Yang, "Analysis of electromechanical stress singularity in piezoelectrics by computed eigensolutions and hybrid-trefftz finite element models," *Computational Mechanics,* vol. 38, no. 6, pp. 551-564, 2006.

[22] Q. H. Qin, "Variational formulations for TFEM of piezoelectricity," *International Journal of Solids and Structures,* vol. 40, no. 23, pp. 6335-6346, 2003.

[23] W. G. Jin, N. Sheng, K. Y. Sze, and J. Li, "Trefftz indirect methods for plane piezoelectricity," *International Journal for Numerical Methods in Engineering,* vol. 63, no. 1, pp. 139-158, 2005.

[24] N. Sheng, K. Sze, and Y. Cheung, "Trefftz solutions for piezoelectricity by Lekhnitskii's formalism and boundary-collocation method," *International journal for numerical methods in engineering,* vol. 65, no. 13, pp. 2113-2138, 2006.

[25] H. Wang and Q.-H. Qin, "Fundamental-solution-based hybrid FEM for plane elasticity with special elements," *Computational Mechanics,* vol. 48, no. 5, pp. 515-528, 2011.

[26] C. Cao, Q.-H. Qin, and A. Yu, "A novel boundary-integral based finite element method for 2D and 3D thermo-elasticity problems," *Journal of Thermal Stresses,* vol. 35, no. 10, pp. 849-876, 2012.

[27] H. Wang and Q. H. Qin, "Fundamental-solution-based finite element model for plane orthotropic elastic bodies," *European Journal of Mechanics, A/Solids,* vol. 29, no. 5, pp. 801-809, 2010.

[28] Q. H. Qin, *The Trefftz finite and boundary element method.* Southampton, WIT Press, 2000.

[29] C. Cao, Q.-H. Qin, and A. Yu, "Hybrid fundamental-solution-based FEM for piezoelectric materials," *Computational Mechanics,* vol. 50, no. 4, pp. 397-412, 2012.

[30] C. Cao, A. Yu, and Q.-H. Qin, "A new hybrid finite element approach for plane piezoelectricity with defects," *Acta Mechanica,* vol. 224, no. 1, pp. 41-61, 2013.

[31] H. Ding, G. Wang, and J. Liang, "General and fundamental solutions of plane piezoelectroelastic problem," *Acta Mechanica Sinica (Chinese Edition),* vol. 28, pp. 441-448, 1996.

[32] Y. Weian and H. Wang, "Virtual boundary element integral method for 2-D piezoelectric media," *Finite elements in analysis and design,* vol. 41, no. 9-10, pp. 875-891, 2005.

[33] K. Sze, X.-M. Yang, and H. Fan, "Electric assumptions for piezoelectric laminate analysis," *International journal of solids and structures,* vol. 41, no. 9-10, pp. 2363-2382, 2004.

SUBJECT INDEX

Qing-Hua Qin

F

G

H

I

K

L

M

N

O

P

Q

R

S

Made in the USA
Las Vegas, NV
03 November 2026

57421601R00212